THE NEOLIBERAL REGIME IN THE AGRI-FOOD SECTOR

For the last three decades, the neoliberal regime, emphasizing economic growth through deregulation, market integration, expansion of the private sector, and contraction of the welfare state has shaped production and consumption processes in agriculture and food. These institutional arrangements emerged from and advanced academic and popular beliefs about the virtues of private, market-based coordination relative to public, state-based problem solving. This book presents an informed, constructive dialogue around the thesis that the neoliberal mode of governance has reached some institutional and material limits. Is neoliberalism exhausted? How should we understand crisis applied to neoliberalism? What are the opportunities and risks linked to the construction of alternatives? The book advances a critical evaluation of the evidence supporting claims of rupture of, or incursions into, the neoliberal model. It also analyzes pragmatic responses to these critiques, including policy initiatives, social mobilization, and experimentation at various scales and points of entry.

The book surveys and synthesizes a range of sociological frames designed to grapple with the concepts of regimes, systemic crisis, and transitions. Contributions include historical analysis, comparative analysis, and case studies of food and agriculture from around the globe. These highlight particular aspects of crisis and responses, including the potential for continued resilience, a neoproductivist return, as well as the emergence and scaling up of alternative models.

Steven A. Wolf is Associate Professor in the Department of Natural Resources at Cornell University, New York, USA. He is also a Senior Lecturer in the Centre for Environmental Policy at Imperial College, University of London, UK.

Alessandro Bonanno is Texas State University System Regents' Professor and Distinguished Professor of Sociology at Sam Houston State University, Texas, USA.

OTHER BOOKS IN THE EARTHSCAN FOOD AND AGRICULTURE SERIES

FOOD SYSTEMS FAILURE
The Global Food Crisis and the Future of Agriculture
Edited by Chris Rosin, Paul Stock and Hugh Campbell

UNDERSTANDING THE COMMON AGRICULTURAL POLICY
Berkeley Hill

THE SOCIOLOGY OF FOOD AND AGRICULTURE
Michael Carolan

COMPETITION AND EFFICIENCY IN INTERNATIONAL FOOD SUPPLY CHAINS
Improving Food Security
John Williams

ORGANIC AGRICULTURE FOR SUSTAINABLE LIVELIHOODS
Edited by Niels Halberg and Adrian Muller

THE POLITICS OF LAND AND FOOD SCARCITY
Paolo De Castro, Felice Adinolfi, Fabian Capitanio, Salvatore Di Falco and Angelo Di Mambro

PRINCIPLES OF SUSTAINABLE AQUACULTURE
Promoting Social, Economic and Environmental Resilience
Stuart Bunting

RECLAIMING FOOD SECURITY
Michael Carolan

FOOD POLICY IN THE UNITED STATES
An Introduction
Parke Wilde

PRECISION AGRICULTURE FOR SUSTAINABILITY AND ENVIRONMENTAL PROTECTION
Edited by Margaret A. Oliver, Thomas F.A. Bishop and Ben P. Marchant

AGRICULTURAL SUPPLY CHAINS AND THE MANAGEMENT OF PRICE RISK
John Williams

SUSTAINABLE FOOD SYSTEMS
Building a New Paradigm
Edited by Terry Marsden and Adrian Morley

For further details, please visit the series page on the Routledge website: http://www.routledge.com/books/series/ECEFA/

THE NEOLIBERAL REGIME IN THE AGRI-FOOD SECTOR

Crisis, Resilience, and Restructuring

Edited by Steven A. Wolf and Alessandro Bonanno

First published 2014
by Routledge
2 Park Square, Milton Park, Abingdon, Oxon OX14 4RN

and by Routledge
711 Third Avenue, New York, NY 10017

Routledge is an imprint of the Taylor & Francis Group, an informa business

British Library Cataloguing-in-Publication Data

A catalogue record for this book is available from the British Library

Library of Congress Cataloging-in-Publication Data

A catalogue record has been requested for this book

ISBN: 978-0-415-81789-9 (hbk)
ISBN: 978-0-203-58311-1 (ebk)

Typeset in Galliard
by Apex CoVantage, LLC

Printed and bound in Great Britain by
CPI Group (UK) Ltd, Croydon, CR0 4YY

CONTENTS

About the Contributors ix

Introduction 1
STEVEN A. WOLF AND ALESSANDRO BONANNO

PART 1
Theoretical Analyses and Key Concepts 11

1 **The Legitimation Crisis of Neoliberal Globalization: Instances from Agriculture and Food** 13
ALESSANDRO BONANNO

2 **How Neoliberal Myths Endanger Democracy and Open New Avenues for Democratic Action** 32
LAWRENCE BUSCH

3 **Policing the New Enclosures: On Violence, Primitive Accumulation, and Crisis in the Neoliberal Food System** 52
MORGAN BUCK

PART 2
Case Studies 71

4 **The Rise and Fall of a Prairie Giant: The Canadian Wheat Board in Food Regime History** 73
ANDRÉ MAGNAN

5 Situating Neoliberalization: Unpacking the Construction of Racially Segregated Workplaces 91
JILL LINDSEY HARRISON

6 Creating Rupture through Policy: Considering the Importance of Ideas in Agri-food Change 112
REBECCA L. SOM CASTELLANO

7 Beyond Farming: Cases of Revitalization of Rural Communities through Social Service Provision by Community Farming Enterprises 129
HARUHIKO IBA AND KIYOHIKO SAKAMOTO

PART 3
Research Opportunities 151

8 To Bt or not to Bt? State, Civil Society, and Firms Debate Transgenic Seeds in Democratic India 153
DEVPARNA ROY

9 Turning of the Tide: Rising Discontent over Transgenic Crops in Brazil 170
KARINE PESCHARD

10 US Agri-environmental Policy: Neoliberalization of Nature Meets Old Public Management 191
STEVEN A. WOLF

11 For Competitiveness' Sake? Material Competition vs. Competitiveness as a National Project 207
ANOUK PATEL-CAMPILLO

12 The Neoliberal Food Regime and its Crisis: State, Agribusiness Transnational Corporations, and Biotechnology 225
GERARDO OTERO

13 "Just Another Asset Class"?: Neoliberalism, Finance, and the Construction of Farmland Investment 245
MADELEINE FAIRBAIRN

14 Neoliberalism in the Antipodes: Understanding the Influence and Limits of the Neoliberal Political Project **263**
GEOFFREY LAWRENCE AND HUGH CAMPBELL

15 Conclusion: The Plasticity and Contested Terrain of Neoliberalism **284**
STEVEN A. WOLF AND ALESSANDRO BONANNO

Index 295

ABOUT THE CONTRIBUTORS

Alessandro Bonanno is Texas State University System Regents' Professor and Distinguished Professor of Sociology at Sam Houston State University. Dr. Bonanno's work focuses on the implications that globalization has for social relations and institutions. In particular, and employing the agro-food sector as an empirical area of concentration, he investigates the impact that globalization has on the state, democracy, and the emancipatory options of subordinate groups.

Morgan Buck is a doctoral student in human geography at the Graduate Center of the City University of New York. Her research interests include the colonial and postcolonial restructuring of African agrarian economies and livelihoods, as well as contemporary processes of informalization and urbanization. She is currently conducting dissertation research in South Africa on social life and political organization among migrants living on Johannesburg's metropolitan fringe. She teaches in the geography department at Hunter College.

Lawrence Busch is University Distinguished Professor of Sociology at Michigan State University. His work over the last decade has focused on the role of standards for persons, products, and processes. He is currently working on two books. The first examines the role of neoliberalism in transforming higher education and research. The second enquires into the outdated myths that continue to inform American society.

Hugh Campbell is Professor of Sociology and Head of the Department of Sociology, Gender and Social Work and Professorial Research Fellow at the Centre for Sustainability (www.csafe.org.nz) at the University of Otago, New Zealand. The consequences of neoliberal reform have provided a major theme to his research interests. These have included the evolution of new forms of private-sector governance over agricultural export industries in New Zealand, particularly around the issue of auditing sustainability and environmental performance in food production and the theorization of privatized governance and neoliberalization within agri-food theory. He has published on these themes in the journals *Rural Sociology, Agriculture and Human Values*, the

International Journal for the Sociology of Agriculture and Food, the *Journal of Rural Studies*, and *Sociologia Ruralis.*

Madeleine Fairbairn is a PhD candidate in community and environmental sociology at the University of Wisconsin–Madison. Her multi-sited dissertation research explores the finance and policy behind the current farmland investment boom, with a focus on the United States and Brazil. Her other research interests include alternative agri-food movements, food security, and land rights.

Jill Lindsey Harrison is Assistant Professor of Sociology at the University of Colorado at Boulder. Her research and teaching areas of focus are environmental sociology, sociology of agriculture and food systems, environmental justice, political theories of justice, and immigration politics, with a regional emphasis on the United States. She has used her research on political conflict over agricultural pesticide poisonings in California and recent escalations in immigration enforcement in rural Wisconsin to identify and explain the persistence of environmental inequalities and workplace inequalities facing Latino immigrants in the United States today.

Haruhiko Iba is Associate Professor of Agricultural Economics at Kobe University, Japan. His principal areas of research and teaching include farm management and farmers' organizations. Iba has been studying public or social goods created by farming organizations through their farming and nonfarming-related activities, such as environmental conservation and agricultural education. His recent research examines Community Farm Enterprises (CFEs), or groups of members of agricultural communities collectively engaged in farming that provide social services in remote and disadvantaged rural areas of Japan. His current research project compares public goods created by British allotment gardens and Japanese CFEs.

Geoffrey Lawrence, PhD, is Professor and Head of Sociology in the School of Social Science at The University of Queensland, Australia. He is coleader of Food Security in the Global Change Institute, at the same university. He is president of the International Rural Sociology Association (2012–16). His current research explores the financialization of agriculture and the governing of food security—nationally and globally. He receives funding from the Australian Research Council, the National Research Foundation of Korea, and the Norwegian Research Council.

André Magnan is Associate Professor in the Department of Sociology and Social Studies at the University of Regina. His research examines the political economy of local and global food systems, with a focus on the history and politics of grain marketing on the Canadian prairies and changing patterns of farm ownership and control in Canada and Australia.

Gerardo Otero is Professor of Sociology and International Studies at Simon Fraser University. Author of *Farewell to the Peasantry? Political Class Formation* *in Rural Mexico* (Westview, 1999), he has published numerous articles and chapters and edited books about political economy of agriculture and food, civil society, and the state in Mexico and Latin America. His latest edited book is *Food for the Few: Neoliberal Globalism and Biotechnology in Latin America* (University of Texas Press, 2008, reissued in paperback in 2010). He is writing a book on empowerment theory or political-class formation from the bottom up. E-mail: otero@sfu.ca. Web page: www.sfu.ca/~otero.

Anouk Patel-Campillo teaches courses on international development and rural social change at The Pennsylvania State University. She studies the interactions between agro-food systems, institutional and multiscalar governance, and social actors. Her current research examines the implications of regulatory change on supply chain (re)organization, the regionalization of food systems, and gendered division of labor.

Karine Peschard holds a PhD in anthropology from McGill University (2010). Her doctoral thesis examines the controversy over agricultural biotechnology in Brazil, looking more specifically at resistance to transgenic seeds among small farmers in Southern Brazil. She is currently a postdoctoral fellow at the Graduate Institute of International and Development Studies in Geneva, where she conducts comparative research on farmers' rights in Brazil and India. Her research interests are centered on global capital, contemporary peasant movements, food sovereignty, agricultural biotechnology, intellectual property rights, and biodiversity.

Devparna Roy is a development sociologist. For her PhD dissertation at Cornell University, she analyzed the experiences of Gujarat (India) farmers with Bt cotton during the years 2002 to 2004. Her current research interests are as follows: the sociology of agro-food systems in India and the United States of America, with an emphasis on comparing different agricultural technologies (e.g., industrial farming using transgenic seeds versus organic farming), the political economy of land seizures in rural settings of India, and social movements related to the development of various natural resources (from seeds to land).

Kiyohiko Sakamoto is an independent translator and lives in Wichita Falls, Texas. He holds a PhD in sociology from the University of Kentucky with emphases on development sociology, sociology of agriculture and food, and science and technology studies. He has been engaged in research projects on science-based risk governance systems, such as food safety and plant quarantine regulations, on sustainable development of farming communities, and on participatory research and development in Latin America, Japan, and the United States.

Rebecca L. Som Castellano completed her doctoral work at The Ohio State University in the summer of 2013 and is Assistant Professor of Sociology at Boise State University. Her research focuses on stratification and agri-food system change. Current projects include examining how the reproduction of gender inequality in alternative agri-food movements limits the achievement of justice in the agri-food system and acts as a social limitation to scaling up alternative agri-food.

Steven A. Wolf is Associate Professor at Cornell University in Ithaca, New York, where he teaches and conducts research on environment and society. He studies environmental governance with a specific focus on the challenges of securing public goods from privatized landscapes. Most of his empirical research is focused on agricultural and forested landscapes of industrialized societies (United States and EU). Current projects address political economy of land resources in Southern India and the Tibetan Plateau of China. He maintains a faculty appointment at Imperial College, London. E-mail: saw44@cornell.edu.

INTRODUCTION

Steven A. Wolf and Alessandro Bonanno

Points of Entry

For the last three decades, neoliberalism has shaped production and consumption processes in agriculture and food. At present, social instability and protest, economic recessions, political uncertainties, and ecological degradation and risks have prompted claims that we now confront a systemic crisis. The regulatory mechanisms and patterns of material flow that constitute contemporary agri-food are implicated in global insecurities, both physical and metaphysical. The capacity to maintain the legitimacy and material coherence of neoliberalism is arguably in doubt. This volume advances dialogue around the thesis that we have reached a number of institutional and material limits. Is neoliberalism exhausted? Are we at the outset or some midpoint of a significant change? And if so, what are the opportunities and risks linked to this change? Of course, in contemplating radical restructuring, we must account for the historical capacity of agri-food governance to deflect critique, co-opt rivals, and sustain the unsustainable.

The book advances a critical evaluation of the evidence supporting claims of rupture of, or incursions into, the neoliberal model. We also seek to analyze proposed responses to these critiques, including policy initiatives, social mobilization, and experimentation at various scales and points of entry. At the level of theory, the book surveys and synthesizes a range of sociological frames that allow us to grapple with the concepts of systemic crisis and transitions. Empirical contributions include case studies from around the globe. These cases highlight particular aspects of crises and responses, including the potential for continued resilience, a neoproductivist return, the emergence of new alternatives, and the scaling up of existing alternative models.

This volume is located in the intellectual tradition of the critical analysis of agriculture and food. Emerging from the now classical critique of post-World War II *Rural Sociology* (Friedland, 1982), critical agriculture and food evolved from the *New Sociology of Agriculture* movement of the 1970s and 1980s (Buttel and Newby, 1980). In the United States, post-World War II sociological studies of agriculture were shaped by the then dominant functionalist theory. Differing from the rest of sociology, works on agriculture did not address hypotheses derived from the grand narrative proposed by Talcott Parsons. Parsons's theory ([1937] 1968) was too abstract to be amenable to testable hypotheses involving microlevel data (i.e., households, individuals, farms). Rather, it was Robert Merton's middle-range theory that inspired and fueled the functionalist turn of agricultural studies in this period (Buttel et al., 1990, pp. 44–5). In this context, sociological works on agriculture took a social-psychological approach, stressing the agency of individual farmers who were assumed to be freely acting individuals responding to external stimuli. Following the established theory of modernization (Parsons, 1971; Rostow, [1960] 1971), studying the evolution of agriculture was reduced to the studying of the diffusion and adoption of new technology and production techniques (Rogers, 1995).

Erroneously defined as atheoretical (Buttel et al., 1990, p. 47), this way of studying social relations in agriculture reinforced the ideology of the stability, desirability, and constant progress of American society. It saw in modernization and functionalism its most decisive theories, in Fordism its socioeconomic regime, and in productivism its strategy for economic development. Denouncing the limits of this view, the group of young scholars that formed the new sociology of agriculture not only overtly applied alternative theoretical frameworks in the analysis of agriculture and food, but also combined empirical analyses with strong normative postures. The neo-Marxist approach and its classical tradition based on the works of Marx, Lenin, Kautsky, and Chayanov offered the context for a great number of studies. Other theoretical positions, such as constructionism (i.e., Busch, 1978; Newby, 1978), world systems (Friedman and McMichael, 1989), and radical pragmatic democracy (Heffernan, 1972; Rodefeld, 1978; Bonanno, 2009), inspired a wealth of contributions that redefined and problematized the evolution of agri-food in mature capitalism.

The various crises of the 1970s, including the domestic farm crisis (i.e., loss of family farms and the concentration of production in fewer larger farms), the articulation of agricultural trade policy of industrialized countries within the failure of the global development project and the emergence of a strong ecological critique of agriculture, and, above all, the end of Fordist socioeconomic arrangements in the following years, opened a new phase in the study of agri-food. Abandoning themes linked to the classical *agrarian question*—that is, the trajectory and conditions of people in agriculture—efforts were progressively directed at understanding the post-Fordist era and the ideology and discourses of neoliberalism that framed it (Bonanno et al., 1994). During the following three decades, a significant number of works proceeded to illustrate the changes brought about by the globalization of society and its neoliberalization. An important aspect of this reframing of the field was an expanded commitment to study of nonfarm components of agri-food systems. In recent decades, this commitment has been extended such that a focus on consumption has in many ways displaced analysis of farm-level dynamics.

Paralleling some of the conditions that ended Fordism in the 1970s, we have entered into a period in which neoliberal globalization and, above all, the tenets and applications of neoliberalism have been progressively questioned. Following recursive and interconnected economic, social, and cultural crises, neoliberalism has emerged as the obstacle to sustainable and just growth and, simultaneously, a formula for its attainment. Drawing on existing debates and original empirical research, this volume probes the legitimacy and instability of the neoliberal agri-food system. Through integrated consideration of discursive, political, economic, ecological, and technological dimensions of agri-food systems, this sociological analysis seeks to evaluate and inform academic, civic, and policy responses to contemporary developments. Method and approaches include grounded analysis, policy analysis, political economy, including commodity systems analysis, industrial organization, trade and food regimes, comparative institutional analysis, discourse analysis, and political philosophy.

The volume aims to synthesize critical perspectives on agriculture and food in line with traditions of sociology, and at the same time, we seek to interrogate the categories and evidence that informs pronouncements that we now confront epochal change. In this sense, the book seeks to extend a line of academic inquiry of over twenty-five years, and at the same time, to challenge conventional assumptions.

New instabilities in models of food provision, the energy sector, poverty alleviation, and land tenure are adding potency to traditional social, economic, and ecological critiques of agriculture and food. Moreover, the contradictions and instabilities of agri-food systems are being recognized by a broader set of actors. Consideration of patterns of food production and consumption now invoke questions of ecological, national, political, and personal security. Framing questions about the social and ecological relations of agri-food in terms of (in)security opens up new vistas and opportunities.

Organizational/Professional Backstory and Acknowledgments

This volume grew out of an effort within the Rural Sociological Society's Sociology of Agrifood Research Interest Group (SAFRIG) to add energy and content to the conference program marking the 75th Anniversary of the Rural Sociological Society, July 26–29, 2012 at the Palmer House Hotel in Chicago, Illinois. SAFRIG hosted a mini-conference featuring twenty-two papers presented over one full day preceding the meeting and in a number of sessions within the conference program. Authors contributed 3,000-word papers that had been shared online in advance with all participants. This format allowed us to invite only brief remarks from authors and to invest a significant amount of time to discussion. The mini-conference brought together sociologists and allied disciplines (e.g., political geography, policy analysis, institutional economics), perspectives from a range of countries, and a large number of young researchers and more senior researchers with shared interests in questions of agri-food dynamics and social problems. The papers that focused most directly on questions of neoliberalism, legitimacy, and restructuring were selected for inclusion in this volume.

The tradition of mini-conferences within SAFRIG and allied groups bears brief mention because this book project is a continuation of a significant intellectual tradition within the sociology of agri-food. Notably, the book *Toward a New Political Economy of Agriculture* (Friedland et al., 1991) marked the fiftieth anniversary of the Rural Sociological Society and its meeting in Madison, Wisconsin. *From Columbus to ConAgra* (Bonanno et al., 1994) was derived from a mini-conference in Columbia, Missouri. A 2003 mini-conference in Austin, Texas at the meeting of the Agriculture, Food & Human Values Society led to the publication of *The Fight over Food* (Wright and Middendorf, 2007). The SAFRIG mini-conferences in Santa Clara, California (2007) and Manchester, New Hampshire (2008) produced, respectively, a special section of *Agriculture and Human Values* (Wolf and Harrison, 2008) and a special section of *Rural Sociology* (Friedland et al., 2010). Mini-conferences organized within the Research Committee on Sociology of Agriculture and Food (RC-40) within the International Sociological Association and within the International Rural Sociological Association also bear mention.

In reflecting on the history recounted above, we want to recognize William Friedland. Bill's name often appears as a leader of these conferences, and this reflects the investment he has made over many years in building this tradition and making the events stimulating and inclusive. With respect to the Chicago mini-conference and this volume, we would like to recognize the leadership of Spencer Wood and Bill Winders, who initiated the effort to organize events within SAFRIG to commemorate RSS's fiftieth year. Wynne Wright, Jessica Goldberger, and Clare Hinrichs served on the SAFRIG planning committee from which the mini-conference emerged. Ralph Brown and Keiko Tanaka were the leaders of the Chicago RSS meeting, and their leadership and administrative efforts were invaluable. Tim Hardwick and Ashley Wright from Earthscan have been encouraging and enabling throughout. Lastly, Meghan Baumer and Winnie Chu provided top-shelf support with the Web site for the mini-conference and with the formatting of the manuscript.

Summary of Chapters

The book is divided into three parts in order to advance our objective to frame, specify, and extend contemporary analysis of political economic relations of agri-food systems. Part 1 is devoted to theoretical analyses that frame key concepts and questions.

Alessandro Bonanno's analysis situates the contemporary claim of legitimation crisis in agri-food within a broader socioeconomic and historical context. Following the crisis of Fordism, the fix represented by neoliberalism has run its course. For Bonanno, the crisis is both ideological and structural. Ideological in that continued belief in market-based coordination is demonstrably naïve. Structural in that we do not seem to have access to a functional alternative. The scope of this analysis emphasizes the value of looking outside of agri-food for the ideas and drivers of some future model. In other words, contemporary problems in agri-food were not specifically borne of agri-food. As we look to identify sources of instability and innovation, we must cast our gaze broadly.

Lawrence Busch surveys the tenets of the market and the government to help us understand neoliberalism and pathways to a more democratic society. In producing a brief intellectual history and drilling down to the core assertions embedded in these concepts as developed in the neoliberal intellectual tradition, this critical examination highlights their abstract and mythic quality. Without denying their power, Busch demonstrates that "markets can be performed in a nearly infinite number of ways" and "the narrowing of the role of government merely allows other less accountable actors to govern." Recognizing that the dominant popular meanings attached to these important ideas are socially constructed and do not produce a democratic surplus creates the space needed to imagine and advance alternatives.

Morgan Buck offers an unabashed condemnation of social relations of agri-food as represented by commitments to enclosure backed up by violence. Based on a historical approach, the chapter identifies the neoliberal agri-food regime as part of a much broader project of accumulation and disciplining of subjects. Buck makes challenging arguments about the stakes of the current socioeconomic and ecological crisis and its relationship to disciplinary function of the state and its corporate and military partners. Looking ahead to what could constitute a response to the current crisis, she identifies creative and critical engagement with "commons" as the core project of analysts and civil society actors.

Part 2 is devoted to case studies. These are theoretically inspired empirical reports that specify relations and dynamics on the ground. This collection of original research offers us a valuable survey of developments through which we can assess the status, trajectory, and meaning of neoliberalism as a mode of governance. The case studies also allow us to reflect on how neoliberalism is variably employed as an analytical frame to make sense of political economic dynamics.

André Magnan analyzes the recent dismantling of the Canadian Wheat Board (CWB) in the context of three decades of neoliberal restructuring in the Western Canadian grains sector. This analysis is timely, as the CWB was effectively dismantled in August 2012, precisely when this collection of papers was presented in Chicago. These recent events testify to the contemporary power of the logic of neoliberalism within national policy making. In the quest for development and legitimacy, entrepreneurial behavior is preferred over relations of solidarity. The case shows neoliberalism to be advancing in significant and stark fashion.

Jill Lindsey Harrison offers a grounded study of how neoliberalization plays out in the agricultural workplace in the United States through analysis of relations of race and national origin in the Wisconsin dairy sector. Harrison argues that "contemporary immigration politics effectively strengthen the neoliberal project—(rationalization)—and thus render it more resilient." The chapter demonstrates how neoliberal immigration policies stabilize labor supply for dairy farms, and at the same time, how the inequity and insecurity that dairy workers confront introduce contradictions that represent points of destabilization for farm owners and the sector at large. In this sense, the issue of crisis is seen as ambiguous. In focusing on racism and economic inequality, the chapter reminds us that we must not try to explain all social problems through reference to neoliberalism. Racism and exploitation of foreigners predate the neoliberal project.

Rebecca L. Som Castellano explores the question of whether subordinate groups can resist and potentially transform the neoliberal food regime via national policy making. The chapter is a critical reflection on processes by which ideas and coalitions can gain a foothold in policy processes to make a positive difference. In her analysis of the role of agency, she examines the recent case of the integration of farm-to-school (FTS) into the National School Lunch Program (NSLP) in the United States. Beyond the modest new national policy commitments to local food, the case brings our attention back to the state as a potentially important domain of struggle and innovation in efforts to respond to social problems.

Haruhiko Iba and **Kiyohiko Sakamoto**'s pragmatic analysis of Japanese rural social service enterprises that have grown out of collective structures supporting small-scale farming draws our attention to devolution as a key dynamic within neoliberal governance. Beyond the value of placing rural communities at the center of the analysis, this chapter advances the important question of how entrepreneurship by local people can represent both accommodation of and resistance to neoliberalism. These authors present a sophisticated theoretical apparatus for analyzing the contemporary moment for rural communities and rural people confronting social problems in an era of declining state support. They also offer a rare opportunity to learn about rural transitions in Japan through a carefully researched case study presented in English.

Part 3 is devoted to a set of broadly scoped, empirically informed analyses that draw our attention to issues on the social science and policy frontier. These chapters extend our horizons by deriving and illustrating programmatic arguments about research opportunities.

Devparna Roy offers a finely tuned analysis of the introduction, social conflict, and social regulation of Bt brinjal (eggplant) and Bt cotton. Her study emphasizes ongoing contests among state, commercial, and civil society actors. The crisis of neoliberalism is analyzed in stark relief. She writes: "The debate over transgenic crops is ultimately a contention over the kind of capitalist structures that Indian citizens wish to live with in the foreseeable future: state-led capitalism or corporate-controlled capitalism (neoliberal globalization), or neither." Real alternatives are at play, and there is contradictory movement.

Karine Peschard analyzes the crisis of neoliberalism quite directly in her study of Brazilian soy growers' confrontations with the corporately controlled transgenics industry. By placing corporate control and transgenics in the center of the debate, Peschard aligns herself with Otero and Roy in this volume. While rupture is not perceived as imminent, farmers' unions have succeeded in shifting the debate to address livelihoods and social dimensions beyond productivism. The withdrawal of large soybean farmers from the advocacy and policy coalition destabilizes the current mode of neoliberal governance and suggests a variety of internal drivers of political economic change. Research on political realignment among elite actors in policy coalitions is an important place to look for trigger points and for accommodation processes that imply resilience.

Steven Wolf analyzes US agri-environmental policy to demonstrate that the rollout of neoliberal policy designs is a highly uneven and contingent process. Efforts to introduce market discipline, accountability through competitiveness, and private funding of conservation in agriculture are, to date, largely stories of frustration for policy entrepreneurs. Despite the flagging legitimacy of bureaucratic traditions of paying farmers to produce public goods, such as water quality, policy reform proposals linked to the concept of payment for ecosystem services remain by and large unrealized. In explaining inertia or stickiness in processes of institutional change, the analysis emphasizes the power of incumbent ideas, practices, and networks. Recognition that market-based rationalization is not unfolding apace in all quarters highlights the open-ended character of neoliberalism. Other modes of governance exist and are possible, specifically modes that combine market mechanisms and democratic process. Understanding and advancing institutional hybridity represents an important challenge on the research and policy frontier.

Anouk Patel-Campillo analyzes national projects of subjecting regions and firms to expanded competition under programs of liberalization through investigation of the cut flower industries of the Netherlands and Colombia. The comparative case analysis highlights unevenness in realization of a core reference within neoliberal thought. Patel shows us how weak legitimacy of national projects of competitiveness at subnational levels "opens windows of opportunity to contain and even contest the neoliberal regime." We learn that ideological projects do not always transfer across scales intact, and this highlights a key governance challenge deserving of attention from researchers.

Gerardo Otero analyzes interplay between the state, agribusiness transnational corporations, and biotechnology as the core of the contemporary neoliberal food regime and attendant crisis signaled most visibly by the 2008 food price spikes. His insistence on the centrality of the state—"both for its deployment and transcendence"—as the hallmark of the neoliberal regime is important, as it helps us appreciate the complex interdependencies between state and market in governance. Further, his analysis of the impotence and exacerbating function of biotechnology within the contemporary food and rural economic crisis highlights the failure of the state. Rather than offer a conclusion emphasizing a failure of technology assessment, Otero focuses our attention on the political economic context of the technology.

Madeleine Fairbairn analyzes the emergence of a market for farmland among institutional investors and the implications for critical engagement with neoliberalism. Through a historically grounded analysis of the practices and narratives through which farmland is being constructed as a new asset class, the chapter provides theoretical and empirical purchase on processes of financialization and related claims of land grabbing. While further incorporation of land into circuits of capital can be seen as fueling a crisis of neoliberalism through introduction of new destabilizing feedbacks, this twenty-first-century land rush extends neoliberal logic by further blurring distinctions between fictitious commodities and material means of production. Quoting Tomaskovic-Devey and Lin (2011, p. 556), Fairbairn remarks, "If neoliberalism is a policy and intellectual movement away from state regulation, financialization is perhaps its most fundamental product." In this sense, future engagement with finance and processes of abstraction constitute important foci of research on agri-food and agri-food politics.

Geoff Lawrence and **Hugh Campbell** critically reflect on their 2003 analysis of the first decade of agri-food neoliberalism in Australia and New Zealand. They "revisit both in terms of how we theorize neoliberalism and how we understand its impacts and limits." The chapter analyzes financialization in Australia as the key contemporary extension of neoliberal thought and practice, reflecting continuity and a deepening of commitments from the previous century. While the analysis of New Zealand demonstrates how cooperative traditions in the dairy and kiwi fruit sectors has served to empower actors confronting the cost-price squeeze of neoliberalism, "there has been overwhelming agreement in both countries, at the political level, for the continuation of neoliberal policy formation including the embedding of economic rationalist settings." Lawrence and Campbell's synthetic analysis points to the value of making sense of contemporary developments and trajectories through reflection grounded in historical, cross-national, and comparative research.

In the concluding chapter, **Steven Wolf** and **Alessandro Bonanno** present the most salient conclusions that can be derived from the contributions contained in this volume. They underscore the complex and multifaceted dimension of neoliberalism, denouncing the reductionist character of views that see it as a unified project. They stress the perils associated with its reification and the adoption of tautological positions that see neoliberalism as a stable regime in the context of a series of regimes in agri-food. Their interpretation of the chapters within the volume indicates that neoliberalism is a system of practices and discourses that generates resistance within the state apparatus and within civil society. The nation-state, they continue, has been transformed through the adoption of neoliberalism. Yet, it is the theater of resistance at various levels. The overall picture that emerges from this examination of neoliberalism is that its characteristics and future trajectories can be better understood in terms of the contested terrain in which opposing social forces and institutions develop, coexist, interact, and struggle.

References

Bonanno, A. (2009) "Sociology of agriculture and food beginning and maturity: The contribution of the Missouri School (1976–1994)," *Southern Rural Sociology*, vol 24, no 2, pp. 29–47.

Bonanno, A., Busch, L., Friedland, W., Gouveia, L., and Mingione, E. (1994) *From Columbus to ConAgra: The Globalization of Agriculture and Food*, University Press of Kansas, Lawrence.
Busch, L. (1978) "On understanding understanding: Two views of communications," *Rural Sociology*, vol 43, no 4, pp. 450–73.
Buttel, F.H., Larson, O.F., and Gillespie, G., Jr. (1990) *The Sociology of Agriculture*, Greenwood Press, Westport, CT.
Buttel, F. and Newby, H. (1980) *The Rural Sociology of Advanced Societies*, Allanheld Osmun, Montclair, NJ.
Friedland, W. H. (1982) "The end of rural society and the future of rural sociology," *Rural Sociology*, vol 47, no 1, pp. 589–608.
Friedland, W.H., Busch, L., Buttel, F.H., and Rudy, A. (1991) *Toward a New Political Economy of Agriculture*, Westview Press, Boulder, CO.
Friedland, W., Ransom, E., and Wolf, S. (2010) "Special segment, agrifood alternatives and reflexivity in academic practice," *Rural Sociology*, vol 75, no 4, pp. 532–7.
Friedmann, H. and McMichael, P. (1989) "Agriculture and the state system: The rise and fall of national agricultures, 1870 to the present," *Sociologia Ruralis*, vol 29, no 2, pp. 93–117.
Heffernan, W. D. (1972) "Sociological dimensions of agricultural structures in the United States," *Sociologia Ruralis*, vol 12, no 2, pp. 481–99.
Newby, H. (1978) "The rural sociology of advanced societies," Pp. 3–30 in H. Newby (ed.) *International Perspectives in Rural Sociology*, Wiley, Chichester.
Parsons, T. ([1937] 1968) *The Structure of Social Action*, Free Press, New York.
Parsons, T. (1971) *The System of Modern Societies*, Prentice Hall, Englewood Cliffs, NJ.
Rodefeld, R. (1978) "Trends in U.S. farm organizational structure and type," Pp. 158–77 in R. Rodefeld, J. Flora, D. Voth, I. Fujimoto, and J. Converse (eds) *Changes in Rural America: Causes, Consequences and Alternatives*, C. V. Mosby Company, St. Louis, MO.
Rogers, E. (1995) *Diffusion of Innovations*, Free Press, New York.
Rostow, W.W. ([1960] 1971) *The Stages of Economic Growth*, Cambridge University Press, New York.
Tomaskovic-Devey, D. and Lin, K. H. (2011) "Income dynamics, economic rents, and the financialization of the U.S. economy," *American Sociological Review*, vol 76, no 4, pp. 538–59.
Wolf, S. A. and Harrison, J. (eds) (2008) "Charting fault lines in US agrifood systems: What can we contribute?," Special section of *Agriculture and Human Values*, vol 25, no 2.
Wright, W. and Middendorf, G. (2007) *The Fight over Food: Producers, Consumers, and Activists Challenge the Global Food System*, Pennsylvania State University Press, University Park, PA.

Part 1

THEORETICAL ANALYSES AND KEY CONCEPTS

1

THE LEGITIMATION CRISIS OF NEOLIBERAL GLOBALIZATION

Instances from Agriculture and Food

Alessandro Bonanno

> The United States has continued to progress: its citizens have become better fed, better clothed, better housed, and better transported; class and social distinctions have narrowed; minority groups have become less disadvantaged, popular culture has advanced by leaps and bounds. All this has been the product of the initiative and drive of individuals co-operating through the free-market. Government measures have hampered not helped this development.
>
> (Friedman, [1962] 1982, p. 200)

> There is a tendency to regard any existing government intervention as desirable, to attribute all evils to the market, and to evaluate new proposals for government control in their ideal form, as they might work if run by able, disinterested men, free from the pressure of special interests.
>
> (Friedman, [1962] 1982, p. 197)

Introduction

In recent years, the issue of the legitimation crisis of neoliberal globalization has become one of the topics of discussion in scientific circles. Pertinent literature underscores that neoliberalism and the process of globalization that it supports has diminished people's democratic participation in decision-making processes (Lupel, 2005; Habermas, 2002, 2012), concentrated power in the hands of the private sector and the super rich (Harvey, 2010; Krippner, 2011), augmented the monopolistic character of markets (Lynn and Longman, 2010), and promoted improbable individualistic solutions to society's problems (Harvey, 2010; Raulet, 2011). These conditions, coupled with the inability of market mechanisms to correct economic and social problems, generated a crisis of legitimation of neoliberal globalization.

The crisis runs deeper than the inability of market mechanisms to address current problems as it involves the concomitant ineffectiveness of state intervention. State actions to address the unwanted consequences of the implementations of neoliberal measures have not corrected problems and have not solved the issues that have been associated with the intervention of the state. Even as the state provides funds to address the crisis, critics from all sides of the intellectual and political spectrum denounce its inadequacy. The state's "steering mechanisms" (Habermas, 1975) have not been able to distribute the wealth generated in the economy and have not created sufficient support for the dominant ideology and norms.[1] In agri-food, this contradictory situation is in full view. Farmers, agri-food corporations, and consumers support the assumed virtues of the market but simultaneously call for state intervention to address its negative outcomes. The assumed ability of market mechanisms to regulate agri-food is contradicted by calls for state intervention.

This chapter addresses the issue of the legitimation crisis by underscoring the ideological and historical limits of neoliberal globalization. Following Habermas's theory of the legitimation crisis, it is argued that neoliberal globalization cannot meet the objectives that it claims. A discussion on the general characteristics of the crisis is presented with instances from the agri-food sector. In particular, the chapter opens with a brief review of the debate on the current legitimation crisis, making references to the legitimation crisis that led to the demise of Fordism in the late 1970s.[2] The second section discusses the ideological and structural components of the legitimation crisis by contrasting theory claims with the historical evolution of socioeconomic arrangements. This segment includes a discussion on the classical work of Milton Friedman and structural contradictions that characterize the economy and society. The following section provides examples of the crisis of neoliberal globalization in agriculture and food. The positions of producers (farmers), corporations, and consumers are illustrated. A brief set of conclusions ends the chapter.

The Historical Dimension of the Legitimation Crisis

Almost forty years ago, the renowned German social theorist Jürgen Habermas wrote his book *Legitimation Crisis* (1975).[3] In that book, Habermas analyzed the conditions that engendered the crisis of legitimation of Fordist capitalism. His thesis centered on the inability of the state to fulfill the claims of economic expansion, social stability, and equality that characterized advanced societies.[4] For Habermas, the legitimation crisis indicated that "the legitimizing system does not succeed in maintaining the required levels of mass loyalty . . . [and] it is not possible by administrative means to maintain or establish effective normative structures to the extent required" (1975, pp. 46–7). For Habermas, the legitimation crisis, therefore, is a destabilizing phenomenon that indicates the historical failure of society to mobilize the means necessary to fulfill immanent social imperatives.[5]

The changes brought about by neoliberal globalization have forcefully reopened discussion on legitimation. Often based on analyses on the "crisis of the nation-state," these contributions call into question the adequacy of existing normative structures to address the consequences of the expansion of transnational social relations. In particular, the adequacy of claims made by neoliberal theorists in regard to enhanced freedom, democracy, and socioeconomic growth are called into question. Employing a definition of legitimation that rests on the gap between the ability of global constituencies to democratically participate in decision-making processes and the very processes with which decisions are made, Lupel (2005) argues that neoliberal globalization has eroded the collective capacity to make "legitimately binding decisions." The current conditions, he concludes, open a crisis of legitimacy as established mechanisms of decision making are ineffective and new and more democratic ones are lacking. Underhill and Zhang (2008) further this argument by contending that current global arrangements undermine the legitimacy of the global system by shifting power overtly in the hands of private actors. Under neoliberal globalization, the public good is increasingly defined and decided upon by private entities that escape collective scrutiny. For these authors, legitimacy is defined as, and refers to, the process of establishing accepted norms of social justice and democracy that allows constituencies to align themselves with the decisions of the rulers (Underhill and Zhang, 2008, p. 537). Focusing on the governance of the financial sector, they conclude that the unchecked dominance of private interests in decision making strips the process of governance of the necessary legitimacy. Discussing the same substantive area, Helleiner (2010) contends that the neoliberal claims of the effectiveness and desirability of market mechanisms clash with calls for state intervention and regulation that followed recent economic crises. These moves unveil a delegitimizing gap between ideology and practice under globalization.[6]

Defining the financial crisis of 2008, as an epiphenomenon of the structural crisis of advanced capitalism, the French philosopher Gérard Raulet (2011) supports the contention that the current situation is reminiscent of the legitimation crisis of Fordist capitalism illustrated by Habermas. As in the case of the 1970s, under the current global regime, the administrative system cannot meet the demands stemming from the economy and the expectations of the masses. The neoliberal withdrawing of the state has magnified the unwanted consequences of the functioning of the market and significantly diminished the ability of large segments of society to participate in decision-making processes. As global capitalism negatively affects the well-being of weaker segments of society, its devastating consequences are addressed though the upsurge and popularity of extreme individualistic interpretations (Harvey, 2010). As society is decollectivized through the individualization of social relations and responsibilities, the regulatory instruments of the state are seen as irrational and inefficient. Individual responsibility placed in the context of the free functioning market is viewed as the most effective replacement of a collective effort to address social needs and inequalities. The result of the resurgence of individualism is the erosion of the instruments and practices that reduced inequality and class polarization and provided opportunities for the lower and middle classes in the past. This system, Raulet concludes, has reached its limits as the market and individualism cannot address the problems emerging from the evolution of contemporary society.

The Ideological and Structural Dimensions of the Legitimation Crisis

The Ideological Dimension

In the classical manifesto of Neoliberalism, *Capitalism and Freedom*, Milton Friedman ([1962] 1982, pp. 196–9) lauds the virtues of the free market and sets the theoretical stage for the argument for the superiority of the neoliberal regime. For Friedman, the virtues of the free functioning of the market go far beyond economic benefits. While the establishment of economic freedom is central to his theory ([1962] 1982, p. 8), the application of neoliberalism solves many of the social and political problems that affect modern advanced societies. Better economic political and social arrangements would emerge from the implementation of neoliberal reforms ([1962] 1982, pp. 8–9). Economically, the key provision of neoliberalism rests on an "affective" freedom of exchange ([1962] 1982, pp. 14–15). It requires that individuals exchange goods and services in ways that do not interfere with the desires and actions of other individuals. Individuals must be free to act, but they should do so without limiting the freedom of action of others. In this context, consumers, Friedman contends, would be protected from coercion from any seller because of the presence of other sellers with whom they can deal. In a free market, consumers are not forced to buy from any one seller. Similarly, sellers are protected from coercion from consumers because of the existence of a multitude of other consumers who can buy their products ([1962] 1982, pp. 14–15). Even workers would be protected from the abuse of employers in a free market, as they can find alternative and more desirable employment. The free market makes unions and their claims in support of the well-being of workers obsolete.

For Friedman, freedom of exchange leads to political freedom. The ability to exchange without coercion gives people "what they want": individuals act freely to achieve independently selected goals. This is significantly different from situations characterized by government intervention in which a relatively small group of political leaders select the course of action for the rest of society ([1962] 1982, pp. 14–15). For Friedman, the essence of political freedom rests on the absence of coercion over individuals. Situations characterized by the existence of guidelines imposed by the government generate limited freedom because individuals are told how to act. Accordingly, the independence of the economic sphere from the political sphere is a condition for freedom ([1962] 1982, pp. 14–15) and a protection against any form of authority ([1962] 1982, p. 21). To be sure, Friedman contemplates some intervention of the state. It should be allowed in a few areas, such as law and order, the definition of property rights, adjudication of disputes, the enforcement of contracts, the provision of a monetary framework, and the implementation of antimonopoly measures. Ultimately, he argues, no government intervention and no entitlements allow people to take care of their interests. People should be left free to act rather than be directed by the state.

The market acts impersonally. No particular group affects its functioning ([1962] 1982, p. 21). This impersonality, Friedman maintains, is fundamental in the creation of conditions that promote the resolution of social problems and a better society ([1962] 1982, pp. 20–2). Employing the instances of the discrimination of minorities, Friedman criticizes the ineffectiveness of government programs designed to alleviate the conditions of minorities. He contends that the best protection against discrimination is the market. Its free functioning allows minorities to earn a living through unconstrained economic activities. Consumers, he explains, do not know the ethnicity of those who produce the bread that they purchase ([1962] 1982, p. 21). They will purchase what is convenient for them regardless of race or ethnicity. Therefore, the market separates economic activities from political views, making the former independent from the latter. The success of economic activities of minorities would depend on their productivity and abilities and not on their "views or their colour" ([1962] 1982, p. 21). Friedman contends that minorities have more to gain from the free market than other groups. Paradoxically, he concludes, these are the groups that claim more often than others that the market generates discrimination.

Addressing the cultural climate of the time, Friedman denounces the existence of a dominant attitude that sees government intervention as desirable while attributing "all evils" to the free market. He maintains that the desirability of government intervention is evaluated in its "ideal form." He argues that government programs are examined in the abstract, as if they were to be executed by individuals without allegiance to any special interest group. After a number of decades of government intervention, Friedman continues, there is historical evidence that it is far removed from its ideal form. It shows not only its ineffectiveness and inefficiency but also its political nature. Government intervention responds to special interests and often creates results that are opposite to its politically established objectives ([1962] 1982, pp. 196–8). The argument made by Friedman and like-minded theorists powered the neoliberal critique of Fordism. Their claims about the neutrality of the market, its impersonality, and the alternative that neoliberalism represents to the ineffectiveness and inefficiency of state intervention found the support of the general public and became instrumental in the establishment of neoliberalism as the dominant ideology.

Today, four decades after the ascendance to dominance of this ideology, Friedman's contention that government intervention is seen in its ideal rather than real form can be turned around and applied to neoliberalism. The social protest of recent years represents the externalization of the discontent with the neoliberal regime and its outcomes (Van Gelden, 2011; Byrne, 2012). Even critics of the

Occupy Movement recognize that, despite the limits of this movement, the dissatisfaction over the power of corporate and financial groups, the concentration of wealth in the hands of the super rich, the impoverishment of the lower and middle classes, high levels of unemployment, and bleak future perspectives for many have created an overall sense of broken and unfulfilled promises (Friedman, 2011; Sorkin, 2012).[7] The mood of the general population in the United States and elsewhere in the world is captured by one of the many popular press commentators who wrote on this issue: "I'll admit, I do not have much patience for this sort of protest . . . but capitalism has been destroyed from within . . . it has been hijacked by corporate interests that are undermining the system that they purport to uphold . . . it reflects a larger broader unrest in which America question why the recovery has benefited a few, the wealthiest 1 percent" (Steffy, 2011, pp. D1–8).

Popular perceptions are backed by scientific research. Studies have demonstrated the inaccuracy of the theoretical argument of neoliberals (Epstein, 2006; Palley, 2007; Stockhammer, 2007; Tabb, 2010). In a study of the concentration of income in the end of the super rich (the so-called 1 percent), Volscho and Kelly (2012) proved that the functioning of the market is affected by political actions, and therefore claims of market neutrality and impartiality are unsubstantiated. Contrary to Friedman's claims, they showed that the last few decades of deregulation and limited state intervention directly impacted market outcomes. This *market conditioning* (Volscho and Kelly, 2012, p. 681) benefited the super rich but also fostered a number of problems for the lower classes. The financial crisis of 2008 was a direct result of the deregulation enacted by neoliberals and the consequent lack of regulatory tools to address the development of new financial instruments and actions (Volscho and Kelly, 2012, pp. 681–2). Additionally, the reduction in union power and influence over wages and labor relations, rather than promoting workers' well-being, engendered conditions that diminished the compensation of members of the working class and increased the concentration of wealth at the top (Volscho and Kelly, 2012, pp. 693–4). The decrease in union power further diminished the bargaining power of labor and increased the market value of companies. This situation resulted in higher stock value that benefitted the rich. The reduction of taxation, and, in particular, the reduction of the capital gains tax rate, was theorized as a way to reduce government intervention and allow the rich to invest more and create additional wealth. This move was said to benefit the entire society, as market mechanisms would redistribute wealth downward to the lower classes. The actual results were quite different. Lower tax rates not only increased the concentration of wealth at the top but prevented the generation of resources to be redistributed to the lower classes through state intervention. In the absence of these redistribution mechanisms, concentration of wealth at the top increased (Volscho and Kelly, 2012, p. 688).

The ideological claims of neoliberalism clash with reality. The functioning of the market is not neutral and impartial, and its outcomes are not equally beneficial to all members of society. This false unity of theory and reality is increasingly visible and fuels arguments for state intervention and the introduction of corrections to market mechanisms. The response of supporters of neoliberalism has been consistently centered on the argument that neoliberal measures have been only partially applied and their functioning has been limited by continuous state intervention and distortions (Firebaugh and Goesling, 2004: Bhagwati, 2004; Steger and Roy, 2010; Crouch, 2011; McNally, 2011). Yet, and contrary to Friedman's recommendation, there is a constant return to an ideal model that is employed to claim the superiority of neoliberalism. The lack of unity of ideology and history is certainly visible in the case of democracy and governance. The withdrawal of the state promoted by neoliberal globalization has engendered the growth in private supervision of many spheres of life. As the economy is conditioned by privately controlled rating programs and consumption is shaped by third party certification, the disconnection between the ruled and the rule makers is clear. The privatization of control and regulation does not mandate nor guarantee the participation of people in decision-making processes. Numerous studies on certification programs in agri-food as well as in other sectors demonstrate this point.[8]

The Structural Dimension

The legitimation crisis of neoliberalism is also structural. Habermas (1975) and left-leaning critics of the Fordist regime (i.e., O'Connor, 1974; Offe, 1985) defined the structural component of that legitimation crisis in terms of the inability of the state to generate the necessary resources to support social programs and capital accumulation. The dominant class's resistances to the cost of the Fordist state social agenda along with its requests for state support in domestic and international affairs were the primary factors that engendered the *fiscal crisis of the state* at the end of the Fordist era. In the case of neoliberal globalization, the structural component of the legitimation crisis rests on two related processes. First, market mechanisms alone are unable to address the crisis, and second, the state intervention is ineffective. As illustrated by the literature reviewed above, the assumed self-corrective market mechanisms never materialized. Instead, state intervention has been sought by all parties. While presented by the US Government as necessary to save jobs and supported by corporate groups and unions alike, state intervention has not been able to solve the crisis (Krippner, 2011; Underhill and Zhang, 2008).

It consisted of two general strategies: austerity measures and increased spending. Austerity measures—the preferred solution of conservative groups—resulted in economic stagnation, high unemployment rates, lack of productive investments, and the deterioration of public services. This state action engendered resistance by the general public,[9] who protested declining socioeconomic well-being and gloomy future perspectives, but also by corporate groups that lamented limited state assistance and the lack of adequate corporate freedom. The economic crisis of the European Union is a case in point (Lapavitsas, 2012; Habermas, 2012). State spending—the preferred solution of the progressive camp (Krugman, 2012)—has achieved similar negative results. US state intervention has not solved the problems affecting the middle and lower classes but has provided support and generated gains for the dominant class. Ultimately, state spending has placed public money back in the hands of those economic elites who engendered the crisis (Underhill and Zhang, 2008; Van Gelden, 2011; Krippner, 2011; Raulet, 2011; Byrne, 2012). The strain of nation-states' finances—state budget deficits in many countries around the world have soared in the last two decades, and in the United States it stands at more than sixteen trillion dollars at the beginning of 2013—rather than be an incentive for more growth, has translated into stagnant employment conditions, lack of productive investment, and, more importantly, economic instability (Tabb, 2010; Krippner, 2011).

The reason for the ineffectiveness of state intervention is not explained by the implementation of inappropriate measures. It is structural. Under neoliberal globalization and through the process of deregulation of the economy, the state actively participated in the generation of those conditions that allowed more freedom of action to global corporations, concentrated wealth in the hands of the upper class, and increased class polarization and the lack of the creation of jobs and economic development (Tabb, 2010; Krippner, 2011). As documented by the abundant literature on globalization, the scope of the nation-state is not adequate to control the global socioeconomic conditions. But, in fact, it was the state's neoliberalization that contributed to the emergence of these conditions (see Bonanno and Constance, 2008). Attempts have been made to create transnational state forms that would address the tasks of regulation and coordination of capital accumulation and social legitimation. Similarly, there have been efforts to create a transnational civil society and market governance. Yet, these attempts have not translated into adequate global institutions that can effectively operate in the global economy (Bonanno and Constance, 2008).

Some brief elaborations on the structural limits of the state are necessary at this point. First, the neoliberal restructuring of the state and the economy has engendered the phenomenon of *financialization*. Financialization made financial capital more powerful and dominant over manufacturing and agricultural capital and more important in the determination of profit expected from investment (Stockhammer, 2007; Erturk et al., 2008; Tabb, 2010).[10] Due to financialization, profit has been transferred from the productive sector to the financial sector (Epstein, 2006; Krippner, 2011). The ensuing inter-capital conflict pits financial companies' short-term strategies and thinking against the strategies preferred by productive companies (Epstein, 2006; Tabb, 2010; Krippner, 2011). Productive sector companies' growth has been traditionally based on profit maximization though the involvement of other stakeholders, such as consumers and workers (Tabb, 2010). Sustained and stable consumption as well as pacified labor relations have been fundamental for the expansion of production and economic growth. Financial corporations' growth, conversely, has been guided by the maximization of shareholder value and the associated rapid creation and trade of financial products, derivatives, and credit arrangements. These strategies created indebtedness and negatively impacted the overall growth rate of the economy and the income of the productive sector's firms, consumers, and workers. Financialization's disincentives for long-term investment—a process known as "short termism"—have been resisted by productive sector companies (Epstein, 2006; Tabb, 2010, p. 163). This situation—along with processes of productive decentralization—has contributed to the growth of corporate profit without a corresponding increase in employment and wages. Defined as the jobless recovery, the years that saw the growth of financialization saw also the stagnation of labor remuneration and rising income inequality and indebtedness. In the United States, wages stagnated as real wages in 2012 were at the same level as 2005 and were below the 1970 point (Hamilton Project, 2012). In Europe, the growth of real wages for the 2000s was 0.45 percent, while in Japan, the real income of workers has decreased significantly in the 1990s and 2000s. These trends occurred in tandem with the exponential growth of the wealth of the richest segments of society (Ikeda, 2004, p. 379; Epstein, 2006; Stockhammer, 2007, p. 7; Tabb, 2010). As wages and salaries stagnated, financial institutions invited consumers to borrow. By providing easy access to credit that members of the working and middle classes could not afford, financial corporations profited, while household debt escalated. As the financial bubble burst, corporations were bailed out, while the working and middle classes were dispossessed and impoverished (Epstein, 2006; Stockhammer, 2007; Tabb, 2010; Harvey, 2010).

Second, while neoliberalism has been characterized by the reduction of state social spending, state intervention in support of corporate interests has continued. With a few exceptions (the United Kingdom and the Netherlands are notable ones in the advanced world), the volume of state spending has not decreased since the 1980s in the advanced countries of North America, Europe, and Asia (Epstein, 2006; Stockhammer, 2007, pp. 18–19; Crouch, 2011). Instead, the state actively redirected resources toward the support of corporate interests, financial capital, and the interests of the upper class (Epstein, 2006; Harvey, 2010; Crouch, 2011). The neoliberal state was one of the principal actors in the restructuring of the economy and the globalization of social relations (Bonanno and Constance, 2008). And these actions established a climate in which it was difficult for the lower classes and their organizations to respond to corporate moves (Epstein, 2006; Harvey, 2010; Steger and Roy, 2010; Crouch, 2011).

Third, the result of these contradictory actions has been a further destabilization of the position of the state. The nation-state appears weak and incapable of addressing emerging problems in the global society. Neoliberal advocates criticize state actions for their attempts to exercise excessive control over the economy and being too lenient in social matters. A globalized world dominated by a free market is inevitable, they contend, and state intervention creates futile attempts to halt a process that, instead, should be accelerated. State actions serve only to retard socioeconomic development (Bhagwati, 2004; Cohen and DeLong, 2010; Greenwood, 2011). Reforms and restructuring of welfare state programs, social control measures (e.g., anti-immigration and surveillance), and further liberalization of the economy are often advocated and practiced. Left-leaning groups denounce the reduced state intervention in social matters and the state overt support of corporate agendas. Theorists of the left also mistakenly framed the development of neoliberal globalization in terms of a polarized debate that stresses the crisis of the power of the state on one side and the growth of state power on the other. The thesis of the "irrelevance/disempowerment of the state" has been countered by the "continuous importance of the state" argument (e.g., Ohmae, 1995; Hirst and Thompson, 1996; Robinson, 2004; Nardone and McDonough, 2010). The inaccuracy of these views rests on their inability to recognize that the crisis of the power of the state is only limited to the social sphere as state support and action in favor of subordinate classes declined (Bonanno and Constance, 2008). The state ceased to support subordinate classes at the same level as it did during the Fordist era (Bonanno and Cavalcanti, 2011). Simultaneously, the neoliberalization of the state and its reregulation resulted in the augmentation of its class dimension (Harvey, 2005). The neoliberal nation-state has become a much stronger class state as its support of the upper class and its control of subordinate classes grew stronger.

Differing from the Fordist period in which the inclusion of the working and middle classes allowed the ruling class not only to expand capital accumulation but also to maintain social legitimation, in the era of neoliberal globalization, class polarization destabilizes society. The financial elites' push for the creation and trade of financial products is destructive to socioeconomic stability and growth. At the outset, the creation of financial products tends to increase collateral value. This expanded value allows more borrowing that finances investment spending and fuels economic expansion. As collateral value decreases, borrowing and investment fall, triggering a downward spiral that results in a crisis. This is a pattern that has occurred in developed and developing countries alike. The uncontrolled inflow and, above all, outflow of capital generated by deregulated financial markets and neoliberal policies undermine productive investment, economic development, and the creation of employment. Communities are left with limited resources and options to counter the crisis (Epstein, 2006; Krippner, 2011; Tabb, 2010).

The Legitimation Crisis in Agriculture and Food

In *Capitalism and Freedom*, Milton Friedman discussed the agri-food sector as one of the most obvious cases of misguided government intervention ([1962] 1982, pp. 181–2). He took issue with commodity price support programs (the section is entitled *Farm Price Supports*) and denounced the distortions that they generated. Introduced under Fordism to stabilize agricultural prices and support the income of farmers—in particular, for those operating small family farms—they were viewed by Friedman as typical instances of the ineffectiveness and inefficiency of government intervention. Not only did these costly programs fail to permit the proper functioning of the market, but they achieved results that were opposite to those that inspired their implementation. For Friedman, price support programs created unwanted surpluses, kept farmers on the land despite low income and limited perspectives for economic growth, placed the United States in difficult diplomatic predicaments, expanded bureaucracy to unnecessary and inflated levels, and, ultimately, discriminated against small, family farmers. These problems were the result of the actions of special interests that could exercise undue power because of the overrepresentation of rural areas in the electoral system and Congress (Friedman [1962] 1982, p. 181). Ultimately, a system guided by political negotiations, compromises, and these powerful special interests was preferred to the unbiased functioning of the market.

Paradoxically, equally critical of price support programs were authors that wrote from the left (Rodefeld et al., 1978; Buttel and Newby, 1980; Mottura and Pugliese, 1980). Both the populist and Marxian left attacked price support programs for their ineffectiveness and inefficiency, waste of resources, and, more importantly, for distributing funds in ways that discriminated against small farmers and poor segments of the rural society. Moreover, and for the case of developing countries, left-leaning critics saw farm and food programs as entities that undermine development. Differing from the neoliberal argument, they contended that state intervention was class-based as it favored corporate interests over those of small farmers and farmworkers. They argued for an intervention of the government that would substantively address the economic problems of subordinate groups in agri-food. Additionally, analyses from the left stressed the social relevance of agri-food in terms of food production, food security, and food sovereignty, whose accomplishment required state intervention. Because of the social importance of agriculture, the state must intervene and guarantee the necessary flow of investment and resources to the sector.[11]

Producers

Despite Friedman's contentions, as the Reagan Administration implemented neoliberal reforms and scaled back farm programs in the early 1980s, the interest for state intervention in the agricultural sector never diminished. Agri-food corporate groups supported liberalization when it involved the opening of new markets for their products and businesses. Yet, they advocated state support when free market competition was adverse to their interests. Simultaneously, farmers and farm communities always searched for "political" ways to protect themselves from the unwanted consequences of the functioning of the market even as they embraced the ideology of the free market. Since the adoption of neoliberalism, the supposed confidence in market mechanisms never materialized as the discourse presented by the farming community remained contradictory.

In the United States, the dominant and recurrent discourse on agriculture legislation, and the "Farm Bill" in particular, has been that of the creation of a "safety net" and the "need of protection" against the consequences of the market in a context in which the free market and the end of state intervention and regulation are preferred. "Who is going to protect us?" farmers ask. A rice farmer in Texas summarized the situation as Farm Bill negotiations continued in the summer of

2012: "We want to do our part in balancing the Federal budget. But if you want to continue to have the most affordable, safest food in the world, we need a *safety net* or folks cannot continue to farm" (Freedman, 2012a, p. A1). Another farmer, interviewed in December 2012, contends: "the Farm Bill is supposed to be there when you need it at the level to keep you in business to farm another day. . . . Can I survive without a *safety net?* Absolutely not . . . I'm more conservative than liberal and I voted republican at the Presidential Elections [of 2012] but Federal support helps agriculture in a positive way" (Freedman, 2012b, pp. D2–4). Discussing the importance of the Farm Bill in an invited column in a major American newspaper, a farmer further articulates the contradictory dimension of farmers' view of neoliberalism (Skalicky, 2012, p. B7). His discourse is centered on the argument that the government should not interfere with business and should not impose regulations. He states, "I do not want the government meddling around in my business . . . I actually do not like farm subsidies." And he adds, "[T]he US has high food safety standards . . . and I can't use certain chemicals in crop production because they are known carcinogens. Yet, the countries we import from . . . can use less expensive outlawed products. And the US imports their products, no questions asked." As the argument for reduced state regulation and control of business is made, calls are also made for intervention of the state. "Why do we need a *safety net?*" the same farmer asked. "Farming is not an occupation you can do when prices are good and get out when prices are bad. . . . Safety nets ensure that after a natural disaster or market collapse, there are still farmers to feed America. . . . Farm subsidies help keep the supermarket prices of our superior US products comparable to inferior products imported from other countries" (Skalicky, 2012, p. B7).

Equally problematic has been the situation at the international level. The assumed opening of markets has been executed with significant partiality. Markets in the North were not opened as much as markets in the South. The latter became "terrains of conquest" for global corporations, while the former remained significantly protected. In the case of the North American Free Trade Agreement (NAFTA), for instance, the protection granted to US products has not been granted to products originating in Mexico. Accordingly, subsidized US agricultural commodities entered Mexico. Mexican farmers, food companies, and political representatives have strongly protested the unequal interpretation of the free trade agreement. More importantly, a significant portion of the Mexican rural population has protested a host of consequences generated by neoliberal policies implemented in the United States (Pechlaner and Otero, 2010).

Consumers

That consumers have expressed concerns over the inability of market mechanisms to control production and distribution processes is well documented (Bonanno and Constance, 2008, pp. 256–60). Similarly well researched are the power and limits of the food consumption movement and the scope of its emancipatory action (Johnston and Szabo, 2011). Consumers in the lower classes have been particularly concerned with the reduced availability of food programs, the increased price of basic food items, and the decline of food sovereignty. Affluent consumers denounced the manner in which the application of neoliberal policies prevented the scrutiny of processes that affect a number of aspects of food production and distribution, including the quality of food and the environmental sustainability of food production. They also questioned the democratic dimension of third party control of food quality. As control of production and distribution processes is increasingly shifted to the private sphere, consumer concerns and dissatisfaction have increased.

Corporations

Significant dissatisfaction has emerged for the unchecked behavior of corporations under neoliberalism. However and surprisingly, concerns about the application of neoliberal policies have been voiced also by corporations. In effect, corporations have been overtly contradictory in their support of neoliberal policies. On the one hand, they have been strong promoters of deregulation and the reduction of barriers to the free circulation of commodities and investment. The elimination of public control over corporate actions and the claim that society will be better served when corporations supervise their own actions are among the most distinctive aspects of corporate affairs in recent decades. On the other hand, the numerous corporate crises of recent years resulted in explicit requests for state assistance. The financial crisis of 2008 is exemplary of this situation in which astronomical corporate losses were covered through state-sponsored bailouts. The issue of financial corporations that "are too big to fail" has become one of the central topics in political and academic circles. Similar instances occurred in other sectors, such as energy. The 2010 BP Gulf of Mexico oil spill is arguably the most visible instance of this decade. The exploitation of labor and natural resources has resulted in escalating corporate profit and sumptuous remunerations for CEOs. According to 2011 data, a CEO earns 244 times the pay of an average worker (AP, 2012). This overwhelming income disparity, along with the greed and arrogance of the corporate class, has been the target of the Occupy Wall Street movement and similar movements around the world (los indignados, etc.). Additionally, economists have demonstrated a growth in monopolies that further increases the concentration of wealth in corporate hands and diminishes the distribution of resources to labor (Lynn and Longman, 2010).

Agriculture and food corporations have also exhibited the contradictory behavior shown in other sectors. The calls for less regulation, the ability to move and invest freely across borders, and the fight against environmental and social concerns have characterized corporate behavior in agri-food. Income disparity is greater in agri-food than in the rest of the economy as most of the ten worst paid jobs in America are in agri-food (Smith, 2012). Simultaneously, agri-food

CEOs earn 352 times the annual salary of farmworkers (Smith, 2012). There is a climate in which the public is increasingly opposed to the freedom that neoliberalism has generated for the corporate world. While corporate supporters continue to call for more deregulations and a "true" free market, mounting popular pressure has resulted in enhanced corporate regulation (such as the Dodd-Frank Wall Street Reform and Consumer Protection Act of 2011, however timid it might be).

Conclusions

The limits of neoliberalism are theoretically clear and empirically evident. Arguably, the crisis of the regime can be seen more as a demonstrated fact rather than a hypothesis. Additionally, existing contradictions make it problematic to argue about the existence of an organized system. Neoliberalism appears more like a project in crisis rather than a regime. Yet, and despite claims of economic unsustainability and lack of substantive democracy, neoliberalism remains the dominant ideology and, in many instances, the preferred political choice of the second decade of the twenty-first century.

This apparent contradiction begs the brief illustration of two points. First, as in the case of the crisis of Fordism, there is a time gap between the existence of visible signs of the regime's crisis and its eventual collapse. Habermas eloquently illustrated the limits of Fordism in the early 1970s. But that regime collapsed years later, at the end of the decade. Moreover, as the case of agri-food demonstrates, components of the Fordist regime were never abandoned as elements of Fordism continued to exist under neoliberalism. The point is that the crisis of legitimation of neoliberalism is a process that has begun but certainly has not ended.

Second, neoliberalism is a regime that has been and remains favorable to the ruling class. Attempts to alter current socioeconomic arrangements need to challenge the power of the dominant elites. As illustrated in recent literature (e.g., Crouch, 2011), the crisis of neoliberalism has strengthened rather than weakened the ruling class. The argument that financial institutions, transnational corporations, and monopolistic conditions have strengthened is compelling. Ideological claims that society cannot allow corporations to fail because of the catastrophic consequences that would follow are widely supported. This is a situation that indicates the level of power that the ruling elites continue to maintain in society. Contrary to the case of the Fordist regime where gains of subordinate classes were institutionalized and supported, the crisis of neoliberalism involves a system that has been created to counter the gains and strength of subordinate classes. As Harvey (2005) has eloquently argued, neoliberalism is a project of the restoration of the power of the capitalist class. As a result of these conditions, the immediate future sees the clash between the powerful ruling class against the continuously dissatisfied masses in a context in which neither the market nor the state can address the crisis. It is the unfolding of this contested terrain that will shape the evolution of society in the years to come. Simultaneously, it represents a historical opportunity to replace social arrangements that have worsened the conditions of numerous groups in society with more just and equitable ones.

Notes

1 According to Habermas (1975), legitimation crises occur when the "steering mechanisms" of a society are ineffective. They fail to provide adaptation and social integration. Dwelling on Parsons, adaptation refers to the ability of the economy to distribute wealth. Social integration refers to the existence of adequate consensus on the dominant ideology and the state actions. Integration depends on the existence of effective mechanisms of domination that employ the normative structures-value systems deployed by the state and the mass media to create norms and identities that foster loyalty to the regime.

2 For a definition of Fordist capitalism, see Aglietta (1979) but also Harvey (1989) and Antonio and Bonanno (1996). To be sure, Habermas does not use the word Fordism in his text. But he speaks about "Advanced Capitalism." In subsequent works, he directly refers to this period as "Fordism." See Habermas (2002, 2009).

3 The original German language version of this book, "Legitimationprobleme im Spätkapitalismus," appeared in 1973. A host of foreign language translations followed, including the 1975 English translation. This version was prepared by Thomas McCarthy, who subsequently translated a significant number of works written by Habermas.

4 Habermas contends that in advanced capitalism, it is the state that is charged with the task of establishing social legitimation. Contrary to competitive liberal capitalism, in advanced capitalism, state intervention is a central condition for socioeconomic growth and the maintenance of social stability as it is the state that intervenes to correct the unwanted consequences of the functioning of the market.

5 Like Habermas (2002), Ulrich Beck (2006) has argued that global transformations have not only weakened the nation-state as a political entity but have created a "methodological revolution" in which the nation-state-centered paradigm of investigation is increasingly inadequate to address issues of democracy and legitimacy.

6 Even among those who see a positive resolution of the 2008 financial crisis (see Porter, 2011), the creation of a regulatory authority that can intervene to control the market appears as a necessary condition for future growth and stability.

7 Responding to criticisms of neoliberalism, social commentator Thomas Friedman (2011) maintains that opposition to neoliberal globalization does not recognize the beneficial components of this process. He contends that huge flows of ideas, innovations, new collaborative possibilities, and new market opportunities are available. While corporations can source the globe for less expensive resources, individuals can benefit from this system as well. Anyone from a village in Africa, for instance, can take a course from Stanford, he contends. The opportunities are as significant as the problems.

8 See the 2011 special issue of the *Journal of Rural Social Sciences*, vol 25, no 3 and the two 2013 special issues of *the International Journal of Sociology of Agriculture*

and Food vol. 20, issues 1 and 2 for summaries of research on the consequences of private certification. Also see Harvey (2005) on the role of nongovernmental organizations (NGOs) in the process of governance. For Harvey, while NGOs often claim that they protect the interests of the disenfranchised, there is no connection between these claims and the actual desires of disenfranchised groups. For the consequences of certification on labor relations, see Bonanno and Cavalcanti, 2012.

9 Political elections in major European countries (France 2012; Italy 2013) as well as in the United States (2012) resulted in the success of candidates and parties that called for the end of austerity measures and the resurrection of state spending as a tool for economic revitalization.

10 More specifically, financialization indicates those processes that reduce all value produced into financial instruments. Under neoliberalism, corporate agents have increasingly operated to produce new financial instruments by creating new assets or by combining existing assets and marketing these repackaged entities to investors.

11 The structuralist argument advanced by the Marxian literature contended that the rate of profit is historically higher in manufacturing than in farming. Accordingly, unless controlled and stimulated by the state, investment would move away from agriculture to be employed in more profitable—yet less strategic—sectors. It was concluded that the intervention of the state in agriculture has been a fundamental aspect for the stability and growth of capitalism.

References

Aglietta, M. (1979) *A Theory of Capitalist Regulation*, New Left Books, London.

Antonio, R. J. and Bonanno, A. (1996) "Post-Fordism in the United States: The poverty of market-centered democracy," *Current Perspectives in Social Theory*, vol 16, pp. 3–32.

AP (Associated Press). (2012) "The typical CEO made $9.6 million last year," Associated Press, New York.

Beck, U. (2006) "The cosmopolitan state: Redefining power in the global age," *International Journal of Politics, Culture and Society*, vol 18, pp. 143–59.

Bhagwati, J. (2004) *In Defense of Globalization*, Oxford University Press, New York.

Bonanno, A. and Cavalcanti, J. S. B. (2011) *Globalization and the Time-Space Reorganization*, Emerald, Bingley, England.

Bonanno, A. and Cavalcanti, J. S. B. (2012) "Globalization, food quality and labor: The case of grape production in northeastern Brazil," *International Journal of Sociology of Agriculture and Food*, vol 19, no 1, pp. 1–19.

Bonanno, A. and Constance, D. H. (2008) *Stories of Globalization: Transnational Corporations, Resistance and the State*, Pennsylvania State University Press, University Park, PA.

Buttel, F. and Newby, H. (eds) (1980) *The Rural Sociology of the Advanced Societies*, Allanheld Osmun, Montclair, NJ.

Byrne, J. (ed.) (2012) *The Occupy Handbook*, Back Bay Books, New York.

Cohen, S. S. and De Long, J. B. (2010) *The End of Influence: What Happens When Other Countries Have the Money*, Basic Books, New York.

Crouch, C. (2011) *The Strange Non-death of Neo-liberalism*, Polity Press, Cambridge, UK.

Epstein, G. A. (ed.) (2006) *Financialization and the World Economy*, Edward Elgar, Northampton, MA.

Erturk, I., Froud, J., Johal, S., Leaver, A., and Williams, K. (eds) (2008) *Financialization at Work: Key Texts and Commentary*, Routledge, London.

Firebaugh, G. and Goesling, B. (2004) "Accounting for the recent decline in global income inequality," *American Journal of Sociology*, vol 110, no 2, pp. 283–312.

Freedman, D. (2012a) "Rice growers fear aid drying up," *Houston Chronicle*, June 18, pp. A1–9.

Freedman, D. (2012b) "Rice support is on the table," *Houston Chronicle*, December 2, pp. D2–4.

Friedman, M. ([1962] 1982) *Capitalism and Freedom*, University of Chicago Press, Chicago, IL.

Friedman, T. (2011) "Which road was path to Wall Street protesters?," *Houston Chronicle*, October 12, p. B7.

Greenwood, D. (2011) "The problem of coordination in politics: What critics of neoliberalism might draw from its advocates," *Polity*, vol 43, no 1, pp. 36–57.

Habermas, J. (1975) *Legitimation Crisis*, translated by Thomas McCarthy, Beacon Press, Boston, MA.

Habermas, J. (2002) "The European nation-state and the pressure of globalization," Pp. 217–34 in P. De Greiff and C. Cronin (eds) *Global Justice and Transnational Politics*, MIT Press, Cambridge, MA.

Habermas, J. (2009). "Life after bankruptcy: Interview with Jürgen Habermas," *Constellations*, vol 16, no 2, pp. 228–34.

Habermas, J. (2012) *The Crisis of the European Union: A Response*, Polity Press, Malden, MA.

Hamilton Project. (2012) "Median annual earnings since 1964," Hamilton Project, www.hamiltonproject.org/multimedia/charts/median_annual_earnings_since_1964/, accessed September 21, 2012.

Harvey, D. (1989) *The Condition of Postmodernity*, Basil Blackwell, Oxford.

Harvey, D. (2005) *A Brief History of Neoliberalism*, Oxford University Press, New York.

Harvey, D. (2010) *The Enigma of Capital: And the Crises of Capitalism*, Oxford University Press, New York.

Helleiner, E. (2010) "A Bretton Woods moment? The 2007–2008 crisis and the future of global finance," *International Affairs*, vol 86, no 3, pp. 619–36.

Hirst, P. and Thompson, G. (1996) *Globalization in Question*, Polity Press, Cambridge, UK.

Ikeda, S. (2004) "Japan and the changing regime of accumulation: A world-system study of Japan's trajectory from miracle to debacle," *Journal of World System Research*, vol 10, no 2, summer, pp. 363–94.

Johnston, J. and Szabo, M. (2011) "Reflexivity and the Whole Foods Market consumers: The living experience of shopping for change," *Agriculture and Human Values*, vol 28, no 3, pp. 303–19.

Krippner, G. R. (2011) *Capitalizing on Crisis: The Political Origins of the Rise of Finance*, Harvard University Press, Cambridge, MA.

Krugman, P. (2012) *End this Depression Now*, Melrose Road Partners, New York.
Lapavitsas, C. (2012) *Crisis in the Eurozone*, Verso, London.
Lupel, A. (2005) "Tasks of a global civil society: Held, Habermas and democratic legitimacy beyond the nation-state," *Globalizations*, vol 2, no 1, pp. 117–33.
Lynn, B. C. and Longman, P. (2010) "Who broke America's jobs machine?," *Washington Monthly*, March 4, www.washingtonmonthly.com/features/2010/1003.lynn-longman.html#Byline, accessed June 10, 2011.
McNally, D. (2011) *Global Slump: The Economics and Politics of Crisis and Resistance*, PM Press, Oakland, CA.
Mottura, G. and Pugliese, E. (1980) "Capitalism in agriculture and capitalist agriculture," Pp. 88–106 in Buttel and Newby, *The Rural Sociology of the Advanced Societies*.
Nardone, E. and McDonough, T. (2010) "Global neoliberalism and the possibility of transnational state structures," Pp. 168–94 in T. McDonough, M. Reich, and D. Kotz (eds) *Contemporary Capitalism and Its Crises*, Cambridge University Press, New York.
O'Connor, J. (1974) *The Fiscal Crisis of the State*, St. Martin's Press, New York.
Offe, C. (1985) *Contradictions of the Welfare State*, MIT Press, Cambridge, MA.
Ohmae, K. (1995) *The End of the Nation State: The Rise of Regional Economies*, Free Press, New York.
Palley, T. I. (2007) "Financialization: What it is and why it matters," *Levy Economic Institute Working Paper Collection*, Working Paper No. 525, December, pp. 1–31.
Pechlaner, G. and Otero, G. (2010) "The neoliberal food regime: Neoregulation and the new division of labor in North America," *Rural Sociology*, vol 75, no 2, pp. 179–208.
Porter, T. (2011) "Public and private authority in the transnational response to the 2008 financial crisis," *Policy and Society*, vol 30, pp. 75–184.
Raulet, G. (2011) "Legitimacy and globalization," *Philosophy Social Criticism*, vol 37, no 3, pp. 313–27.
Robinson, W. I. (2004) *A Theory of Global Capitalism: Production, Class and State in a Transnational World*, Johns Hopkins University Press, Baltimore, MD.
Rodefeld, R., Flora, J., Voth, D., Fujimoto, I., and Converse, J. (1978) *Change in Rural America: Causes, Consequences, and Alternatives*, C. V. Mosby Company, St. Louis, MO.
Skalicky, M. (2012) "A rice farmers' view of the U.S. fiscal cliff. U.S. agriculture needs level field to compete globally," *Houston Chronicle*, December 12, p. B7.
Smith, J. (2012) "The best and worst paying jobs in America," *Forbes*, May 15, www.forbes.com/sites/jacquelynsmith/2012/05/15/the-best-and-worst-paying-jobs-in-america/, accessed May 16, 2012.
Sorkin, A. R. (2012) "Occupy Wall Street: A frenzy that fizzled," *New York Times*, September 17, dealbook.nytimes.com/2012/09/17/occupy-wall-street-a-frenzy-that-fizzled/, accessed September 23, 2012.
Steffy, L. (2011) "Behind the protest lie some very mainstream concerns," *Houston Chronicle*, October 21, pp. D1–8.
Steger, M. B. and Roy, K. R. (2010) *Neoliberalism: A Very Short Introduction*, Oxford University Press, Oxford.

Stockhammer, E. (2007) "Some stylized facts on the finance-dominated accumulation regime," *Political Economy Research Institute*, Working Paper series 142, University of Massachusetts, Amherst.

Tabb, W. K. (2010) "Financialization in the contemporary social structure of accumulation," Pp. 148–67 in T. McDonough, M. Reich, and D. Kotz (eds) *Contemporary Capitalism and Its Crises*, Cambridge University Press, New York.

Underhill, G. and Zhang, X. (2008) "Setting the rules: Private power, political underpinnings, and legitimacy in global monetary and financial governance," *International Affairs*, vol 84, no 3, pp. 535–54.

Van Gelden, S. (ed.) (2011) *This Changes Everything: The Occupy Wall Street and the 99% Movement*, The Positive Futures Network, New York.

Volscho, T. M. and Kelly, N. J. (2012) "The rise of the super-rich: Power resources, taxes, financial markets, and the dynamics of the top 1 percent, 1949 to 2008," *American Sociological Review*, vol 77, no 5, pp. 679–99.

2

HOW NEOLIBERAL MYTHS ENDANGER DEMOCRACY AND OPEN NEW AVENUES FOR DEMOCRATIC ACTION

Lawrence Busch

Introduction

Although I daresay that much of the world's population is largely unaware of it, we live in a world that has been made in large part by neoliberalism. Whether we find this a good or a bad state of affairs, we must admit that this is an extraordinary feat of social engineering. Put in its simplest form, over the last sixty to seventy years, and certainly building on the shoulders of giants, neoliberals have been able to create two interlocking myths: the myth of the market and the myth of government. The former myth asserts that markets always lead to the best, freest, and most efficient outcomes in human affairs. The latter asserts that to promote freedom, government must be radically limited to the promotion of "free" markets and regulation from a distance. Both myths have been widely applied in the agri-food sector and elsewhere, although their application has been uneven across nations, has often met with resistance (e.g., protests against Walmart's destruction of small town businesses), and has resulted in sometimes contradictory positions (e.g., farmers supporting "free" markets even as they lobby for continuing price supports). Moreover, the application of these myths has not only led to the rapid expansion of trade but also to the financial crisis, widespread unemployment, growing hunger, hollowing out of the state, and weakening of democratic governments. But it is important to emphasize that (1) we cannot do without myths; all societies have them, and (2) as W.I. Thomas wrote, "If men define situations as real, they are real in their consequences" (1928, p. 572); thus, the enactment of neoliberal myths have consequences that are independent of their truth values.

Origins

Although it has roots earlier, neoliberalism can be traced back to an invitational colloquium held in Paris in 1938.[1] Organized by the philosopher of logic and mathematics Paul Rougier, it honored Walter Lippmann (1938), who had recently published *The Good Society*. Participants included sociologists Alfred Schütz and Raymond Aron, Austrian economists F.A. Hayek and his mentor Ludwig von Mises, philosopher Michael Polanyi, several American and German liberal economists, Lippmann himself, and others from the banking and business community. At the colloquium, Rougier suggested that a new liberalism was needed, a *neo*liberalism. Only by reinventing liberalism would the world be able to withstand the onslaught of Nazism, fascism, communism, and the growing state-led economies in the United States, France, and Britain. Unlike the older laissez-faire approach to liberalism, which emphasized the creation of a space from which the state was largely excluded, the new liberalism would be aided by direct market-making interventions from the state. The end result, it was argued, would be greater individual liberty. It was agreed to form a Centre for the Renewal of Liberalism in Paris to develop these ideas. However, Paris was occupied soon after, and the participants at the colloquium, including those from France, were scattered across the world.

Hayek joined the faculty of the London School of Economics and while there, in the midst of World War II, he wrote *The Road to Serfdom* ([1944] 2007). Although he initially had difficulty in finding a suitable publisher, *Reader's Digest* eventually serialized it, and it became a best-seller. In it, Hayek made his case for a market-based society. In particular, Hayek argued that everyone's knowledge was incomplete, but that the market—a human invention—could collect all the knowledge of its participants and summarize it in a price. Prices could thus provide signals to both buyers and sellers and allow coordination without perfect knowledge on anyone's part. In contrast, according to Hayek, since knowledge was incomplete, government planners could never have sufficient knowledge to plan adequately. Hence, means were needed to prevent governments from engaging in such planning. What would do this job was an international organization that would have as its goal the facilitation of trade, even as it barred governments from intervening in the marketplace. Moreover, when the war ended in 1945, he led the establishment of the Mont Pelerin Society (2006), an organization devoted to debate among neoliberals as to how society should be restructured so as to make the neoliberal model become reality. For several years, the Mont Pelerin Society stumbled, even to the point of approaching dissolution. But despite disputes among its members over what neoliberalism was to be, it eventually stabilized and became one of many venues where neoliberal ideas were promoted.

In addition, after the war, Hayek moved to the Law School at the University of Chicago where he and Milton Friedman, among others, worked out some of the more technical details of neoliberalism. With generous support of several foundations, they produced a justification for a world dominated by markets. Somewhat later, the Heritage Foundation and the American Enterprise Institute were created, thereby furthering the neoliberal agenda.[2]

Neoliberalisms

Older philosophies had assumed that complete, absolute knowledge was possible. Descartes ([1637] 1956), for example, believed that by following his rules of method, complete knowledge of the world would be available to all within a century. Neoliberals, in contrast, were among the first postmodernists in that they argued that one's knowledge was always incomplete. But they went far beyond that, arguing that the market could solve the problem of incomplete knowledge, even as it protected individual liberty. Specifically, the market would determine the correct price of every good and service by virtue of aggregating supply and demand. Goods would thereby be distributed based on their relative scarcity. Moreover, prices would send signals to producers telling them to produce more or less of a given good or service. Importantly, no buyer or seller would ever need to know how the price came to be such as it was. Indeed, regardless of whether an agricultural commodity, such as wheat, brought a high price as a result of drought, floods, disease, insect pests, a change in consumer preferences, or new uses for the grain, the market price would act as a signal, telling producers to produce more. Conversely, regardless of whether excellent weather, new pesticides, declining consumer demand, or new substitutes for wheat caused prices to decline, the market price would act as a signal telling producers to plant less the following season.

Moreover, they developed a model of society that differed considerably from that of the nineteenth-century liberals, even as it accepted the assumptions of *homo œconomicus*, that is, of methodological individualism combined with an entrepreneurial self. That model abandoned the sharp division between state and market enshrined in the principles of laissez-faire and replaced it with a state that was activist in its promotion of markets, yet restricted in its provision of non-market goods and services.

That said, in citing the basic tenets of neoliberalism, one needs to be careful not to reify them. As Mirowski and Plehwe (2009) have argued, neoliberalism is best understood as a "thought collective"—that is, a group of like-minded people contributing to an ongoing debate about what a market society should look like and not a set of principles that are inviolate. Moreover, the diversity of neoliberalism as practiced in various nations and even regions reflects its theoretical diversity. In short, one can say that there are *neoliberalisms*, each linked to the others in some ways, but each distinctive in how it is theorized and applied.

Despite this theoretical and practical variation, there do appear to be some key points shared widely among neoliberals. These consist of:

1 Recognition of the limits of human knowledge: *no one person, organization, or government can know enough to plan adequately.* Hence, neoliberals argue, for example, that government programs for agriculture—from crop insurance to soil conservation to commodity payments—can never be designed in such a manner as to ensure the desired outcomes. There is simply too much missing data, too many variables that cannot be measured or of which government officials are simply unaware to design such programs adequately.

2 However, despite the limits of human knowledge, neoliberals claim to have found an Archimedean point, based not on empirical but on mathematical and logical knowledge. They argue that *only the market, as demonstrated by mathematical or logical models*, can arrive at an optimum solution in which freedom is equated with free exchange. Importantly, for Hayek (1937, 1943, 1969), mathematical truths need not be verified by empirical observation. As his mentor, Ludwig von Mises ([1933] 1978, p. 13), explained it:

> The science of human action that strives for universally valid knowledge is the theoretical system whose hitherto best elaborated branch is economics. In all of its branches this science is a priori, not empirical. Like logic and mathematics, it is not derived from experience; it is prior to experience. It is, as it were, the logic of action and deed.

Hayek (1943, p. 11), far and away von Mises's star student, made much the same point several years later:

> All that the theory of the social sciences attempts is to provide a technique of reasoning which assists us in connecting individual facts, but which, like logic or mathematics, is not about the facts. It can, therefore, and this is the second point, never be verified or falsified by reference to facts.

Murry Rothbard (1957) made this argument even more forcefully in an article entitled "In Defense of 'Extreme Apriorism,'" while Milton Friedman ([1962] 2002, p. 120) has put the argument in somewhat weaker terms, likening competition to Euclidean geometry.

The price mechanism is used as an example. Producers and consumers will respond to price without knowledge as to why the price is whatever it is. Moreover, even if they are aware of the reasons for price changes, their awareness has little or no effect on their behavior in the marketplace. Hence, if the price of tomatoes increases due to drought, disease, or tax increases, I will similarly restrict my purchases of tomatoes accordingly.

3 To make society fit the logical model, governments must be reshaped so as to make them positively support the formation of (quasi)markets, encourage competitions of all sorts and govern indirectly through markets or market-like mechanisms. From the neoliberal perspective, there is a need to develop what Simons ([1934] 1948) called "a positive programme for laissez-faire." This goes far beyond the liberalism of the eighteenth and nineteenth centuries in that it actively involves governments in market formation and even turns otherwise nonmarket institutions into (quasi)markets by introducing competition wherever possible.

Thus, under the banner of public choice theory, government has been reconceptualized as a market (e.g., Buchanan, 1968). Legislators and other elected officials are said to be vying solely for their own interests (including that of remaining in office) rather than any notion whatever of the public good. Note that once one assumes the stance of methodological individualism, such an argument makes perfect sense. Perhaps Hobbes ([1642] 1991, p. 112–13) said it best: “All society therefore is either for gain, or for glory; that is not so much for love of our fellows, as for love of ourselves.” Similarly, unelected officials are said to be interested more in promoting their personal interests than in carrying out the functions of government. Hence, given this dismal view of government, little good can come from governmental action, so the best policy is to limit direct government action to those functions that are indispensable—for example, national defense and maintenance of law and order.

What is claimed for elected officials and bureaucrats is also said to hold for courts, although in somewhat different form. Here neoliberals disagree among themselves. On the one hand, Ronald Coase (1988) argued that most legal conflicts are about conflicting property rights. Among other things, Coase used a conflict between ranchers and farmers to illustrate the problem. If wandering cattle cause damage to a neighbor’s crops, then the question for the courts is whether the rancher’s rights trump those of the farmer, or the farmer’s rights trump those of the rancher. Coase argued that if there are no transactions costs (i.e., no costs incurred from participating in the market), then there is no reason for government to be involved at all. One only need specify the rights of the parties and then let the market internalize the externality. Of course, Coase understood that often there are transactions costs, in which case government intervention might be needed.

On the other hand, US (and to a lesser extent other nations’) civil court decisions have been heavily influenced by neoliberal theory in the form of the compensation principle, also known as Kaldor-Hicks efficiency or wealth maximization. In keeping with the general neoliberal view that markets maximize efficiency, its proponents argue that compensation between the plaintiff and defendant should be resolved in a manner such that overall wealth is maximized. However, as Mercuro and Medema (2006, p. 27, authors’ emphasis) explain:

> The concept of the compensation principle was consciously developed without any requirement that the compensation actually be paid. In an attempt to maintain a positivist stance with respect to the evaluation of legal-economic policy, the idea was that it would be enough to show the policy generated gains sufficiently large to *potentially* compensate the losers.

Actual compensation may or may not take place but is excluded from discussion since (it is claimed) doing so would involve normative decisions, while economics is (allegedly) only about the facts. In practice, the courts have increasingly moved toward awarding compensation based on the wealth maximization approach. Cost-benefit analysis also employs this rule; the costs and potential benefits are calculated, but the compensation of the losers is left to the politicians (see, e.g., Porter, 1995).

Gary Becker expanded this general view by arguing that all behavior can be subjected to economic analysis. As he put it, "[T]he economic approach is clearly not restricted to material goods and wants, nor even to the market sector" (Becker, 1978, p. 6). In recent years, judge and legal theorist Richard Posner (2011), very much influenced by Becker, has become arguably the foremost proponent of this view of the relations between law and economics. Posner rejects the Coasian distinction between economic and noneconomic actions, arguing for an economic analysis of law. Put differently, for Posner, not only are the consequences of law relevant to economics, but economics can be used as a tool for analyzing and improving the legal system itself.

Given this view of government, it is hardly surprising that neoliberals should see government as something that must be minimized in order to protect liberty. As such, they have encouraged a variety of government policies designed to restrict government action or to make government operate only at a distance. These include:

- Diminishing and eventually eliminating government planning on the grounds that such planning (1) assumes perfect knowledge on the part of planners, (2) imposes the planners' ultimately arbitrary or self-interested values on the liberties of otherwise free actors, and (3) sends the wrong signals to markets leading to under- or overproduction of goods and services.
- Contracting out of as many government services as possible to the private sector on the grounds that the competition among private firms will result in doing the job better and more efficiently. For example, the USDA is currently seeking to reduce the number of government inspectors at poultry processing plants and shift that activity to the companies themselves (Food Safety and Inspection Service, 2012).
- Repealing banking regulations of various kinds (leading in part to the crash of 2008) so as to allow the market to do its magic.
- Governing at a distance by changing the roles of many government employees to that of (often underfunded) oversight of external contractors.
- Eliminating or reducing the scope of coverage of many government programs (e.g., welfare, public housing) as these are seen to provide unfair (subsidized) competition or to interfere with market competition among private firms.
- Promoting choice in government programs wherever possible as a means of promoting freedom. Hence, we are now faced with ever more complex choices that affect nearly every aspect of our lives: To which school shall I send my children? Which pension plan shall I invest in? What

health insurance policy should I purchase? Which social networks should I join so as to further my career? Which certifications should I look for in purchasing food? Fair trade? Organic? Rainforest friendly? Which company's recommendations should I (as a farmer) follow when deciding about agricultural practices?

4 *International bodies must be created that serve simultaneously to promote markets internationally and to limit the power of States* outside the realm of promoting markets (e.g., the World Trade Organization [WTO], World Intellectual Property Organization, International Monetary Fund, World Bank). This takes several forms. We have the so-called Washington Consensus on "structural adjustment" (see, by way of comparison, Williamson, 2004–5) whereby state-led development is blocked by lending institutions. We also have severe limits on what the WTO can consider in making decisions about trade; in particular, environmental and labor issues are excluded. Finally, we have decisions made by the World Bank in providing loans to poor countries, usually putting numerous strings on those loans so as to ensure that non-neoliberal objectives are excluded (Goldman, 2005).
5 *The notion of "social justice" is rejected* as insufficiently clear as a basis for either projects of positive law or court decisions (Hayek, 1976). Hence, "freedom to" is promoted, while "freedom from" is demoted.[3] Hence, one has the freedom to choose to buy an increasingly large range of fresh produce at the local supermarket, but freedom from poverty and hunger is harshly limited.
6 *The notion of monopoly is reinterpreted by focusing not on the concentration of power (as Theodore Roosevelt did), but on the lack of competition.* As Hayek (1979, p. 83) put it, "it is not monopoly as such but only the prevention of competition which is harmful." Hence, starting in the 1960s, US prosecutors and courts (Mercuro and Medema, 2006) (as well as those in Britain [Competition Commission, 2000] and other nations) have tended to treat monopoly as problematic only when a clear increase in consumer prices can be demonstrated. The mere fact of market dominance (e.g., Walmart's dominance of US food retailing) is generally seen as fleeting and irrelevant as long as it does not result in rising consumer prices, which would suggest inefficiencies had been introduced into the market. As a consequence, we have witnessed a rapid global concentration of ownership and control in the farm input, output processing, and food retailing industries.
7 *Reconstruction of the self as an isolated, entrepreneurial self.* As Mead ([1934] 1962), Schutz (1970) and others argued many years ago, the self is fundamentally social. Moreover, the self is multiple; the self that we display to others varies markedly depending on the situation in which we find ourselves. Thus, a farmer will display a different self to a fertilizer salesperson than to a family member. However, the neoliberal project aims to make hegemonic the entrepreneurial, calculating, isolated self. This has been accomplished through (1) the promotion of choice in all things, (2) a strong emphasis on

competition in all human activities, and (3) the growing necessity to ceaselessly *calculate and invest* in one's future (through individually supported educational activities, pensions, health care, career decisions, and the like) (Rose, 1996; Foucault, 2008). Thus, for example, small farmers must weigh the benefits of growing higher value perishable produce for a global market against the risks that they may not meet the standards imposed by supermarket chains, that the weather might not cooperate, that the market might collapse, that transport to a packing facility might not be available, and so on. As Langley (2007, p. 72) puts it, "[o]n the one hand, (neo)liberal government respects the formal freedom and autonomy of subjects. On the other hand, it governs within and through those independent actions by promoting the very disciplinary technologies deemed necessary for a successful autonomous life."

***

It is also important to emphasize that the neoliberal program is a normative program described in positive terms. Put differently, a mathematical model is built, which is (claimed to be) sufficient unto itself. Then, society is to be "corrected" so that it looks more and more like the model. This has been done in numerous subfields of economics. For example, Gary Becker (1964, 1993; Becker and Becker, 1997) has introduced the model of human capital—that is, of each person as an entrepreneur of her- or himself. Each of us is to spend our life striving to acquire the right education, the right friends, even the right marriage partner. Similarly, we are to evaluate engaging in criminal behavior as merely another risk to be taken (or not) in the achievement of our individual goals and preferences. Put differently, the neoliberal program is designed to make hegemonic calculating selves who are incessantly involved in weighing risks, costs, and benefits of various investments in their individual futures.

Consequences

Overall, these policies and practices have had a dual effect on democratic governance and on specific policies: On the one hand, this has been accomplished largely by redefining freedom narrowly as the ability to act in the marketplace; governments are defined as institutions that limit freedom instead of letting individuals decide for themselves. Indeed, neoliberals have attempted to transform all social relations into market relations by actively destroying social solidarities (Bourdieu, 1998) and by enacting policies that demand that we all become more entrepreneurial. In so doing, they have created the illusion that *by definition* government is largely an impediment to freedom.

On the other hand, governments have been set up to be incompetent, as they rely more and more on private contractors to do much of their business while deskilling and discouraging many government officials. Thus, government officials have found themselves overseeing the work of myriad contractors without the necessary skilled staff to ensure that contractors carry out their contracts in accordance with the law; in short, government officials become auditors of the paper trail left by contractors. Moreover, using the principles of New Public Management, they are required to design ever more specific accountability regimes with ever more clearly specified deliverables (see Steven A. Wolf, this volume).

Together, the enactment of these strategies to transform markets and states has had profound effects on the agri-food sector over the last several decades. As I illustrate below, this transformation has extended from research to production to processing and retailing. For example, with respect to research, we have seen a marked shift from the so-called formula funds provided to researchers in the United States at State Agricultural Experiment Stations under the Hatch Act and similar support provided to agricultural researchers elsewhere, toward the promotion of competitive funding. The former involves planning under bureaucratic rules, while the latter involves introducing competition and market-like forms of organization. (Importantly, competitive funding means that researchers spend vast amounts of their time in writing grant applications.) In some nations, such as the United Kingdom, much of the public research infrastructure (e.g., the Cambridge Plant Breeding Institute) has been sold to private companies on the grounds that they can engage in this kind of research more efficiently, effectively, and competitively than can governmental research institutions.

Furthermore, in the United States and many other nations, public agricultural research has been moved upstream. Put differently, where once public researchers created products, techniques, and management strategies that were of direct use to farmers and ranchers, today their research is largely confined to "upstream" innovations that can only be used if filtered through the private input and/or processing sectors (Wolf and Zilberman, 2001). This has the obvious consequence of allowing private firms to compete to market products that have their origins in public research institutions. Less obvious is how it filters public research by putting control of the research agenda firmly in the hands of private sector actors. Only those findings of public research that provide opportunities for profit-seeking firms will likely be implemented. Hence, promising alternatives such as apomixes are for all practical purposes abandoned (Charles, 2003).[4] And, users will pay twice in effect—once for the public research and later for the private product.

Similarly, agricultural extension services worldwide have been downsized or privatized and thereby subjected to market discipline (Wolf, 1998). Hence, many functions previously provided free of charge by extension agencies are now performed for a fee by private companies. In New Zealand, extension activities have been completely privatized, such that individual farmers must now pay for those formerly freely available services. In some poor nations, extension services have simply collapsed due to a lack of funds. In short, after decades of slow but steady growth, the public agricultural research and extension agencies of many nations have been radically weakened under structural adjustment.

Also of note is that the last several decades have seen the transformation of numerous state-run marketing boards, such as the Canadian Wheat Board (CWB). The CWB was established in 1935 as the sole marketing system for wheat and barley in Canada; however, in 2012, marketing through the CWB became a voluntary activity for farmers (see Magnan chapter in this volume). Numerous other collective marketing agencies have been eliminated or radically modified, leaving farmers at the mercy of volatile global markets.

Lucrative markets for smallholders that were linked to particular trading partners, such as the preferential marketing of African bananas to Western Europe, have weakened or disintegrated. Public seed multiplication and distribution agencies have been eliminated or replaced by private companies.

In poor nations, attempts have been made to replace the freely available and open-pollinated cultivars of the Green Revolution (themselves not without their own problems) with hybrids that must be bought annually from private companies at whatever price the market will bear. Often, when crops fail for whatever reasons, poor farmers find that they no longer have the means to grow a crop the following year. Nor have the seeds available through the market always been subject to quality standards that would ensure that what is on the package label is actually inside the package (Stone, 2004).

Finally, the neoliberal interpretation of monopoly combined with the decline in tariffs and quotas has opened much of the world to a few upstream and downstream corporations that now dominate the agri-food and other sectors. Since capital is heavily concentrated in the industrialized nations of the North and is easily transferred globally, reducing or eliminating tariffs and quotas has the effect of allowing northern firms to purchase promising southern firms and/or to simply extend their reach far beyond national boundaries. With few exceptions, the reverse is rarely the case. Therefore, on the input side, Monsanto now dominates the global seed industry, while a handful of firms dominate the agrochemical industry (e.g., Dow, DuPont, BASF). On the downstream side, a few large supermarket chains (e.g., Walmart, Tesco, Carrefour, Royal Ahold) now dominate not only in the wealthier nations but increasingly even in poorer nations (Busch and Bain, 2004; Michelson et al., 2011; Reardon and Timmer, 2012).

In short, the neoliberal perspective has become the dominant one to explain and transform much of the world.

A Critique and an Appreciation

A Critique

Since countless others have critiqued neoliberalism (e.g., Harvey, 2005; Apple, 2006; Klein, 2007; Foucault, 2008; Guthman, 2008; Lazzarato, 2009; Mirowski and Plehwe, 2009; Pestre, 2009; Stanfield and Carroll, 2009; Dean, 2010;

Giroux, 2011), I shall only briefly note some of the more obvious issues posed by neoliberalism:

- Neoliberals mistakenly presume in Cold War terms that there are only two ways to organize society—via the market or via the state. But, in point of fact, markets are themselves creatures of the state. At the very least, they depend on (1) property law (including intellectual property law), (2) corporate (limited liability) law, (3) contract law, and (4) antifraud law. Moreover, they depend on the active and consistent enforcement of these laws.
- Neoliberals mistakenly assume that all markets are the same. Yet, the market is merely a useful fiction. Farmers' markets, the real estate market, the stock market, and commodities markets all have different dynamics and are necessarily governed by specialized laws that apply only to these markets (e.g., laws for food safety, land purchase, application of farm chemicals, purchase of securities, etc.).
- For neoliberals, government is the only source of governance. But governance is part of the human condition. Among the many institutions that engage in governance are families, businesses, corporations, private voluntary organizations, foundations, educational institutions, religious groups, markets, and governments (although differently through legislative, judicial, executive, and regulatory agencies). Hence, farmers are governed by Monsanto every time that they sign a licensing agreement for the use of genetically modified seeds. They are governed by supermarkets when they agree to abide by certifications required by those chains. Similarly, consumers are governed by supermarkets every time they enter a store. In addition, the world of technical artifacts governs us by allowing certain actions and disallowing others. For example, corn harvesters allow rapid harvest of vast acreages, but only if the corn is grown in rows and grows only to a certain height. Robotic dairies allow farmers to escape the discipline of milking cows, but only if the cows are such that they learn how to use the system and the system itself is maintained. Canned foods have a long shelf life but must be purchased based on what is written on the label and can only be easily opened with a can opener (Cochoy, 2002).
- Focusing on reducing the scope of government has had the unintended effect of promoting an explosion of another form of governance: private standards (Busch, 2011). In the agri-food sector alone, we now see individual firms (e.g., Walmart), industry associations (e.g., The Consumer Goods Forum), groups of supermarket chains (e.g., GlobalGAP), and nongovernmental organizations (e.g., The Rainforest Alliance) all involved in standards development so as to govern various actors in agri-food supply chains. These standards (1) often conflict with one another, (2) may confuse both producers and purchasers, and (3) are usually developed with only minimal attention to democratic governance.

- Farming out essential government services has several unintended consequences. On the one hand, it makes it necessary for government employees to oversee technical activities even as it reduces their competence to understand the details of those activities. Thus, the Agency for International Development, which once had a large staff of agricultural experts, now must oversee contractors, but lacks much technical expertise of its own. Instead, it relies on formal paper evaluations (with occasional well-orchestrated field trips) to evaluate the effectiveness of its programs. On the other hand, it encourages contractors to lobby the state for higher profits and larger contracts.
- The putting into practice of neoliberals' methodological individualism has numerous undesirable consequences, including (1) declining support for democratic government (Brown, 2006; Revault d'Allonnes, 2010) and (2) the rise of various forms of antidemocratic extremism. (The current unrest and rise of an overtly fascist party in Greece is a possible vision of things to come.)
- Perhaps most importantly, neoliberals assume that the market world can be extended to all aspects of human activity. In so doing, they attempt to force the multidimensional world in which we live into the box provided by the market. As Boltanski and Thévenot ([1991] 2006) have noted, in everyday life, we rely on numerous worlds of justification to explain our actions—among them the civic, opinion, industrial, inspirational, environmental, and domestic worlds, in addition to the market world. And, as Walzer (1983) argues, a society in which markets invade all spheres of social life approaches a totalitarian society.
- Market failure is far more commonplace than neoliberals would like to admit. Markets do a poor job of addressing many of the major issues of our day, including global warming, declining species diversity, and the rising costs of food in poor nations. Addressing these issues requires governments and institutional changes as they have consequences for the necessities of human life and on future generations who cannot be represented in contemporary markets (Bromley, 1998).

An Appreciation

Most of the critics of neoliberalism content themselves with noting the flaws in the neoliberal position as well as the sometimes catastrophic consequences of the implementation of neoliberal policies (e.g., Klein, 2007). However, this amounts to throwing out the baby with the bathwater. In point of fact, there are several things we can learn from the neoliberal experiment:

- The neoliberals are certainly correct in their view that knowledge is always incomplete. The very incompleteness of knowledge means that we can never be absolutely sure that we are doing the right thing. (This, of course, includes both the policies of neoliberals and of their critics!)

- The neoliberals are constructionists, as are many, perhaps most, sociologists. They believe that the social world—including markets—is not given but made. This implies that it might be made differently.
- Moreover, the neoliberals take their constructionism seriously. They actively engage in changing the world to make it fit the neoliberal model of the good society. Through the Mont Pelerin Society, the American Enterprise Institute, the Heritage Foundation, the American Legislative Exchange Council, and other neoliberal organizations, they have been able to build an extraordinary edifice of neoliberal governance.
- Neoliberalism is about the government of government (Dean, 2010). That is, neoliberals are reflexive about how government should be performed. In this respect, they differ substantially from neoclassical economists for whom governments as well as firms are often reified. And, they also distinguish themselves from those social scientists sometimes described as "dust-bin empiricists," who fail to recognize the constant tension between what is and what might be (Castoriadis, [1975] 1998).
- Even as neoliberals wish to establish a singular market order that would replace or become hegemonic over other social orders, they recognize the importance of customs, traditions, and laws in creating any form of social order. They also recognize that different laws and customs create different social orders.
- Finally, neoliberals appear to have accidentally stumbled on something first hinted at by symbolic interactionists (e.g., Mead, [1934] 1964) and phenomenologists (e.g., Schutz, [1932] 1967): If the self is social, then changes in institutions and laws not only change social relations (by promoting or inhibiting certain actions) but also transform selves.

Conclusions: Performing a Better Future

Ironically, the neoliberals have shown us (at least part of) the way toward enacting a better future. Although they have, knowingly or unknowingly, done this in a manner that has created an entrenched global elite while producing massive inequalities in the name of freedom, there is no particular reason why the world needs to be performed in this way. Other futures are both possible and more desirable.

Moreover, there is a growing literature on performativity and enactment. Austin (1962) has shown how language itself is performed through speech acts. Foucault (2008) has shown how neoliberal selves are constructed. More recently, Rose (1996) has shown how we participate in the enactment of our very selves and how the "psy" disciplines (e.g., psychology, psychiatry) are involved in those enactments. Butler (2007) has illustrated the complex ways in which gender is performed, as well as emphasizing how those who perform inadequately or "wrongly" may be disciplined. Callon (1998; Callon et al., 2007) and his colleagues have shown how economies are performed through the actions of economists and market participants. Mol (2002) has shown how even bodies are enacted. And Jasanoff (2005) and Latour (2005), among others, have shown how technoscience is performed in certain ways and not others.

These various approaches to the performance of society open new avenues for social change. On the one hand, they replace the structure/agency debate that has plagued the social sciences for decades with a perspective that shows how structures and agents only exist through their performance. Put differently, only when people—social beings—perform the world in certain ways does that particular world come into existence. Thus, only when we perform in certain ways as consumers, supermarket staff, and management do supermarkets come into existence. Moreover, when some choose to reject (all or part of) supermarkets, performing instead in farmers' markets, community gardens, or community supported agriculture, these come into existence as alternatives. On the other hand, performative approaches emphasize the constant tension between what is and what could be, between the now dominant myths of the market and government and new, transformed, reenacted myths that may prove to be better, more just, more democratic, and more effective in achieving our collective ends.

Hence, it appears that the key is to ask specifically what kinds of actions need to be taken to reenact society—actions not too different in form although very different in substance from those taken by the neoliberals. Among the tasks to be performed with respect to agri-food are:

1. Creation of multiple opportunities for deliberative democracy wherein ordinary citizens can identify shared goals, establish priorities, and discuss how to resolve issues of concern (e.g., Callon et al., 2009). Such opportunities might employ Habermas's (1981) notion of a communications community as an ideal towards which to strive. Indeed, the Cooperative Extension Service previously served to create those opportunities (Kirkendall, 1966) and might do so again.
2. Development of discussions and debates on goals, priorities, and policies in specific areas of human endeavor, such as agriculture and food. The key here is to involve all of the relevant publics. The difficulty in doing so is evident in the recent International Assessment of Agricultural Knowledge, Science and Technology for Development (IAAKSTD, 2008); the large seed/chemical companies participated initially but pulled out after it became apparent that their views would not be the dominant ones.
3. Promotion of forms of social solidarity among various constituent groups. For example, we should not forget that at one time farmers were in the lead in creating cooperative institutions to serve their mutual interests (Gray and

Mooney, 2009; Gray et al., 2001). Cooperatives may also serve the interests of consumers (e.g., Kimura, 2010).

4 Design of policy proposals as follows:

   a Constitutional changes. Sociotechnical changes have not been matched by constitutional changes. Hence, many constitutions are inadequate for grappling with these newly emerging problems. For example, innovations from genetic engineering and genomics have created considerable problems with respect to intellectual property protection. In many cases, this has left farmers responsible for the consequences of pollen drift across their property lines. Likely, these problems will only be resolved constitutionally.
   b Statutory changes. Many food safety problems of recent origin appear to be the result of inadequate practices and conflicts of interest among certifiers. Statutory changes are likely needed to determine just what role (if any) certifiers should have with respect to food safety, as well as what oversight might be needed by government agencies.
   c Actions to be taken by private voluntary associations. Voluntary associations can be effective means of social change or can inhibit such changes. At least two issues need to be raised here. On the one hand, one might ask to what degree voluntary associations adhere to democratic principles, both among their members and among the people they claim to serve. On the other hand, one might ask when it is more appropriate for such associations to lobby the state and when it is more appropriate to lobby firms. Guthman (2008) and Brown and Getz (2008) argue that lobbying firms puts such organizations squarely in the neoliberal camp even as they attempt to resist it.
   d Individual changes that can be taken to further already identified and tentatively agreed upon goals. Making such changes involves enhancing self-reflexivity by better understanding how institutions and laws make selves, as well as how selves are disciplined by such institutions and laws. For example, as consumers, individuals can decide to purchase most food at farmers' markets and can avoid certain stores and products to protest undesirable policies. And as citizens, individuals can join with others in protesting injustices, in demanding greater attention to the collective rights and wants of farmworkers and small farmers, and in supporting policies that are alternatives to those of neoliberals. However, they can only do this effectively if it is part of a larger social movement, in solidarity with others.

5 Preparation for the moment when the neoliberal policies once again create havoc. Given the fiscal chaos in Greece, and to a lesser extent Spain, Portugal, Ireland, and Italy, as well as the weaknesses in the Chinese economy and the weak "recovery" in the United States, as well as the various environmental and food crises that confront us, that moment may not be far off. (Some

might argue that it is already upon us.) Grappling with these issues will not be simply a matter of returning to the pro-state policies of the past; it will require debate and deliberation to develop alternatives now, not at some remote point in the future.

In closing, let me note that I am aware that this is hardly an easy task. But let us remember that the neoliberals have been quietly at work for seventy-five years or more. In the past, revolutionaries wished to change the world overnight. In nearly every instance, their attempts met with disaster, with terror, with oppression, and with death. Through greater understanding of the enacted character of human society, we can have insights into better, more effective, and more democratic means of transformation. As Slavoj Žižek (2008, p. 474) writes, "[w]e do not obey and fear power because it is in itself so powerful; on the contrary, the power appears powerful because we treat it as such."

Notes

1 For the proceedings of the colloquium, see Centre International d'Etudes Pour la Rénovation du Libéralisme (1939). For a summary of these events relative to the agri-food sector, see Busch (2010).
2 For more on the history of neoliberalism, see Foucault (2008), Mitchell (2005), Mirowski and Plehwe (2009), and Wapshott (2011).
3 For an extended analysis of how neoliberalism has changed the commonly accepted notions of justice, see Mercuro and Medema (2006) and Davies (2010).
4 Apomixis is the asexual reproduction of plants through seeds. Like hybrids, apomicts could incorporate desirable traits, but since, unlike hybrids, apomicts are clones of the parent plant, they would not have to be purchased anew each year. Ironically, apomixis appears most easily achieved using genetic engineering, but its widespread use would undercut the current and growing dominance of large seed companies.

References

Apple, M.W. (2006) *Educating the 'Right' Way: Markets, Standards, God, and Inequality*, Routledge, New York.

Austin, J. (1962) *How to Do Things with Words*, Cambridge University Press, Cambridge.

Becker, G.S. (1964) *Human Capital: A Theoretical and Empirical Analysis*, University of Chicago Press, Chicago, IL.

Becker, G.S. (1978) *The Economic Approach to Human Behavior*, University of Chicago Press, Chicago, IL.

Becker, G.S. (1993) "Nobel lecture: The economic way of looking at behavior," *Journal of Political Economy*, vol 101, no 3, pp. 385–409.

Becker, G.S. and Becker, G.N. (1997) *The Economics of Life: From Baseball to Affirmative Action to Immigration, How Real-World Issues Affect Our Everyday Life*, McGraw-Hill, New York.

Boltanski, L. and Thévenot, L. ([1991] 2006) *On Justification: Economies of Worth*, Princeton University Press, Princeton, NJ.
Bourdieu, P. (1998) "The essence of neoliberalism," *Le Monde Diplomatique* (English Edition), available at mondediplo.com/1998/12/08bourdieu, accessed November 4, 2007.
Bromley, D.W. (1998) "Searching for sustainability: The poverty of spontaneous order," *Ecological Economics*, vol 24, no 2–3, pp. 231–40.
Brown, S. and Getz, C. (2008) "Privatizing farm worker justice: Regulating labor through voluntary certification and labeling," *Geoforum*, vol 39, no 3, pp. 1184–96.
Brown, W. (2006) "American nightmare: Neoliberalism, neoconservatism, and de-democratization," *Political Theory*, vol 34, no 6, pp. 690–714.
Buchanan, J.M. (1968) *The Demand and Supply of Public Goods*, Rand McNally, Chicago, IL.
Busch, L. (2010) "Can fairy tales come true? The surprising story of neoliberalism and world agriculture," *Sociologia Ruralis*, vol 50, no 4, pp. 331–51.
Busch, L. (2011) *Standards: Recipes for Reality*, MIT Press, Cambridge, MA.
Busch, L. and Bain, C. (2004) "New! Improved? The transformation of the global agri-food system," *Rural Sociology*, vol 69, no 3, pp. 321–46.
Butler, J. (2007) *Gender Trouble: Feminism and the Subversion of Identity*, Routledge, New York.
Callon, M. (ed.) (1998) *The Laws of the Markets*, Basil Blackwell, Oxford.
Callon, M., Lascoumbes, P., and Barthe, Y. (2009) *Acting in an Uncertain World: An Essay on Technical Democracy*, MIT Press, Cambridge, MA.
Callon, M., Millo, Y., and Muniesa, F. (eds) (2007) *Market Devices*, Blackwell, Oxford.
Castoriadis, C. ([1975] 1998) *The Imaginary Institution of Society*, MIT Press, Cambridge, MA.
Centre International d'Etudes Pour la Rénovation du Libéralisme (ed.) (1939) *Le Colloque Walter Lippmann*, Librairie de Médicis, Cahier no. 1, Paris.
Charles, D. (2003) "Corn that clones itself," *Technology Review*, vol 106, pp. 33–5, 38, 40–1.
Coase, R.H. (1988) *The Firm, the Market and the Law*, University of Chicago Press, Chicago, IL.
Cochoy, F. (2002) *Une Sociologie du Packaging ou l'Âne de Buridan Face au Marché*, Presses Universitaires de France, Paris.
Competition Commission. (2000) "Supermarkets: A report on the supply of groceries from multiple stores in the United Kingdom," presented to Parliament by the Secretary of State for Trade and Industry by Command of Her Majesty, London, available at www.competition-commission.org.uk/rep_pub/reports/2000/446super.htm, accessed July 28, 2005.
Davies, W. (2010) "Economics and the 'nonsense' of law: The case of the Chicago antitrust revolution," *Economy and Society*, vol 39, no 1, pp. 64–83.
Dean, M. (2010) *Governmentality: Power and Rule in Modern Society*, Sage, London.
Descartes, R. ([1637] 1956) *Discourse on Method*, Bobbs-Merrill, New York.
Food Safety and Inspection Service, USDA. (2012) "Modernization of poultry slaughter inspection: Proposed rule," *Federal Register*, vol 77, pp. 4408–56.
Foucault, M. (2008) *The Birth of Biopolitics: Lectures at the Collège de France, 1978–79*, Palgrave Macmillan, New York.

Friedman, M. ([1962] 2002) *Capitalism and Freedom*, University of Chicago Press, Chicago, IL.

Giroux, H.A. (2011) "Neoliberalism and the death of the social state: Remembering Walter Benjamin's Angel of History," *Social Identities*, vol 17, no 4, pp. 587–601.

Goldman, M. (2005) *Imperial Nature: The World Bank and Struggles for Justice in the Age of Globalization*, Yale University Press, New Haven, CT.

Gray, T.W., Heffernan, W.D., and Hendrickson, M.K. (2001) "Agricultural cooperatives and dilemmas of survival," *Journal of Rural Cooperation*, vol 29, no 2, pp. 167–92.

Gray, T.W., and Mooney, P. (2009) "Tillamook Co-operative, Monsanto, and rBGH: Discourse struggle and common sense," *Journal of Co-operative Studies*, vol 42, no 2, pp. 24–37.

Guthman, J. (2008) "Thinking inside the neoliberal box: The Micro-politics of agro-food philanthropy," *Geoforum*, vol 39, no 3, pp. 1241–53.

Habermas, J. (1981) *The Theory of Communicative Action*, Beacon Press, Boston, MA.

Harvey, D. (2005) *A Brief History of Neoliberalism*, Oxford University Press, Oxford.

Hayek, F.A. (1937) "Economics and knowledge," *Economica, New Series*, vol 4, no 13, pp. 33–54.

Hayek, F.A. (1943) "The facts of the social sciences," *Ethics*, vol 54, no 1, pp. 1–13.

Hayek, F.A. (1969) "The primacy of the abstract," Pp. 309–23 in A. Koestler and J.R. Smythies (eds) *Beyond Reductionism*, Macmillan, New York.

Hayek, F.A. (1976) *The Mirage of Social Justice*, University of Chicago Press, Chicago, IL.

Hayek, F.A. (1979) *The Political Order of a Free People*, University of Chicago Press, Chicago, IL.

Hayek, F.A. ([1944] 2007) *The Road to Serfdom*, University of Chicago Press, Chicago, IL.

Hobbes, T. ([1642] 1991) *Man and Citizen (De Homine and De Cive)*, Hackett, Indianapolis, IN.

IAAKSTD. (2008) International Assessment of Agricultural Knowledge, Science and Technology for Development, IAAKSTD, Washington, DC.

Jasanoff, S. (2005) *Designs on Nature: Science and Democracy in Europe and the United States*, Princeton University Press, Princeton, NJ.

Kimura, A.H. (2010) "Between technocracy and democracy: An experimental approach to certification of food products by Japanese consumer cooperative women," *Journal of Rural Studies*, vol 26, no 2, pp. 130–40.

Kirkendall, R.S. (1966) *Social Scientists and Farm Politics in the Age of Roosevelt*, University of Missouri Press, Columbia.

Klein, N. (2007) *The Shock Doctrine: The Rise of Disaster Capitalism*, Picador, New York.

Langley, P. (2007) "Uncertain subjects of Anglo-American financialization," *Cultural Critique*, vol 65, pp. 67–91.

Latour, B. (2005) *Reassembling the Social: An Introduction to Actor-Network-Theory*, Oxford University Press, Oxford.

Lazzarato, M. (2009) "Neoliberalism in action: Inequality, insecurity and the reconstitution of the social," *Theory, Culture & Society*, vol 26, no 6, pp. 109–33.

Lippmann, W. (1938) *The Good Society*, Allen & Unwin, London.

Mead, G.H. ([1934] 1962) *Mind, Self, and Society from the Standpoint of a Social Behaviorist*, University of Chicago Press, Chicago, IL.

Mercuro, N. and Medema, S.G. (2006) *Economics and the Law: From Posner to Post-Modernism and Beyond*, Princeton University Press, Princeton, NJ.

Michelson, H., Reardon, T., and Perez, F. (2011) "Small farmers and big retail: Trade-offs of supplying supermarkets in Nicaragua," *World Development*, vol 40, no 2, pp. 342–54.

Mirowski, P. and Plehwe, D. (eds) (2009) *The Road from Mont Pèlerin: The Making of the Neoliberal Thought Collective*, Harvard University Press, Cambridge, MA.

Mitchell, T. (2005) "The work of economics: How a discipline makes its world," *Archives of European Sociology*, vol 46, pp. 297–320.

Mol, A. (2002) *The Body Multiple*, Duke University Press, Durham, NC.

Mont Pelerin Society. (2006) The Mont Pelerin Society, https://www.montpelerin.org, accessed September 2, 2013.

Pestre, D. (2009) "Understanding the forms of government in today's liberal and democratic societies: An introduction," *Minerva*, vol 47, no 3, pp. 243–60.

Porter, T.M. (1995) *Trust in Numbers: The Pursuit of Objectivity in Science and Public Life*, Princeton University Press, Princeton, NJ.

Posner, R.A. (2011) *Economic Analysis of Law*, 8th edn, Aspen Publishers, New York.

Reardon, T. and Timmer, C.P. (2012) "The economics of the food system revolution," *Annual Review of Resource Economics*, vol 4, pp. 225–64.

Revault d'Allonnes, M. (2010) *Pourquoi nous n'aimons pas la démocratie*, Seuil, Paris.

Rose, N. (1996) *Inventing Our Selves: Psychology, Power, and Personhood*, Cambridge University Press, Cambridge.

Rothbard, M.N. (1957) "In defense of 'extreme Apriorism,'" *Southern Economic Journal*, vol 23, no 3, pp. 314–20.

Schutz, A. ([1932] 1967) *The Phenomenology of the Social World*, Northwestern University Press, Evanston, IL.

Schutz, A. (1970) *On Phenomenology and Social Relations*, University of Chicago Press, Chicago, IL.

Simons, H.C. ([1934] 1948) "A positive program for laissez faire: Some proposals for a liberal economic policy," Pp. 40–77 in *Economic Policy for a Free Society*, University of Chicago Press, Chicago, IL.

Stanfield, J.R. and Carroll, M.C. (2009) "The social economics of neoliberal globalization," *Social Economics*, vol 38, no 1, pp. 1–18.

Stone, G.D. (2004) "Biotechnology and the political ecology of information in India," *Human Organization*, vol 63, no 2, pp. 127–40.

Thomas, W.I. (1928) *The Child in America: Behavior Problems and Programs*, Alfred A. Knopf, New York.

von Mises, L. ([1933] 1978) *Epistemological Problems of Economics*, Ludwig von Mises Institute, Auburn, AL.

Walzer, M. (1983) *Spheres of Justice: A Defense of Pluralism and Equality*, Basic Books, New York.

Wapshott, N. (2011) *Keynes Hayek: The Clash that Defined Modern Economics*, W. W. Norton, New York.

Williamson, J. (2004–5) "The strange history of the Washington Consensus," *Journal of Post Keynesian Economics*, vol 27, no 2, pp. 195–206.

Wolf, S.A. (ed) (1998) *Privatization of Information and Agricultural Industrialization*, CRC Press, Boca Raton, FL.
Wolf, S.A. and Zilberman, D. (eds) (2001) *Knowledge Generation and Technical Change: Institutional Innovation in Agriculture*, Kluwer Academic Publishers, Boston, MA.
Žižek, S. (2008) *In Defense of Lost Causes*, Verso, London.

3

POLICING THE NEW ENCLOSURES

On Violence, Primitive Accumulation, and Crisis in the Neoliberal Food System

Morgan Buck

> As I see it, history moves from one conjuncture to another rather than being an evolutionary flow. And what drives it forward is usually a crisis . . . Crises are moments of potential change, but the nature of their resolution is not given.
>
> (Hall and Massey, 2010, p. 57)

Introduction

Since the 1970s, the contours of the global agri-food system have been shaped by an unprecedented concentration of corporate control that has aimed for a wholesale reorientation toward the market as the only pathway for food. An extensive disruption and penetration by the logics of industrial production into previously unintegrated sectors through successive waves of global enclosure has inaugurated an unforeseeably networked world of food production, distribution, and consumption that is both denser and more rigorously disciplined by the principles of the free market than any previous period. However fundamentally the neoliberal food regime has transformed global relations of production, it is currently facing a swath of crises—a series of global economic fissures and political eruptions that threaten to fundamentally undermine the regime's legitimacy, as well as its ability to turn a profit. What will emerge out of this regime is, as Hall's above quote suggests, certainly "not given." Rather, it will be the end result of a struggle between a viable alternative to neoliberal capitalism and the reorganization and recapitulation of existing power relations and imperatives.

The aim of this contribution is to interrogate the everyday and extraordinary forms of violence that buttress the neoliberal food system and, moreover, to emphasize primitive accumulation and policing as the main principles through which these forms of violence are organized and deployed. My intention is to make visible some of the complex relations of real and symbolic violence,[1] the machinations of which continue to underwrite the contemporary food system, while drawing together disparate temporal and spatial locations, histories, and subjectivities under the rubric of enclosure and ongoing primitive accumulation.

In exploring questions of violence and discipline, I ultimately aim to open up a wider space within agri-food scholarship, and food regime literature specifically, in which to think more critically and analytically about the conditions of global struggles against the neoliberal food regime both within and against its ongoing crises.

As the neoliberal food regime's internal faltering compounds the cracks produced by mounting political resistance, capital has tended to respond by unleashing new rounds of privatization and enclosure buttressed by new and more violent extensions of law and order. Examples of the everyday ferocity of enclosures abound in the contemporary food system, as do instances of resistance, particularly in the name of reasserting the role of the commons and the reinvention of autonomous modes of production and reproduction. While it is impossible to adequately capture such a complex global struggle, I employ some specific instances throughout this chapter as a means of highlighting and giving material form to the intersecting relations of the neoliberal regime. In other words, I present some geographically specific glimpses into a larger reality. In this way, I aim to underscore the fact that while each iteration of neoliberal policing is in its own way increasingly local, embedded in the history of a particular geographic space as well as in narratives of isolation and of intimacy, each is also bound to impacts and phenomena occurring at a number of broader scales, including the regional, the national, and the global.

I am interested not only in what a series of locations, histories, and subjectivities have in common but ultimately what their relationships can tell us about the future of an agro-food system that is facing down the barrel of recurring global crises, which, perhaps today more than ever, seem to have opened up a new space for potentially radical change. Borrowing from feminist geographer Cindi Katz (2001),[2] I intend to contribute to this volume with a "non-innocent topography," albeit a preliminary one, of primitive accumulation, violence, policing, crisis, and resistance within the neoliberal food regime.

The chapter will proceed in four parts, organized loosely along temporal and thematic lines. First, I provide some theoretical and historical context by exploring the contemporary processes of primitive accumulation through which the neoliberal food system has been wrought, particularly by identifying the intersection between food regime theory and contemporary themes in Marxist political economy. Building on the terms laid out in the first section, I next examine in greater detail the question of material and symbolic violence and policing in relation to contemporary forms of primitive accumulation. This is followed by a discussion and a response of the neoliberal food system to contemporary forms of crisis. The final section of this contribution will be located in the spaces of potential transformation, which have been opened up in this moment of crisis, specifically addressing contemporary movements around the commonalities and issues emerging within these struggles.

MORGAN BUCK

Enclosures in the Neoliberal Food Regime

Food regime scholarship has been paramount in describing the global shift in agricultural production and consumption on a global scale for over two decades. Food regime theory aims to inscribe international relations of food production and consumption within a historical analysis of periods of capitalist development, offering a periodization of the strategic role of agriculture and agri-food relationships in the world economy (Buttel and Goodman, 1989; McMichael, 2009). As a lens for historical analysis, food regime theory is premised on the existence of relatively stable periods punctuated by moments of large-scale transition. Transition is the product of crisis, the result of gradual accumulation of contradictions that together rupture relations of hegemony and causes their necessary reorganization according to shifts in power over production. Methodologically, food regime theorists tend to observe the organization of social forces and relations of production in the global food system through their expression in various locations, contexts, and scales (Araghi, 2003; McMichael, 2009). This materialist methodology makes the food regime concept a particularly useful tool for analyzing ongoing changes in global value relations and the articulation of struggle on transnational, national, and local levels (Wolf et al., 2001; Araghi, 2003; Pritchard, 2009; Friedmann, 2009).

Food regime theorists have described the current food regime as the product of the global economic and political crisis of the early 1970s (Friedmann, 1993). As a response to the crisis of the Keynesian state and Fordist production, the current regime is generally characterized by the receding of the state and empire and their replacement by the market as the primary organizing principle of global power relations. To summarize a more involved and nuanced literature, the contours of the neoliberal turn in the food system have been defined by a twofold global process aimed at the deterritorialization of capital on one hand and the disaggregation of the state on the other (McMichael and Mhyre, 1991; Le Heron, 1993; Friedland, 1994; Bonanno et al., 1994; McMichael, 2005). The neoliberal food regime is thus predicated primarily on the expansion of global markets, facilitated by rapid and increasingly volatile global flows of capital and by the reorientation of state functions toward accelerating rates of production and of capital accumulation. As a result, power over the food system is today concentrated in the hands of a relatively small number of corporate actors. These actors—integrating control over all elements of the food chain, from agricultural inputs to farming technology to marketing and distribution—operate largely with impunity as the state's role in regulating national and transnational agri-food policies are redeployed from concentrating nation-state power through control over global food flows toward creating a unified, global agricultural marketplace (McMichael, 2009).

Just how does the market become the organizing principle of the food system? What social relations are at the root of this kind of global transition toward mass privatization and concentrated corporate control? These questions are central to a growing dialogue on enclosure and primitive accumulation in relation to the current agri-food system. Enclosure, and the history of capitalism in which the term is rooted, is used today to describe the mechanisms by which goods, both material and immaterial, are transferred by noneconomic force from the nonmarket sphere into the market sphere as both commodities and raw materials for capital accumulation.

In the seventeenth, eighteenth, and nineteenth centuries, the enclosure of land and the expulsion of British commoners from the grounds of their subsistence laid the foundation for the original accumulation of capital via agricultural and industrial revolution (see Linebaugh, 2001, 2009). In the early 1990s, as the neoliberal model was first taking shape, the Midnight Notes (1990) presciently introduced the term "the new enclosures" in their argument for primitive accumulation as a recurrent historical phenomenon as opposed to a one-time antecedent to original capitalist expansion. Especially over the last decade or so, this formulation of primitive accumulation as an ongoing phenomenon has become increasingly widespread, mainly in the discipline of geography, and most especially in relation to the agri-food system (Vasudevan et al., 2008; see also McCarthy, 2001; Akram-Lodhi, 2007; Harvey, 2005; Glassman, 2005, 2006; Hart, 2006; Prudham, 2007; Wolford, 2007).

Today's brand of (distinctly neoliberal) enclosures takes on a number of forms, which together enable the expansion of market relations across more and more of the socio-natural world. They work by way of simultaneous and ongoing revolutions—in science and technology, finance, and across the world political scene—which provide the raw materials as well as the social relations necessary for the ongoing accumulation of wealth (Midnight Notes, 1990; De Angelis, 2001). This is particularly the case for the contemporary agri-food system, wherein rapid advances in mechanical, biological, and financial technologies are at the forefront in opening new grounds for capitalist control over production and circulation of agricultural commodities.

Though not as clearly rooted in a project of nation-state hegemony, contemporary enclosures, like their historical counterparts, are central to the maintenance of global class forces. They are necessary in managing, both combating and pre-empting, crises of the capitalist system—including over/underproduction, as well as crises created by various forms of resistance to and rebellion against the inequalities produced by the market-based system. By separating people from their sources of livelihood, enclosures are also a means of attacking the reproductive base of social struggles, both agrarian and otherwise. Whether imposing relations of market-dependence or deepening those that already exist, enclosure does the work of systematically breaking down nonmarket forms of social solidarity, reordering social life instead according to the logics of market discipline and competition.

Within a bourgeoning literature on agrarian enclosures, these two forms are discussed most centrally: land grabbing and biotechnology. Advertised as promoting fair and "distortion-free" land transactions, policies of market-led agrarian reform are overwhelmingly embedded in an environment of increasing agro-industrial competition, predominantly favoring the circulation of land directly into the hands of large-scale agribusiness, according to Lahiff et al. (2007). Massive waves of land privatization over the last two decades have been integral to the expansion of biofuels production in response to the global energy crisis of the 1970s (Houtart, 2010; Goodman et al., 1987). The privatization of land, particularly in African and Latin American countries, further concentrates control over production as a new conglomeration of "energy, chemical, auto, and biotechnology capitals join the rush for cheap land" (McMichael, 2012a, p. 692). While the so-called original enclosures described by Marx were predominantly the purview of the state writing violence into law, today's land grabbing involves a complex organization of apparatuses, legal and extralegal, state and nonstate. According to World Bank studies, total documented land acquisitions between 2008 and 2009 alone indicated the transfer of 56.6 million hectares of land worldwide (Deininger et al., 2011).

It has been convincingly argued that biotechnology is by and large a response to the myriad crises faced by the Green-Revolution era agro-industrial industry in the 1970s (see Kloppenburg, 2005). Where the Green Revolution was tied into a hegemonic state project by being embedded in American foreign policy, biotechnology inverts these relations in the context of the neoliberal regime by allowing for regimes of private property and corporate control to be embedded directly into organisms themselves (McMichael, 2004; Kloppenburg, 2005; Prudham, 2007). Corporate-driven genetic modification represents what Neil Smith (2007, p. 25) refers to as the capitalist production of nature "all the way down"—wherein natural entities are physically altered to suit the requirements of profitability for those who manufacture them (see also Castree, 1995; Boyd et al., 2001). Much like the commodification of land, the process of biological enclosure involves the "privatization and fragmentation of a commons into a series of dissassociable parts, but also their enmeshment with within the logic of exchange value rather than use value" (Bridge et al., 2002, p. 168; see also Rifkin, 1998; Mannion, 1992; McAfee, 2003a; Rajan, 2006).

In addition to land and biotechnology, financialization has been described as a third kind of contemporary enclosure, although this literature is at present somewhat less elaborated in relation to the agri-food system. While land-based speculation and commercial development undercuts subsistence food production, the manipulation of the market by commodity speculation deepens the reach of overarching logics of speculative commodity investment (Burch and Lawrence, 2009; Ghosh, 2009; McMichael, 2012b). In this sense, new economic technologies have had an important role in shaping the agri-food system as a new and distinctly neoliberal response to the various constraints imposed by the material relations of production. For example, by transferring the locus of accumulation away from the point of production, financial capital is far less susceptible to the power of workers' struggles than its industrial counterpart (Midnight Notes Collective, 2009).

Overall, the concept of new enclosures—along with its historical, political, and theoretical underpinnings—continues to contribute a great deal to food regime theory and indeed to understandings of the agri-food system in general. Its main contributions, I argue, are twofold. First, the concept of enclosure provides a framework through which to articulate the social relations underlying the rise of the neoliberal food system and through which to interpret these social relations at multiple scales. Second, I argue that this literature contributes to the project of identifying and analyzing the role of the state, particularly in relation to corporate and other nonstate actors, in facilitating this global market system.

Based on the literature and examples presented above, I argue that the neoliberal food regime hinges on three broad and deeply intertwined modes of enclosure: land, biotechnology, and finance. In using these categories, I wish to emphasize their flexibility and multifunctionality in interpreting the issues facing the current agri-food system. Most evidently, they represent three categories of enclosure, each distinct and yet in many ways materially overlapping. I also use them as a way of envisioning more broadly the multiplicitous nature of enclosure in general; enclosure *as a social relation* is at once spatial, biopolitical, and economic. The real and multiple overlaps between the three forms of enclosure give material weight to the interpenetration of the various scales at which enclosure and primitive accumulation operate: at the scale of the seed, of the region and nation-state, and the scale of finance-capital, the flows of which are decidedly global. I expect this triptych of enclosures-as-social-relations to demonstrate a quintessential point about the global neoliberal food system, derived from Katz's (2004, 2005) broader analysis of globalization. Namely, I wish to emphasize that globalization is both local *and* global and that, in fact, one is unintelligible without the other.

Policing the New Enclosures

Carol Rose (1994) reminds us that "property" is never a static, pregiven entity. Rather, it depends on a continual, active process of "doing." This "doing," which subsumes elements of the socio-natural world into the order of the market, is essentially rooted in a logic of separation and exclusion; it entails the drawing and regulating the boundaries—both material and discursive—of private property. Enclosures in this sense rely on the intervention of a hegemonic power, one that can generate a measure of consent, as well as enacting measures of coercive force. With this in mind, I employ the notion of *policing* as an inherent element in the production of enclosures—namely, the twinned elements of material and symbolic violence—fundamental to both the creation and circulation of private property, and thus in the recurrent process of primitive accumulation in the agri-food system.

Who is responsible for policing the new enclosures? According to Marx's tale of the ejection of the English commoners, what differentiates enclosure from theft is the material and ideological backing of the state, whereby drawing boundaries becomes the expressed purview of the state, and whereby "*the law itself* becomes now the instrument of the theft of the people's land" (1992, p. 796, emphasis added). The state's central role in policing extended to the colonies as well—emphasizing the relationship between enclosure and the rise of British hegemony in the agri-food industry throughout the first food regime (Friedmann, 1987, 1993), as well as the United States' rise to dominance through the reinstatement of neo-imperial relations (McMichael, 2009).

Questions surrounding the impacts of neoliberal tendencies toward deregulation and deterritorialization have figured prominently in food regime literature. As Bonnano et al. (1994) have clearly elaborated, through neoliberal tendencies toward deregulation and deterritorialization, state power seems to have been effectively disaggregated, with many functions of the state delegated to a variety of nonstate actors operating at both supra- and subnational scales. This notably includes supranational and transnational governance organizations, such as the United Nations, World Trade Organization, the International Monetary Fund (IMF), and the World Bank, as well as nongovernmental organizations (NGOs), and various other civil-society organizations. Overall, this observation has led to the conclusion that the neoliberal state's actual functions currently revolve around the legitimizing behavior of transnational agents across the global socioeconomic scene, as well as the management of acute social and political contradictions (crises) that emerge as a result of the accumulation process (Bonanno, 2004; Bonanno et al., 1994; see also O'Connor, 1988).

More recently, interrogating the impact of deregulation and disaggregation of state capacities in the agri-food system, Pechlaner and Otero (2008, 2010) have argued against the concept of deregulation—rather a reorganization of state capacities toward new modes of regulation, or *neoregulation*—within the global agri-food system. Despite a seemingly fragmented or diffuse form, the state is still responsible for the formation and maintenance of new regulatory institutions through which contemporary mechanisms of global regulation take shape and through which the state aids the proliferation of new technologies of accumulation, such as biotechnology (Pistorius and van Wyk, 1999; Pechlaner and Otero, 2008) and financialization (Burch and Lawrence, 2009).

While these formulations demonstrate the ongoing relevance of the state within relations of the neoliberal food system—through new regulatory formations and through the production of legitimacy and the management of crises—they present a somewhat murkier picture of the material and discursive relationship between these three categories. While food regime theorists do in many ways acknowledge the myriad forms of violence within the food system, attention to the dynamics of violence *as discipline* tend to be more descriptive than analytical. I therefore assert policing specifically as a means of emphasizing the structural role of violence within neoliberal modes of primitive accumulation.

It is undoubtedly obvious that the restructuring and rescaling of state functions under the guise of deregulation and deterritorialization has made it more and more difficult to articulate the composition of contemporary disciplinary mechanisms and the organization of the powers and logics by which they are deployed. The dismantling of state protection for agriculture under the banner of liberalization and integrated global markets does not, however, mean that the state has withdrawn from the organization of policing mechanisms. Rather, punitive practices are the cornerstone in the state's ability to manage the contradictions and crises wrought by neoliberal accumulation while also establishing and enabling new and deepening forms of enclosure. Combined with the production and maintenance of conditions propitious to ongoing economic expansion, conjoined policies of state retrenchment in some spheres with increasingly forceful regulation of others, represent the splintering of a Keynesian model of centralized redistribution toward more repressive, if also more complex and diffuse, modes of discipline and governance.

Relations of order and disorder within the agri-food system have taken on both biopolitical and, as Ferguson (2006, p. 207) notes, distinctly "private-patchwork" characteristics. In other words, the reorganization of class forces per the neoliberal model has been managed through the reorganization of policing toward modes of biopolitical forms of discipline accompanied by more repressive—though complex and fragmented—expressions of coercive force involving both state and private actors. Some have argued, however, that despite the diffusion of disciplinary mechanisms and coercive force, these policing strategies are nevertheless part of a current project of "proactive statecraft." Peck (2003, p. 225) argues, for instance, the following:

> in the context of pervasive—if unevenly realized and contested—processes of neoliberalization, these diverse manifestations of accelerated state restructuring-cum-rescaling are often being premised on an uneasy marriage between economic liberalization and authoritarian governance.

The production of genetically modified crops exemplifies policing apparatus' embedded directly into the flow of biological matter as both commodity and raw material for commercial production (McAfee, 2003a; Prudham, 2007). Given the elimination of state management in the agricultural sector, the economic impacts of liberalization are felt much more at the level of individual and community, making market discipline a distinctly biopolitical mode of regulating productive relations. The rubric of "personal responsibility," for instance, is espoused by credit and micro-credit schemes (in various ways and with highly uneven results) as a means of paradoxically producing the farmer as self-disciplined entrepreneurial subject while hinging the production cycle to increasingly volatile cycles of debt-fueled financial speculation (Guthman, 2008; Trostle, 2008).

The spread of biotechnology necessitates constant policing—from the ideological and legal projects of redefining private property and repartitioning biological life (McAfee, 2003b), to the global dissemination of intellectual property rights and the prosecution of infringement cases (Prudham, 2007), to the private security corporations hired to guard corporate fields against populations bent on protecting their access to biological commons of non-GM seed (Shiva, 1992).

Similarly, though the contemporary scramble for land is guided by the kinds of market logic and economic rationality promoted with mottos like "willing buyer–willing seller," it is predominantly noneconomic disciplinary interventions—often piecemeal strategies—that are responsible for removing barriers to the privatization and commodification of land. This is often achieved through the forcible eviction of peasants and the repression of organized resistance (Borras and Ross, 2007; Lahiff et al., 2007). In many of these cases of violence and intimidation around evictions, the state is indeed not the primary actor. Instead, paramilitaries, hired thugs, and militias are mobilized to extricate peasants who do not willingly abdicate. The interactions between private and public violence in relation to land-grabs fundamentally challenges conventional understandings of the relationship between "criminality" and the law-and-order state and between forms of state and nonstate violence:

> Armed violence becomes institutional violence when public institutions of property rights enforcement recognize grabbed land. Public institutions then become a vector of social exclusion, consolidating new forms of authority intimately related to private violence and crime. (Grajales, 2011, p. 772)

For both state and corporate actors, the 2007–8 food riots and protests represented a systemic challenge to the legitimacy of finance-based accumulation and capitalism in general. The repressive force demonstrated against the protests was part of a broader strategy of coordinated displacement through both discursive and repressive apparatuses orchestrated predominantly by the state. By declaring the current moment one of ultimate threat to national security, stability, and peace, any movement can be made a site for targeted repression. This in many ways explains why control over television and radio airwaves accompanied applications of military force—that is, to manipulate the narrative of political protest from one of rebellion and popular resistance to a basal reaction of the hungry, desperate, and dangerous (Patel and McMichael, 2009; Holt-Giménez and Patel, 2010). The invention of the "violent protester" as a pressing threat in this sense grants the state authority to respond—*not* to the actual causes of a crisis but rather with an overinvestment in repressive technologies targeted at some of the most vulnerable most affected by conditions of crisis.

These instances further demonstrate a complex political relation between extralegal property regimes and the contemporary constitution of state authority. While the actual work of eviction is delegated to the realm of extralegal force, both agribusiness and the state are able to maintain a facade of legitimacy by distancing themselves from the violence. With its legitimacy ostensibly intact, the state itself is able to politically neutralize the whole process through legal recognition. This is fundamental to what Lund and Sikor (2009) call a *contract of mutual authorization*, whereby legal recognition of extralegal violence in turn reinforces the legitimate influence of the authorizing agencies themselves.

Policing and the Neoliberal Legitimation Crisis

As Beverly Silver (2003) points out, there are two distinct sides to the crisis of capitalism. These sides are manifestations of the tension between the drive toward maximum accumulation and the need to maintain legitimacy, particularly to those whom it most oppresses and exploits. In other words, the crisis of accumulation can only be solved by means that bring about a crisis of legitimation (namely, ramping up violence and exploitation), and vice versa. This was indeed the case in the 1970s transition to the neoliberal food regime, as capitalism reorganized itself by attacking global working-class struggles and diffusing responsibility through the private sector. By regaining profitability, neoliberalism also laid the groundwork for its own subversion, setting the stage for an impending legitimation crisis.

This crisis of legitimacy is rooted both in neoliberalism's basic inability to deliver on its promises of global prosperity and in the increasingly harsh modes of discipline that have accompanied its broken promises around the word. The legitimacy crisis is particularly observable in the mounting strength of agrarian social movements—from anti-GMO (genetically modified organisms) activism in India as well as the West, to the Zapatistas and the MST (Movimento dos Trabalhadores Rurais Sem Terra, Brazil's Landless Peoples' Movement), as well as innumerable smaller struggles around the world, and to organizations of global solidarity (see Patel, 2006; Borras et al., 2008; Borras, 2010).

Neoliberalism's response to crisis is complex and ultimately difficult to characterize. However, there are two particularly relevant trends that I wish to highlight, which are particularly relevant to the framework of enclosure and primitive accumulation. I characterize the first strategy as the invention of the capitalist commons, a strategy geared toward the weak placation, co-optation, and manipulation of struggles against enclosures and for the reinstatement of common wealth. The second can be summarized as the integration of crisis as an accumulation strategy, one through which the maneuvering, aggravation, and even invention of instability plays a central role in enabling capital accumulation. By exploring these strategies, I ultimately aim to emphasize neoliberalism as a unique regime with the capacity to metabolize myriad crises, both internal and external, through the rapid reorganization of structures of accumulation and mechanisms of policing.

After two decades of a "there is no alternative" approach to neoliberal economic reform, agrarian resistance to enclosure met the self-defeating realities of a model based on the privatization of everything. As Federici (2012, p. 140) points out, since "capitalist accumulation is structurally dependent on the free appropriation of immense quantities of labour and resources that must appear as externalities to the market," the actual marketization of all spheres of life is ultimately detrimental to the functioning of the capitalist economic system. As a response to this, as well as to mounting resistance from local and transnational movements, development specialists and policy makers from the UN to the World Bank started crediting various kinds of micro-social relations and noncapitalist forms of organization as a "naturally efficient" way of managing resources. This shift is based on the new argument that given the right conditions, the collective management of common resources, including agricultural land, can indeed "produce very well for the market" (Federici, 2012, p. 140; see also Juma and Ojwang, 1996; Caffentzis, 2010).

This strategy accomplishes two important ends in relation to neoliberalism's economic and legitimacy crises. First, the implementation of various state-led policies of communal tenure or private-public partnerships, which—what De Angelis (2010a) calls a "distorted commons"—serve capital politically "at least as a stop-gap, transitional institution when the revolts of the landless or the devastation of the forests become destabilising to the general exploitation of a territory or population" (Caffentzis, 2010, p. 29). At the same time, both the state and private sector are relieved from the responsibility, cost, and risk of managing land, water, and infrastructure. Moreover, the proliferation of a notion of the "global commons" has served as a means of effectively eschewing corporate and state responsibility for environmental degradation caused by capitalist production. The notion of air and sea (see Denmark and Mulvenon, 2010) or soil as global resource commons, for instance, promotes a conception of global responsibility that shifts the burden of proof and ultimately of responsibility from the worst polluters to an abstract interpretation of the global community.

In addition to manipulating the commons toward capitalist ends, crisis itself has been integrated into a strategy of deepening repressive and authoritarian modes of governance. For Hall et al. (1978), policing represents a state response to general conditions of crisis operating through the creation of a "moral panic" to which the state is granted the authority to respond with an outlay of disciplinary technologies. This represents an *exceptional* moment in that it is the increased reliance on coercive mechanisms or apparatuses already in the state's repertoire of power, mobilizing an "authoritarian consensus," by which these powers of coercion are revealed (Hall et al., 1978, p. 217). In other words, through a combination of neoliberal restructuring processes and exclusionary policies, the capacities of the state have effectively been organized to leverage, deploy, and redistribute general contradictions and crises of the state and capitalist class onto target territories and populations. Once displaced, these crises are confronted anew with an emergent concatenation of punitive practices and militaristic technologies.

Federici (2012), for instance, emphasizes the role of war and civil conflict as an integral element of neoliberal expansion and capital accumulation. While the state withdraws social protections for peasants, it reenters actively in the form of military-led efforts to quell the civil unrest—manifested largely as religious and territorial skirmishes—caused by neoliberal forms of economic, political, and territorial marginalization of the poor. In this sense, the disorder caused by neoliberal reorganization serves as a justification for prescribing more neoliberal reforms along with new kinds of law and order in the countryside. Often in collaboration with NGOs and international relief/aid organizations, state intervention is provided with the opportunity to rebuild decimated physical and social infrastructure in such a way as to gain access to resources and to strengthen dependence of societies on "international capital and the powers that represent it" (Federici, 2012, p. 77).

In this sense, acute and protracted socioeconomic and environmental crises finish the work that neoliberal policy started—destabilizing societies so as to pave the way for neoliberal strategies of restabilization on multiple—individual, community, geopolitical—scales. This mutually assuring relationship between neoliberal economic reform and political violence is surprisingly *not* anathema to contemporary notions of liberal statehood (Ballvé, 2011). The reality in fact is quite the opposite, as today's popular tropes of institution building, good governance, and integrated human security actually hinge on the ability of states to attract capital investment by cleansing rural spaces of their ungovernable attributes.

From Crisis and Enclosure, Toward the Reclamation and Reinvention of the Commons

"It has become increasingly clear that we are poised between an old world that no longer works and a new one struggling to be born" (Bollier and Helfrich, 2012, p. 1). Indeed, crisis in many ways signifies a political and material opening. However, the nature of this "new world" is yet unknown. Especially given the central role of crisis in deepening regimes of enclosure and dispossession, as well as advancing more organized forms of everyday coercion and violence, the crisis of neoliberalism cannot be presumed to be a necessarily liberatory phenomenon. Rather, as Hall suggests, crises are moments of *potential* change, the resolution of which is not predetermined.

While the contemporary legitimation crisis may have indeed caused a refraction large enough to open generative spaces for the organization of transformative movements, such spaces may also present an opportunity for the forces of capital and the logics of policing, discipline, and control to be recapitulated, expanded, and reinforced. This chapter draws on existing theoretical and empirical scholarship on primitive accumulation and enclosure as a means of clarifying and emphasizing the central role of disciplinary economic and social policies in relation to the question of crisis in the neoliberal agri-food industry. I argue that in order to analyze the contemporary food system, it is crucial to develop an analysis of the manifold mechanisms of enclosure and forms of organized coercion, repression, and violence that undergird the fundamental separation of people from their means of livelihood and the basis of organizational power.

As the neoliberal food regime faces the accumulation of manifold global crises, the practices of subversion that are otherwise repressed or dwell in its interstices are brought to the fore as both topplers of the regime and creators of a new order. Over the last decade, the notion of the commons has gained unprecedented popularity in social movements as well as academic literature. The commons is a powerful lens—both novel and embedded in an ongoing history of commoners' struggles against enclosure in the global North and South alike—through which to organize and interpret the many and various existing struggles (sometimes minor) that assert some notion of common wealth against forces of capitalist enclosure and privatization (De Angelis, 2010a). Despite its popularity and impact, the commons is nevertheless a barbed concept with many, often incongruous, definitions and political implications.

Where the notion of the commons has emerged in the discourses of struggles, it has more often than not retained a categorical abstractness in its political articulations. This kind of soft focus risks lending the commons usage in service, both intentionally and unintentionally, of the status quo of neoliberal development. As the example of the IMF's neoliberal commons would suggest, the commons can easily be construed as a convenient fix to various internal and external roadblocks to capital accumulation through the exploitation of unwaged labor and the externalization of negative impacts of capitalist production. However, even in its better-intentioned forms, the commons can also be deployed therapeutically as a means of easing tensions and soothing the harsh effects of capitalism while fundamentally maintaining existing relations of production.

One prevalent notion of reinventing the commons is that which posits the commons as a version of "capitalism 3.0" (Barnes, 2006). This notion of re-embedding economy in an ethical community is the basis of a just society in which the workings of the state and capital are mediated by a network of commons. This sentiment of a "return to small" is valuable for its ability to reveal the disastrous scale and tempo of contemporary global capitalism. However, the employment of commons as remedial tonic while maintaining an adherence to principles of free market entrepreneurialism also undermines to a significant extent the root causes of crisis. By offering a formulation of the commons that is compatible with, and in many ways potentiates ongoing capitalist accumulation, this version of the commons also precludes the more radically transformative potential of the commons as a terrain of global struggle.

In order to be a viable alternative to neoliberal enclosure and the violent social relations that underwrite it, I argue that the commons must be conceived of as an active process that is both outside of and resistant to the capitalist economic system. As Linebaugh's (2009, 2013) work emphasizes in great detail, the commons has historically served as a source of both physical and organizational strength for planetary commoners against the conditions of their exploitation. By providing a source of livelihood that relies on neither the state nor the market, the act of protecting the old and producing the new commons should be an act of resistance against exploitation as well as the privatization and monopolization of resources. Building on these assertions, De Angelis (2010b, p. 955) emphasizes the active nature of commons as being most fundamentally about:

> the (re)production of/through commons . . . there are no commons without incessant activities of *commoning*, of (re)producing in common. But it is through (re)production in common that communities of producers decide for themselves the norms, values and measures of

> things . . . there is no commons without commoning, there is no commons without communities of producers and particular flows and modes of relations. [emphasis added]

For the purpose of this chapter, I hold that the act of commoning is enabled at least in part by a politically precise understanding of the forms of policing, forms of both material and symbolic violence, which support and enable it. The aim of this chapter is thus to enable the advancement of existing struggles against enclosure and for autonomous social reproduction and the fashioning of new modes of life in common, while also highlighting some of the existing lacunae within these struggles. In North America especially, the success of present and future struggles will require reinventing and reconstructing forms of community that are largely lost to us, as well as constructing new forms of community that can address the ethnic, racial, and geopolitical hierarchies that have thus far served as the basis of expropriation and exploitation. Especially in the context of the global agri-food system, the transformative power of the commons is located in its fundamental role as predicate of resistance and therefore the production of—from the shell of the neoliberal regime—a new and more socially just world.

Notes

1 This emphasis on mapping the duality of real and symbolic violence is inspired by Michael Watts's (2004, p. 64) study of extractive industries in Africa.

2 In an exploration of globalization as a phenomenon through which the histories and geographies of global and local are increasingly intertwined, Katz (2001) uses the notion of a "noninnocent topography" in order to "get at the specific ways globalization works on particular grounds in order to work out a situated, but at the same time scale-jumping and geography-crossing, political response to it" (p. 1216).

References

Akram-Lodhi, A. (2007) "Land, markets and neoliberal enclosure: An agrarian political economy perspective," *Third World Quarterly*, vol 28, no 8, pp. 1437–56.

Araghi, F. (2003) "Food regimes and the production of value: Some methodological issues," *Journal of Peasant Studies*, vol 30, no 2, pp. 41–70.

Ballvé, T. (2011) "Territory by dispossession: Decentralization, statehood, and the Narco Landgrab in Colombia," International Conference on Global Land Grabbing, University of Sussex, April 6, 2011.

Barnes, P. (2006) *Capitalism 3.0: A Guide to Reclaiming the Commons*, Berrett-Koehler Publishers, San Francisco.

Bollier, D. and Helfrich, S. (2012) "Introduction," in D. Bollier and S. Helfrich (eds) *The Wealth of the Commons: A World Beyond Market and State*, Levellers Press, New York, available at http://wealthofthecommons.org/essay/introduction-commons-transformative-vision, accessed April 12, 2013.

Bonanno, A. (2004) "Globalization, transnational corporations, the state and democracy," *International Journal of Sociology of Agriculture and Food*, vol 12, pp. 37–48.

Bonanno, A., Friedland, W. H., Llambi, L., Marsden, T., Belo Moreira, M., and Schaeffer, R. (1994) "Global post-Fordism and concepts of the state," *International Journal of Sociology of Agriculture and Food*, vol 4, pp. 11–29.

Borras, S. (2010) "The politics of transnational agrarian movements," *Development and Change*, vol 41, no 5, pp. 771–803.

Borras, S., Edelman, M., and Kay, C. (2008). "Transnational agrarian movements: Origins and politics, campaigns and impact," *Journal of Agrarian Change*, vol 8, no 2–3, pp. 169–204.

Borras, S. and Ross, E. B. (2007) "Land rights, conflict, and violence amid neo-liberal globalization," *Peace Review: A Journal of Social Justice*, vol 19, pp. 1–4.

Boyd, W., Prudham, S., and Schurman, R. (2001) "Industrial dynamics and the problem of nature," *Society and Natural Resources*, vol 14, no 7, pp. 555–70.

Bridge, G., McManus, P., and Marsden, T. (2002) "The next new thing? Biotechnology and its discontents," *Geoforum*, vol 34, no 2, pp. 165–74.

Burch, D. and Lawrence, G. (2009) "Towards a third food regime: Behind the transformation," *Agriculture and Human Values*, vol 26, no 4, pp. 267–79.

Buttel, F. and Goodman, D. (1989) "Class, state, technology and international food regimes," *Sociologia Ruralis*, vol 29, no 2, pp. 86–92.

Caffentzis, G. (2010) "The future of 'The Commons': Neoliberalism's 'Plan B' or the original disaccumulation of capital?," *New Formations*, no 69, pp. 23–41.

Castree, N. (1995) "The nature of produced nature: Materiality and knowledge construction in Marxism," *Antipode*, vol 27, no 1, pp. 12–48.

De Angelis, M. (2001) "Marx and primitive accumulation: The continuous character of capital's 'enclosures,'" *The Commoner*, no 2, pp. 1–22.

De Angelis, M. (2010a) "On the commons: A public interview with Massimo De Angelis and Stavros Stavrides," available at www.e-flux.com/journal/on-the-commons-a-public-interview-with-massimo-de-angelis-and-stavros-stavrides/, accessed April 29, 2013.

De Angelis, M. (2010b) "The production of the commons and the 'explosion' of the middle class," *Antipode*, vol 42, no 4, pp. 954–77.

Deininger, K. and Byerlee, D., with Lindsay, J., Norton, A., Selod, H., and Stickler, M. (2011) "Rising global interest in farmland: Can it yield sustainable and equitable benefits?," World Bank, Washington, DC.

Denmark, A.M. and Mulvenon, J. (eds) (2010) *Contested Commons: The Future of American Power in a Multipolar World*, Center for a New American Security, Washington, DC.

Federici, S. (2012) *Revolution at Point Zero: Housework, Reproduction, and Feminist Struggle*, Common Notions, Brooklyn, NY.

Ferguson, J. (2006) *Global Shadows: Africa in the Neoliberal World Order*, Duke University Press, Durham, NC.

Friedland, W. (1994) "Globalization, the state, and the labor process," *International Journal of Sociology of Agriculture and Food*, no 6, pp. 30–46.

Friedmann, H. (1987) "International regimes of food and agriculture since 1870," in T. Shanin (ed.) *Peasants and Peasant Societies*, Basil Blackwell, Oxford.

Friedmann, H. (1993) "The political economy of food: A global crisis," *New Left Review*, no I/197, pp. 29–57.

Friedmann, H. (2009) "Discussion: Moving food regimes forward: Reflections on symposium essays," *Agriculture and Human Values*, vol 26, no 4, pp. 335–44.

Ghosh, J. (2009) "The unnatural coupling," *The IDEAs Working Paper Series*, pp. 1–24.

Glassman, J. (2005) "The new imperialism? On continuity and change in US foreign policy," *Environment and Planning*, vol 37, no 9, pp. 1527–44.

Glassman, J. (2006) "Primitive accumulation, accumulation by dispossession, accumulation by extra-economic means," *Progress in Human Geography*, vol 30, no 5, pp. 608–25.

Goodman, D., Sorj, B., and Wilkinson, J. (1987) *From Farming to Biotechnology: A Theory of Agro-Industrial Development*, Basil Blackwell, Oxford.

Grajales, J. (2011) "The rifle and the title: Paramilitary violence, land grab and land control in Colombia," *Journal of Peasant Studies*, vol 38, no 4, pp. 771–92.

Guthman, J. (2008) "Neoliberalism and the making of food politics in California," *Geoforum*, vol 39, no 3, pp. 13–24.

Hall, S., Critcher, C., Jefferson, T., Clarke, J., and Roberts, B. (1978) *Policing the Crisis: Mugging, the State and Law and Order*, Palgrave Macmillan, London.

Hall, S. and Massey, D. (2010) "Interpreting the crisis," *Soundings*, vol 44, pp. 57–72.

Hart, G. (2006) "Denaturalising dispossession: Critical ethnography in the age of resurgent imperialism," *Antipode*, vol 38, no 5, pp. 977–1004.

Harvey, D. (2005) *A Brief History of Neoliberalism*, Oxford University Press, Oxford.

Holt-Giménez, E. and Patel, R. (2010) *Food Rebellions: Crisis and the Hunger for Justice*, Food First Books, Oakland, CA.

Houtart, F. (2010) *Agrofuels: Big Profits, Ruined Lives and Ecological Destruction*, Pluto Press, London.

Juma, C. and Ojwang, J. (1996). *In Land We Trust: Environment, Private Property and Constitutional Change*. Zed Books, London.

Katz, C. (2001) "On the grounds of globalization: A topography for feminist political engagement," *Signs*, vol 26, pp. 1213–34.

Katz, C. (2004) *Growing Up Global: Economic Restructuring and Children's Everyday Lives*, University of Minnesota Press, Minneapolis.

Katz, C. (2005) "Partners in crime? Neoliberalism and the production of new political subjectivities," *Antipode*, vol 37, pp. 623–31.

Kloppenburg, J. R. (2005) *First the Seed: The Political Economy of Plant Biotechnology, 1492–2000*, 2nd edn, University of Wisconsin Press, Madison.

Lahiff, E., Borras, S. M., and Kay, C. (2007) "Market-led agrarian reform: Policies, performance and prospects," *Third World Quarterly*, vol 28, no 8, pp. 1417–36.

Le Heron, R. (1993) *Globalised Agriculture*, Pergamon, Oxford.

Linebaugh, P. (2001) *The Many-Headed Hydra: Sailors, Slaves, Commoners, and the Hidden History of the Revolutionary Atlantic*, Beacon Press, Boston, MA.

Linebaugh, P. (2009) *The Magna Carta Manifesto: Liberties and Commons for All*, University of California Press, Berkeley.

Linebaugh, P. (2013) *Stop, Thief!: The Commons, Enclosures, and Resistance*, PM Press, Oakland, CA.

Lund, C. and Sikor, T. (2009) "Access and property: A question of power and authority," *Development and Change*, vol 40, no 1, pp. 1–22.

Mannion, A. M. (1992) "Biotechnology and genetic engineering: New environmental issues," Pp. 147–60 in A. M. Mannion and S. Bowlby (eds) *Environmental Issues in the 1990s*, Wiley, Chichester.

Marx, K. (1992) *Capital vol. 1: A Critique of Political Economy*, Penguin Classics, London.

McAfee, K. (2003a) "Neoliberalism on the molecular scale. Economic and genetic reductionism in biotechnology battles," *Geoforum*, no 34, pp. 203–19.

McAfee, K. (2003b) "Biotech battles: Plants, power, and intellectual property in the new global governance regimes," Pp. 174–94 in R. A. Schurman and D. D. Kelso (eds) *Engineering Trouble: Biotechnology and Its Discontents*, University of California Press, Berkeley.

McCarthy, J. (2001) "Environmental enclosures and the state of nature in the American west," Pp. 117–45 in N. Peluso and M. Watts (eds) *Violent Environments*, Cornell University Press, Ithaca, NY.

McMichael, P. (2004) "Biotechnology and food security: Profiting from insecurity?," Pp. 137–53 in L. Benería and S. Bisnath (eds) *Global Tensions: Challenges and Opportunities in the World Economy*, Routledge, London.

McMichael, P. (2005) "Global development and the corporate food regime," *New Directions in the Sociology of Global Development*, vol 11, pp. 269–303.

McMichael, P. (2009) "A food regime genealogy," *Journal of Peasant Studies*, vol 36, no 1, pp. 139–69.

McMichael, P. (2012a) "The land grab and corporate food regime restructuring," *The Journal of Peasant Studies*, vol 39, no 3–4, pp. 681–701.

McMichael, P. (2012b) "Food regime crisis and revaluing the agrarian question," Pp. 99–122 in R. Almås and H. Campbell (eds) *Rethinking Agricultural Policy Regimes: Food Security, Climate Change and the Future Resilience of Global Agriculture (Research in Rural Sociology and Development, Volume 18)*, Emerald, Bingley, UK.

McMichael, P. and Myhre, D. (1991) "Global regulation vs. the nation-state: Agro-food systems and the new politics of capital," *Capital & Class*, vol 15, no 1, pp. 83–105.

Midnight Notes. (1990) "The new enclosures," *Midnight Notes*, no 10, available at midnightnotes.org, accessed April 29, 2013.

Midnight Notes Collective. (2009) "Promissory notes: From crisis to commons," available at midnightnotes.org, accessed April 29, 2013.

O'Connor, J. (1988) "Capitalism, nature, socialism: A theoretical introduction," *Capital, Nature Socialism*, vol 1, no 1, pp. 11–38.

Patel, R. (2006) "International agrarian restructuring and the practical ethics of peasant movement solidarity," *Journal of Asian and African Studies*, vol 41, no 1–2, pp. 71–93.

Patel, R. and McMichael, P. (2009) "A political economy of the food riot," *Review*, vol 35, no 1, pp. 9–35.

Pechlaner, G. and Otero, G. (2008) "The third food regime: Neoliberal globalism and agricultural biotechnology in North America," *Sociologia Ruralis*, vol 48, no 4, pp. 1–21.

Pechlaner, G. and Otero, G. (2010) "The neoliberal food regime: Neoregulation and the new division of labor in North America," *Rural Sociology*, vol 75, no 2, pp. 179–208.

Peck, J. (2003) "Geography and public policy: Mapping the penal state," *Progress in Human Geography*, vol 27, no 2, pp. 222–32.

Pistorius, R. and van Wyk, J. (1999) *The Exploitation of Plant Genetic Information: Political Strategies in Crop Development*, CABI Publishing, Oxon.

Pritchard, B. (2009) "The long hangover from the second food regime: A world-historical interpretation of the collapse of the WTO Doha Round," *Agriculture and Human Values*, vol 26, no 4, pp. 297–307.

Prudham, S. (2007) "The fictions of autonomous invention: Accumulation by dispossession, commodification and life patents in Canada," *Antipode*, vol 39, no 3, pp. 406–29.

Rajan, K. (2006) *Biocapital: The Constitution of Postgenomic Life*, Duke University Press, Durham, NC.

Rifkin, J. (1998) *The Biotech Century*, Victor Gollancz, New York.

Rose, C. M. (1994) *Property as Persuasion: Essays on the History, Theory, and Rhetoric of Ownership*, Westview Press, Boulder, CO.

Shiva, V. (1992) *The Violence of Green Revolution: Third World Agriculture, Ecology and Politics*, Zed Books, London.

Silver, B. (2003) *Forces of Labor: Workers' Movements and Globalization Since 1870*, Cambridge University Press, Cambridge.

Smith, N. (2007) "Nature as accumulation strategy," *Socialist Register*, vol 43, pp. 19–40.

Trostle, R. (2008) "Global agricultural supply and demand: Factors contributing to the recent increase in food commodity prices," USDA Economic Research Service, Washington, DC.

Vasudevan, A., McFarlane, C., and Jeffrey, A. (2008) "Spaces of enclosure," *Geoforum*, vol 39, pp. 1641–6.

Watts, M. (2004) "Resource curse? Governmentality, oil and power in the Niger Delta, Nigeria," *Geopolitics*, vol 9, no 1, pp. 50–80.

Wolf, S., Hueth, B., and Ligon, E. (2001) "Policing mechanisms in agricultural contracts," *Rural Sociology*, vol 66, no 3, pp. 359–81.

Wolford, W. (2007) "Land reform in the time of neoliberalism: A many-splendored thing," *Antipode*, vol 39, no 3, pp. 550–70.

Part 2

CASE STUDIES

4

THE RISE AND FALL OF A PRAIRIE GIANT

The Canadian Wheat Board in Food Regime History

André Magnan

Introduction

This chapter analyzes the recent dismantling of the Canadian Wheat Board (CWB) in the context of three decades of neoliberal restructuring in the Western Canadian grains sector. The prairie region of Western Canada has been a grain exporting "bread basket" since the 1870s. Over this long history, relations of conflict and cooperation among farmers, the state, and agribusiness have given rise to different institutional configurations for regulating the sector and integrating it into world markets. In the postwar period, the CWB became the linchpin for regulating relations among state, market, and society in the grains sector. Since the 1980s, Canadian governments, agribusiness, and some farm groups have pursued a neoliberalizing agenda that has substantially restructured the sector. These actors translated the neoliberal impulse for "liberating individual entrepreneurial freedoms and skills" (Harvey, 2005, p. 2) into a program for scaling back state support for agriculture, deepening the sector's export orientation and pursuing free trade. Yet, restructuring of the prairie grain sector has unfolded unevenly in a process shaped by political contestation and institutional contexts.

The case of the CWB is particularly significant. For one, its dismantling represents the end of nearly seventy years of state-sponsored collective grain marketing, representing a major change for the industry. Yet, although the CWB was emblematic of a state-led form of agri-food regulation, its fate in the neoliberal era was not a foregone conclusion. As I argue below, its fate was shaped by political struggles among agri-food and state actors with diverging ideological commitments and interests. This case therefore sheds light on the processes of contestation through which neoliberalization is transforming agri-food sectors. I situate these conflicts in the larger context of changing food regime relations (Friedmann and

McMichael, 1989; Pritchard, 2009), the norms, rules, and institutional frameworks regulating food production, distribution, and consumption internationally in periods of stability and change. The key insight from the food regimes framework, as it applies to this case, is that more than simply a conflict over agricultural policy, struggles over the CWB question involved competing strategies for integrating the prairie grains sector into world markets under shifting conditions. My analysis therefore takes up Pelchaner and Otero's (2010) enjoinder to look within nation-states to examine how instances of contention affect the way in which agri-food sectors respond to and shape international food regimes.

From 1943 until 2012, the CWB served as a "single-desk" marketing agency for grain produced on the Canadian prairies.[1] Farmers delivered their grain to the CWB, which pooled it on the basis of quality and end-use, sold it to domestic and foreign markets, and returned all of the proceeds (minus administrative costs) to farmers. Over the course of three food regimes, collective marketing through the CWB served as a means of integrating the prairie region into world markets and for mediating relations among farmers, the state, and capital. The Canadian prairie region became integrated into the first food regime (1870–1914), defined principally by the transatlantic trade in basic food staples underpinned by British hegemony (Friedmann and McMichael, 1989), as an exporting breadbasket serving the massive demand for wheat in industrializing Europe. During this period, prairie farmers began to demand collective grain marketing as a way of addressing the structural inequalities of the grain trade and achieving orderly marketing (Irwin, 2001; Skogstad, 2008, pp. 110–11). During the crisis of the first food regime, marked by the dislocations of world war and the Great Depression, organized farmers and the state experimented with different versions of collective marketing. It wasn't until World War II, however, that the state finally implemented the "single-desk" marketing system that became the hallmark of the prairie wheat economy.

The second food regime (1945–73) emerged out of the restructuring of domestic agricultures and world trade under US hegemony. Here, governments took a stronger role in regulating agricultural prices and markets to achieve domestic rural development and food security goals. Internationally, the United States and other powerful actors structured agricultural trade around "food aid," international commodity agreements, and other mercantilist practices. In Canada, the CWB served several goals that complemented the agendas of farmers and the state, including increasing and stabilizing farm incomes and establishing new export outlets. The crisis of the second food regime, precipitated by the oil and food shocks of the 1970s, ushered in a period of intensified international competition, price volatility, and the unraveling of many postwar institutions. These changes challenged the CWB's traditional role in Canadian agricultural and trade policy, opening the door to political conflicts over its marketing role. Some farm groups began to argue that the CWB system was inefficient and bureaucratic, holding back the prairie grain sector from its true potential, a view that gained the sympathy of conservative political parties.

These tensions intensified in the 1990s, as domestic political debates and the restructuring of international agricultural markets and trade eroded the CWB's legitimacy (Skogstad, 2008, p. 107). Partly, the challenge was ideological, as the neoliberal ethos of agricultural restructuring and free trade painted that CWB as a statist anachronism. However, changing commercial conditions also posed a challenge to the CWB's centralized marketing structure. CWB critics suggested that centralized marketing of bulk commodities was ill suited to an increasingly privatized and differentiated world import market. Over the 1990s, the CWB embarked on a process of strategic reorientation, including governance changes that gave farmers direct control over the agency.[2] These changes helped defuse conflicts over the CWB for a time and provided a reasonably successful strategy for adapting to commercial changes in world wheat markets. Beginning in 2006, however, the Canadian government, seeking to complete the neoliberal transformation of the prairie grains sector, undermined the farmer-controlled CWB and eventually stripped it of the "single-desk."

I focus here on the processes of conflict and change that shaped the CWB from the 1990s to 2012, arguing that debates over its future reflected competing strategies for integrating the prairie grains sector into the neoliberal food regime. The CWB's strategy of centralized marketing, branding, and quality differentiation sought to leverage Canada's traditional strengths while responding to new forms of demand. The deregulated environment implemented by the federal government has abandoned this coordinated approach in favor of one in which farmers navigate highly concentrated global markets as isolated units, competing with one another to make grain sales. While the current food regime is subject to instabilities and contradictions, including the impasse at the World Trade Organization (WTO) (Pritchard, 2009), the global "food crisis" (Magdoff and Tokar, 2010), and long-run ecological and social contradictions (Weis, 2010; McMichael, 2010), the restructuring of the prairie grains sector suggests a deepening of the neoliberal paradigm in Canada. The prairie grains sector is likely to become further integrated into global agri-food capital, to see the further concentration of farming resources, and to shift towards production of high-yielding varieties for biofuels, feed, and industrial markets.

Towards a Third Food Regime

Since the food regimes perspective was first articulated, there has been a great deal of debate over whether, and under what conditions, a third regime has emerged. In order to understand the contours of this debate, I review some of the events that flowed from the collapse of the second food regime. The crisis of the second food regime triggered a new era of intense international competition and market volatility (Friedmann, 1993). By the early 1980s, a virtual trade war between the United States and the European Union, waged through escalating export subsidies, severely depressed world grain prices. Facing low prices, high interest rates, and rising production costs, farmers in many industrialized states experienced a farm income crisis. Meanwhile, severe debt crises in the global South caused governments to scale back or eliminate farm supports and food subsidies, often under International Monetary Fund (IMF) or World Bank-mandated structural adjustment plans (Weis, 2007). International commercial conflicts and the rising costs of farm subsidies called into question the postwar compact between farmers and the state in many countries, though the path taken in response to these conditions has varied. All of these conditions laid the groundwork for the global restructuring of agri-food sectors under neoliberal hegemony (Bonanno et al., 1994; McMichael, 1994; Goodman and Watts, 1997).

Beginning in 1986, the key players in international grain markets sought a new set of multilateral trade rules that would liberalize the agri-food sector. For the United States and the European Union, this was intended to defuse international trade conflicts arising from the export subsidy war. Unable to compete with the agro-subsidies of the European Union and the United States, smaller exporters, such as Canada, New Zealand, Australia, Thailand, and Argentina, hoped that a set of multilateral trade rules for agriculture would level the global playing field. The pursuit of free trade in agri-food was thus a return to international regulation, but this time under a neoliberal framework giving priority to free markets over international cooperation and rural development goals. The Agreement on Agriculture, a sub-agreement of the WTO signed in 1995, institutionalized these principles (Weis, 2007).

Scholars continue to debate whether conditions have stabilized sufficiently to warrant the recognition of a third food regime (Friedmann, 2009; Pritchard, 2009). Nevertheless, several recognize either an emerging (Friedmann, 2005) or actual, though crisis-laden (McMichael, 2005; McMichael, 2012) and contested (Pelchaner and Otero, 2008), regime based on several important trends. The third food regime is characterized by the consolidation of agri-food capital, the accelerating global integration of the agri-food sector, and the uneven incorporation of farmers and agricultural sectors into the global food economy. Spurred by agri-food liberalization, transnational corporations, including firms from emerging economies (Wilkinson, 2010), continue to integrate their international production and distribution networks, seeking access to cheap raw materials, advantageous labor and environmental regimes, and new markets. Many agri-food sectors are more concentrated than ever, including the global grain-trading complex, dominated by four firms: ADM, Bunge, Cargill, and Louis Dreyfus (Clapp, 2012, pp. 96–102). Agri-food firms, especially retailers, also play an increasingly prominent role in developing private quality assurance schemes (Busch and Bain, 2004; Campbell, 2005; Friedmann, 2005).

Neoliberal governments have weakened or dismantled many of the pillars of postwar agricultural policy, including price supports, supply management, collective marketing schemes, and other farm programs. This has involved a process of neoregulation, where governments actively impose market forces on the agri-food sector in line with neoliberal principles and the class projects of influential actors (Pelchaner and Otero, 2010). In many industrialized countries, farm sector rationalization involved the elimination of smaller and medium-sized operations in favor of more competitive—that is, more highly capitalized and larger—operations. Governments have attempted, with mixed results, to cut or reorient farm sector spending. In the United States, some important forms of state intervention, such as supply management, have been rolled back (Winders, 2009), although overall agricultural spending has remained essentially stable. In the international arena, the United States, along with other major agro-exporters and to a lesser extent the European Union, have promoted neoliberal reforms and trade liberalization, but with contradictory results. The European Union and the United States have been criticized for failing to significantly decrease their agricultural subsidies, even as many countries in the global South have opened their sectors to increasing imports. Rather than resolving the international tensions in the global food trade, the "free trade" agenda under the WTO has served to "lock in" and institutionalize gross North-South imbalances (Pritchard, 2009).

In Canada, neoliberal agricultural restructuring has involved three key components. First, the federal government pursued free trade agreements, both at the continental scale as well as through the WTO, with important implications for agriculture (Knuttila, 2003). The prairie grains sector would henceforth be exposed to more direct competition from other major exporters. Second, as of the early 1990s, the federal government initiated a major restructuring of farm income stabilization programs, replacing them with a new cost-shared model (Skogstad, 2008, p. 78). Third, the government deregulated statutory freight rates for grain, put in place in the late 1890s in order to mitigate the high cost of shipping grain out of the geographically isolated prairie region. Following partial deregulation in 1984, the government completely eliminated grain shipping subsidies in 1995, drastically increasing the cost to ship grain for prairie farmers (Schmitz and Furtan, 2000, p. 166). Reforms to farm programs and transportation subsidies served two goals: to reduce government spending, under pressure from mounting budget deficits, and to meet new obligations under free trade agreements.

The restructuring of the prairie grains sector had major implications for the CWB. While the CWB stood to benefit in some ways from trade liberalization, especially the reduction of competitors' price-depressing export subsidies, free trade paradoxically increased conflicts between Canada and the United States. In the North American free trade environment, Canada's exports to the United States increased sharply, prompting US farmers and government trade officials to allege that the CWB was an unfair trader (Schmitz and Furtan, 2000, pp. 111–12). Between 1989 and 2005, the United States launched fourteen separate trade challenges, none of which succeeded in proving the alleged trade distorting practices (Skogstad, 2008, p. 129). After 2005, the United States adopted a new tack, seeking instead to have state-trading enterprises, such as the CWB, prohibited under proposed rules for a new WTO agreement. From the early 1990s onward, domestic pressure on the CWB also increased. The discontent among some prairie farmers was illustrated by a civil disobedience campaign intended to discredit the single-desk marketing system. Seeking to sell grain directly to US buyers, some Canadian farmers made illegal cross-border deliveries to American elevators.[3] Both international and domestic conflicts over the single-desk heightened the tensions around the CWB's marketing role.

The restructuring of the 1990s also undermined the social basis for the CWB's political support. Farming operations became increasingly diversified by size, commodity, and political orientation, continuing a longstanding trend. By the early 2000s, a small number of very large farms accounted for the majority of total farm production and receipts for the sector. In 2005, for instance, farm operations with over C$250,000 in total farm receipts represented only 14.3 percent of the number of farms, but 57.8 percent of Saskatchewan's total farm receipts (Statistics Canada, 2006). Even as the overall number of farms continued to decline, the rate of decline was highest for small and medium operations. These changes eroded one of the traditional bases of solidarity within the prairie farming class: the relative uniformity of family farm operations. Indeed, the CWB's own producer surveys indicated that support for the CWB and for the "single-desk" system was weakest among large operations (CWB, 2006a; Cross, 2011a). The consolidation and privatization of the prairies' farmer-owned grain-handling cooperatives, the "Pools," also undermined political support for the CWB. For decades, the Pools had mobilized their membership and political influence in support of the single-desk system (Fairbairn, 1984). While the Pools provided farmer clout in the grain-handling sector, the single-desk CWB provided clout in the grain market. Beginning in the late 1990s, however, a series of mergers and acquisitions among the prairie Pools led to their eventual consolidation and privatization (Painter, 2004). By 2007, all that remained of the four prairie Pools was the multinational grain company, Viterra, now a publicly traded corporation rather than a cooperative.

With the transformation of the prairie grains sector from the mid-1980s onward, the single-desk marketing system took center stage in political debates over the sector's future. In line with the growing influence of neoliberal ideals, some economic and political actors questioned the CWB's legitimacy. For their part, farmers became increasingly divided over the question, leading to an increasingly bitter debate between "pro" and "anti" CWB factions. In response to these political challenges, the CWB embarked on a process of "institutional adaptation" (Skogstad, 2005) that drew strength from a coalition of pro-single-desk actors, including farm organizations and a sympathetic government. At the same time, the CWB faced commercial challenges resulting from a restructuring of the world grain market. This sparked a commercial reorientation that, I argue, helped the prairie grains sector respond to new forms of demand in the corporate food regime.

Commercial Reorientation

At the beginning of the 1990s, the CWB's leadership recognized the need to respond to commercial changes that posed a challenge to some aspects of the agency's marketing system (Steers, 1990). One change was the growing importance of more demanding quality specifications among wheat buyers. Partly, this change resulted from the rapid deregulation of monopoly state-importing agencies in the Former Soviet Union as well as in other key importers (Abbott and Young, 1999). Henceforth, the CWB would deal with a much larger number of private wheat buyers (e.g., millers), who, compared to state importers, are less concerned with high-volume purchases at low prices and more concerned with quality characteristics. At the same time, technological and commercial changes in milling and baking industries meant that food manufacturers developed increasingly stringent quality specifications for grain. Some wheat buyers, especially those developing market niches for premium industrial bread, developed new sourcing requirements going beyond the traditional quality assurances of the Canadian bulk handling and grading system (Kennett, 1997). These changes reflected the increasing importance of quality claims—as a basis for competition and market differentiation—more generally in the reconfiguration of agri-food sectors (Busch and Bain, 2004; Friedmann, 2005).

The CWB responded to these new circumstances, first, by capitalizing on the deregulation of state-importing agencies through a strategy of quality- and price-differentiation. Dealing directly with private buyers, the CWB sought to maximize its returns by meeting their specific quality demands and charging different prices accordingly. In this way, the CWB actually succeeded in increasing its market share among some Latin American markets following deregulation (Kraft et al., 1996). Second, the CWB helped develop new commercial relations based on tailor-made sourcing arrangements with certain high-profile buyers. This differed from the broader quality-differentiation strategy explained above in that it involved identity-preserved (IP) shipments of particular wheat varieties rather than traditional bulk shipments. Beginning in 1995, the CWB developed an IP sourcing arrangement for the British bakery Warburtons, which has pursued a successful strategy of market premiumization in the UK bread sector (Magnan, 2011a). The CWB-Warburtons contract was significant for showing how the niche demands of important buyers could be accommodated within the centralized marketing system of the CWB (Oleson, 1999). Third, and related, the CWB developed a more comprehensive branding strategy for Canadian wheat, seeking to capitalize on new opportunities for market differentiation and to consolidate its reputation for sales service and product quality and consistency.

In sum, the CWB responded to the commercial challenges of the 1990s by positioning itself as a provider of high-quality grain for premium markets. In fact, this strategy built upon a decades-old set of relationships that had defined Canadian wheat as "high-quality," a function both of geographic factors and the set of public institutions for regulating grain quality, which worked in tandem with the CWB's marketing role (Sinclair and Grieshaber-Otto, 2009). It must be pointed out that high-quality grain makes up a small proportion of the total world market. The CWB traditionally sold into premium markets, such as the UK, Japan, and North America, as well as into other markets with generally lower quality standards, such as Iran, Saudi Arabia, and Indonesia. Competitors such as the United States also sell some high-quality grain into premium markets. The difference is that by virtue of its market information and centralized control over sales, branding, and market development, the CWB was able to develop a sector-wide quality-differentiation strategy. CWB leaders believed this to the most appropriate strategy for extracting rents from the increasingly deregulated and quality-differentiated wheat market. In turn, this commercial strategy was deployed in the service of its public mandate to maximize returns for farmers.

The modernizing CWB of the 1990s therefore mediated state-market-society relations in a very particular way. The agency derived its powers from government statute, yet operated at arm's length from government policy as the agent of industrial grain farmers. Although its single-desk structure contradicted "free market" ideals, the CWB pursued a commercial strategy quite compatible with a neoliberalizing agri-food environment. The wheat and barley sector continued, nevertheless, to face major challenges, including chronically depressed prices and continual trade harassment from the United States. While the commodity price spikes of recent years have benefited grain farmers, some of this has been offset by the rising value of the Canadian dollar.

Political Evolution

Though some farmers and farm organizations began calling for the end of the CWB's marketing monopoly in the late 1970s, the debate over the single-desk intensified considerably in the 1990s. In 1993, the Conservative government attempted to remove the CWB's monopoly power over barley sales, a move intended to satisfy political constituencies convinced that the single-desk system was holding back the growth of value-added and livestock industries. This prompted a legal battle between the government and CWB supporters over the terms under which the agency's mandate could be changed. The courts ruled that major changes to the CWB system required legislative approval and could not be made by simple executive order (Skogstad, 2005, p. 538). Following a federal election in 1993, the incoming Liberal government launched a federal inquiry that would deal with the controversial CWB issue.

The Western Grain Marketing Panel report of 1996 proposed several changes to the marketing system, but the government decided not to act on the recommendations without first consulting farmers. To this end, it conducted a farmer plebiscite, in 1997, on the CWB's marketing powers over barley. Over 60 percent of prairie farmers voted to retain the single-desk selling system (Skogstad, 2005). Following the plebiscite, the government changed the CWB's governing legislation in order to institutionalize two key principles: (1) The CWB would henceforth be governed by farmers, through a new board of directors controlled by elected farmer-members, and (2) No changes to the CWB's marketing mandate could be made without first consulting farmers (e.g., through a plebiscite), second, consulting the CWB's leadership, and finally, obtaining parliamentary approval. By devolving governance of the CWB in this way, the government avoided making a politically difficult decision and made prairie farmers masters of their own destiny on the marketing question. The CWB's legitimacy would henceforth depend on the level of farmer buy in for the farmer-controlled agency.

Under farmer control, the CWB launched a number of new initiatives. It introduced new, more flexible payment and delivery options for farmers, called Producer Payment Options, which allowed farmers to price their grain using instruments that mimicked contracts available on the open market. The farmer-elected board of directors also led the CWB to take on a more prominent role as a farmer advocate in the grain industry. Beginning in the late 1990s, the CWB repeatedly challenged the prairies' rail duopoly in complaints over rail service and freight rates (Schmitz and Furtan, 2000, pp. 171–2). The CWB also led a broad-based coalition of actors in opposing the commercialization of Monsanto's Roundup Ready wheat in Canada, something it believed necessary for protecting lucrative export markets in GM-wary countries, such as Japan and the European Union (Magnan, 2007). At a cost of tens of millions of dollars, the CWB defended the prairie grain industry against a series of US trade complaints and attempts to delegitimize it at the WTO. Here, the CWB successfully demonstrated that it did not engage in unfair trade practices (Skogstad, 2008, p. 133).[4]

The significance of the CWB's political evolution after 1998 was twofold. First, it allowed the CWB, through its advocacy role, to partially fill the political void left by the decline of the prairie Pools. Second, it empowered farmers, through the farmer-elected board of directors and the plebiscite mechanism, to take direct responsibility for running the agency and for the future of the "single-desk" system. This governance structure reflected neoliberal ideals to the extent that it sought to empower farmers as economic actors. Here, the farmer-elected CWB leadership framed the benefits of the "single-desk" system in terms of its "value proposition"—that is, its ability to manage risk and extract rents from the market. Although farmers committed to free market ideals continued to oppose the "single-desk," the CWB could draw legitimacy from the fact that it was now a farmer-controlled agency. Acute conflicts over its marketing role therefore declined considerably between 1998 and 2006.

Dismantling the Single-Desk: 2006–12

This situation changed with the election of a Conservative government in 2006. Vocal opponents of collective marketing, the new government announced its intention to dismantle the single-desk for barley. The Conservative's opposition was to some degree philosophical, following a libertarian argument that considered the monopoly powers of the CWB an infringement of farmers' property rights. This argument was accompanied by claims that the monopoly marketing system was inefficient, market distorting, excessively bureaucratic, and outdated. The government therefore proposed to replace the single-desk with a "dual marketing" system under which farmers would be free to sell their grain through the CWB or to any other buyer. The plausibility of a dual market is contested by many single-desk supporters and some agricultural economists who dispute the idea that a voluntary CWB could survive in an environment dominated by large, multinational grain companies (Fulton, 2006).[5]

In pursuing its goal of implementing a "dual market" for barley between 2006 and 2008, the government's actions undermined farmer control of the CWB (Magnan, 2011b). Among other things, the government refused to consult with promonopoly farm groups, replaced four government-appointed members of the CWB board of directors with supporters of Conservative policy,[6] refused to commit to consulting prairie farmers by plebiscite, and fired the CWB's president and CEO for speaking out against its actions. Under pressure from farm organizations, who generally refused to side with either the pro- or anti-single-desk position but supported farmers' right to decide, the government eventually agreed to conduct a second plebiscite on the barley question. The plebiscite presented farmers with three options: the status quo, the elimination of the CWB, or the "dual market." While 38 percent of farmers voted to retain the single-desk, 48 percent voted for "choice," and only 14 percent voted for the CWB's abolition. Combining the votes for the latter two options, the government concluded that 62 percent of farmers favored a change to the system (AAFC, 2007). Many farm groups disputed this interpretation, however, since none of the options received a majority of votes and because the "dual market" option was vague and misleading. The government proceeded nevertheless to introduce an executive order to strip the CWB of its monopoly over barley. As in 1993, however, the change was blocked by a court ruling that upheld the requirement for the government to obtain parliamentary approval for major changes to the CWB (Hansen, 2007). Without a parliamentary majority, the Conservatives were prevented from following through on their policy.

The Conservative's electoral victory in 2011, which yielded a parliamentary majority, prompted the government to relaunch its campaign to dismantle the single-desk, this time for both wheat and barley. The government argued that the federal election results, in which it received a large share of the prairie rural vote, gave it a clear mandate to proceed with its goal. The CWB and its supporters continued to insist that any major change to the agency had to be decided by farmers. As the president of the CWB put it:

> The Wheat Board belongs to us. As farmers, we pay for its operations from the sale of our grain. We run it, through our elected representatives on its board of directors. But we are not being allowed to decide its future. (2011a)

Faced with the government's refusal to conduct a plebiscite, the CWB organized its own vote in August 2011. Of the 38,000 producers who participated (a participation rate of 56 percent), 62 percent voted to retain the single-desk for wheat and 51 percent voted to retain the single-desk for barley (CWB, 2011b). The Minister of Agriculture disregarded the results, declaring that "No expensive survey can trump the individual right of farmers to market their own grain" (Cross, 2011b, p. 2). The government proceeded to introduce legislation that dissolved the farmer-controlled board of directors, brought the agency under direct government control, and stripped it of its "single-desk" powers. As of August 1, 2012, the CWB began to compete directly with grain companies for farmers' business.

Throughout this episode, the CWB's farmer-elected leaders, "single-desk" supporters, and opposition parties resisted the government's actions. Farm groups engaged both in mobilization and court challenges. The Canadian Wheat Board Alliance, founded in 2006, launched a series of rallies and public information meetings across the prairies. As for legal challenges, a group called the Friends of the CWB, together with the CWB's elected farmer-directors, contested the government's right to eliminate the single-desk without adhering to the democratic principles enshrined in the CWB Act. The challengers achieved a partial victory when, in December 2011, the Federal Court of Canada ruled that the government had acted illegally in introducing the legislation to dismantle the CWB without first having conducted a plebiscite (Campbell, 2011). This decision did not have the legal force to prevent the legislation to come into effect, however, and was later overturned on appeal.

The political struggle over the single-desk from 2006 to 2012 pitted a government committed to a neoliberalizing agenda against the coalition of CWB supporters. Though CWB supporters argued forcefully for the government to allow farmers to determine the fate of the agency, the governing party used its state powers to overrule the principles of farmer control put in place by a previous government. Here, the government acted on the desires of some CWB opponents who have argued that the "single-desk" system was illegitimate whether or not a majority of farmers supported it. In the government's political calculation, its tactics, though heavy-handed, were unlikely to upset its core constituency of conservative rural voters. In dismantling the "single-desk," the Conservatives realized a long-standing policy goal, one that would significantly reorient the prairie grains sector.

Discussion

Government actors and their supporters have advanced a narrative based on claims that eliminating the single-desk system will bring many benefits to the prairie grains sector. This narrative is emblematic of neoliberal claims about the power of individual entrepreneurialism and market efficiency. The federal agriculture ministry, for instance, suggested that the open market would lead to more investment, innovation, and value-added jobs in the sector, in turn unleashing "the true economic potential and entrepreneurial energy of the Western Canadian grain sector" (AAFC, 2011). Under this scenario, it is understood that the end of the single-desk will differentially affect individual farm operations. In line with the neoliberal narrative, however, this is not problematic since competition will help weed out the least efficient farms.

Opposing this logic is the argument that the end of the single-desk will harm the collective interests of prairie farmers as an increasing share of the value produced by farmers flows to the large grain companies that dominate the sector. If, as some economic analyses suggest, the single-desk allowed farmers to capture price premiums from the market (Furtan et al., 1999), those premiums will evaporate. Estimated to be several hundred million dollars per year, the premiums will instead be captured by grain companies. Key grain trading firms, such as Richardson International, Bunge, Louis Dreyfus, and ADM, have anticipated greater participation in the sector in the new environment (Waldie, 2012; Cross, 2012). Viterra, the prairies' largest grain company, foresees increasing its profitability by forty to fifty million Canadian dollars per year with the end of the single-desk (Waldie, 2012). These circumstances will encourage further consolidation of the prairie region's grain-handling and transportation system. Two independent studies predicted that without the single-desk CWB, many smaller, independent grain-handling companies will disappear (*Western Producer*, 2011). Indeed, some of the sector's key players are in the process of significant restructuring in anticipation of economic gains they expect to see materialize with the demise of the single-desk. In March 2012, global commodities giant Glencore International PLC announced its intention to buy Viterra for C$6.1 billion (Briere, 2012). This will replace the prairies' own multinational grain firm with a foreign-owned entity.

Stripped of its monopoly powers, the CWB is seeking to develop a viable commercial model. It continues to offer pooling options as well a number of other contracts that farmers can sign up for on a voluntary basis. Because it has no grain-handling infrastructure of its own, however, it must depend on its competitors for access delivery points in order to make grain sales. To this end, the CWB has signed strategic partnerships with all of the major grain-handling companies (CWB, 2012). The prospects for success of this model are uncertain, however, since grain companies are likely to give priority to handling grain related to its own sales rather than to the CWB's contracts. One possibility over the medium term is that the CWB—now a small grain company—will be absorbed by one of its much larger competitors. This is what happened in Australia, when Cargill acquired the deregulated Australian Wheat Board in 2011.

Finally, there are a number of indirect consequences the end of the single-desk may have on the prairie grains sector. The most significant are:

- the increased likelihood that genetically engineered wheat will be introduced given that the CWB was among its most influential opponents;
- the erosion of Canada's quality assurance system for wheat and barley. The single-desk structure of the CWB allowed farmers to capture a quality premium flowing from a set of public institutions for market development, grading, branding, and marketing. Without the single-desk CWB, there is less incentive for public investment in Canada's quality assurance system. Private grain companies do not have the same incentive to preserve Canada's

overall reputation for grain quality, since they are focused on differentiating themselves from their competitors. Furthermore, the business model of the large grain companies depends on maximizing the volume of grain passing through their facilities more than on grain quality (McCann, 2011);
- the threat posed to farmers' right to load their own grain for shipment in "producer cars." This right, which is enshrined in Canadian law, allows farmers to bypass the corporate grain-handling industry by giving grain producers direct access to rail cars. Historically, the vast majority of producer car usage has been for transporting wheat and barley and was coordinated by the CWB. With its large sales program and predictable supplies, the CWB was able to coordinate producer car deliveries to coincide with sales opportunities. Given the relatively small quantity of grain in question, it is doubtful that private grain companies and railways will be prepared to accommodate the needs of producer car shippers. If producer car shipping declines, this will harm the short-line, independent railways that depend on this cargo for a large share of their business.

These changes could lead to a significant reorientation of the prairie grain economy. The sector is likely to become more closely integrated to the US industry, sharing a continental network for grain handling, storage, and transportation. The historic tendency for the prairie grains sector to emphasize sales of high-quality wheat and barley could also be undermined. Individual farmers will, of course, pursue their own quality-differentiation strategies, seeking sales to niche buyers and markets, but the sector will no longer have the CWB as a champion of the Canadian brand. Instead, a variety of actors—whether farmers or grain companies—will compete with one another to differentiate their products and make sales. The end of Canada's centralized approach to quality-differentiation, branding, and marketing could open the door to more high-yielding, lower quality varieties destined for feed, biofuels, and industrial uses, an outcome promoted by some of the CWB's longtime opponents (e.g., Carter et al., 2006). This is a risky strategy, however, since Canada is ill suited to compete with low-cost producers. Prairie farmers are much further from port than their peers, have relatively high production costs, and have shorter, harsher growing seasons.

As for farmers, the open market environment will favor larger operations that are better able to achieve the economies of scale necessary to compete in a high-volume, low-margin environment. Larger operations, especially those able to devote significant resources to sales and marketing, may also be better positioned to capture higher prices than smaller farms with few options for selling their grain other than at the local elevator. Finally, many farmers will have lost price pooling as a low-cost form of risk management. Farmers selling in the open market will have greater responsibility for protecting against sudden market fluctuations. In an environment of generally high commodity prices, such as has existed between 2007 and 2012, the risk-management advantages of price pooling may seem less salient. This could change in a period of depressed or volatile prices.

The dismantling of the CWB was a political project driven by the Conservative government, farmer allies, and the grain industry. Given the uneven effects the change will have on agricultural sector players—increasing risks and diminishing returns for family farmers while expanding accumulation opportunities for capital—the project can be read in class terms. Single-desk supporters and their allies contested the change on the basis of defending farmers' collective economic interests and their right to decide the future of the CWB's marketing system. The government's actions, inspired both by ideology and political calculation, provide a vivid example of the state imposing market logic in the name of neoliberal ideals. Yet, given that previous neoliberal governments (i.e., the Liberals) chose to devolve the CWB's governance to farmers rather than simply dismantle it, the agency's trajectory of change was relatively open. For this reason, the CWB case confirms the importance of paying attention to the contested processes by which social actors in national contexts shape the contours of the neoliberal food regime (Pelchaner and Otero, 2010).

The case also provides an instance in which competing coalitions of actors struggle over the terms under which an agricultural sector will be integrated into broader market relations in the neoliberal food regime. The anti-CWB coalition, including the current government, envisions a reorientation of the prairie grains sector towards low-margin, low-cost production controlled by a shrinking number of highly industrialized and vertically integrated farms. The pro-CWB coalition, by contrast, had staked the future of the wheat and barley sector on a strategy for targeting premium markets, including both niche markets and bulk shipments, under the centralized coordination of a farmer-controlled entity. Far from being incompatible with the neoliberal food regime, there are successful examples of monopoly-based farmer-owned agri-food entities, notably in the New Zealand kiwi and dairy sectors (Campbell, 2005; Stringer et al., 2008). The dismantling of the single-desk system was therefore not a failure of the CWB's commercial strategy, per se, but the outcome of a political contest over different ways of integrating the prairie grain sector into world markets.

Conclusion

The fate of the CWB in the third food regime is somewhat paradoxical. Between 1990 and 2006, the CWB developed a strategy for integrating the prairie grains sector to the new commercial and political conditions of the time. The idea was to preserve the benefits of single-desk marketing while pursuing the new opportunities presented by changing quality demands and new market participants. When farmers gained control of the agency in 1998, the political debate over the single-desk became a question to be resolved by farmers, not the state. Inspired by neoliberal discourses of market freedom, entrepreneurialism, and choice, the Canadian government undermined and replaced this model with one based on competition among farmers and further penetration of nonfarm capital into the sector.

The path towards neoliberalizing the prairie grains sector has been far from straightforward, involving contestation among competing coalitions of social actors at every stage. The contest over the CWB specifically suggests that there is no mechanical or necessary link between the neoliberal food regime and a given configuration of state-market-society relations. It is conceivable that under a different political climate, the modernizing CWB of the 1990s and early 2000s would have continued to carve out a successful commercial strategy in the global agri-food sector. Indeed, in 2006, the CWB farmer leadership proposed a major restructuring that would have further entrenched farmer control, rendered it more independent from government, and allowed the CWB to invest in other parts of the grain value chain as a means of providing farmers more clout in the farm inputs and food processing sectors (CWB, 2006b). These possibilities were foreclosed by the government's unraveling of farmer-control and the single-desk system. In the process, the prairie wheat economy has lost one of its most distinctive features—centralized grain marketing in the name of farmers.

Notes

1 The CWB's jurisdiction covered the provinces of Manitoba, Saskatchewan, and Alberta, as well as the grain-growing region of northwestern British Columbia. The CWB had the exclusive right to sell prairie wheat and barley, with the exception of animal feed for Canadian use.

2 Until 1998, the CWB's leadership consisted of government-appointed commissioners.

3 The farmers' actions violated the Customs Act, which required that grain sellers obtain an export permit from the CWB for any foreign grain sales.

4 After losing its case against the CWB at the WTO tribunal, the United States lobbied for language prohibiting state-trade enterprises in the text for the Doha Round negotiations.

5 One of the key arguments is that since the CWB owns no grain elevators, it must depend on its direct competitors, grain-handling companies such as Cargill and Viterra, for access to the grain-handling system. Grain-handling companies are likely to give priority to moving grain related to its own sales rather than the CWB's.

6 After the 1998 governance reforms, the CWB was governed by a board of directors composed of ten elected farmers and five government appointees.

References

AAFC (Agriculture and Agrifood Canada). (2007) "Barley producers choose marketing choice," news release, March 28.

AAFC (Agriculture and Agrifood Canada). (2011) "Marketing freedom for Grain Farmers Act gives farmers and economy more opportunities," press release, November 2, available at www.agr.gc.ca/cb/index_e.php?s1=n&s2=2011&page=n111102, accessed April 26, 2013.

Abbott, P. C. and Young, L. M. (1999) "Wheat-importing state trading enterprises: Impacts on the world wheat market," *Canadian Journal of Agricultural Economics/ Revue Canadienne d'Agroeconomie*, vol 47, no 2, pp. 119–36.

Bonanno, A., Busch, L., Friedland, W. H., Gouveia, L., and Mingione, E. (eds) (1994) *From Columbus to ConAgra: The Globalization of Agriculture and Food*, University Press of Kansas, Lawrence.

Briere, K. (2012) "Viterra sale in hands of regulator," *Western Producer*, March 29, p. 3.

Busch, L. and Bain, C. (2004) "New! Improved? The transformation of the global agrifood system," *Rural Sociology*, vol 69, no 3, pp. 321–46.

Campbell, H. (2005) "The rise and rise of EurepGAP: European (re)invention of colonial food relations?," *International Journal for the Sociology of Agriculture and Food*, vol 13, no 2, pp. 1–19.

Campbell, J. (2011) "Federal Court of Canada decision, Docket T-1057–11," Federal Court Canada, decisions.fct-cf.gc.ca/en/2011/2011fc1432/2011fc1432.html, accessed April 26, 2013.

Carter, C., Berwald, D., and Loyns, A. (2006) *The Economics of Genetically Modified Wheat*, University of Toronto Centre for Public Management, Toronto.

Clapp, J. (2012) *Food*, Polity Press, Malden, MA.

Cross, B. (2011a) "Both sides use CWB survey results as fodder," *Western Producer*, July 21, www.producer.com/2011/07/both-sides-use-cwb-survey-results-as-fodder/, accessed April 26, 2013.

Cross, B. (2011b) "Majority favour CWB: Plebiscite," *Western Producer*, September 15, pp. 1–2.

Cross, B. (2012) "Grain firm paints rosy picture for marketing system," *Western Producer*, January 26, p. 25.

CWB. (2006a) *Canadian Wheat Board Annual Producers' Survey*, Winnipeg.

CWB. (2006b) *Harvesting Opportunity: Strengthening Farmers' Competitive Advantage*, Winnipeg.

CWB. (2011a) "Farmers urged to face future with eyes wide open: CWB chair," news release, June 16.

CWB. (2011b) "Farmers vote to keep Canadian Wheat Board," news release, September 12.

CWB. (2012) "CWB completes agreements with all Prairie grain handlers," news release, available at www.cwb.ca/news/16/cwb-completes-agreements-with-all-prairie-grain-handlers, accessed April 26, 2013.

Fairbairn, G. (1984) *From Prairie Roots: The Remarkable Story of Saskatchewan Wheat Pool*, Western Producer Prairie Books, Saskatoon.

Friedmann, H. (1993) "The political economy of food: A global crisis," *New Left Review*, vol 197, pp. 29–57.

Friedmann, H. (2005) "From colonialism to green capitalism: Social movements and emergence of food regimes," Pp. 227–64 in F. Buttel and P. McMichael (eds) *New Directions in the Sociology of Global Development, Research in Rural Sociology and Development*, Elsevier, Oxford.

Friedmann, H. (2009) "Discussion: Moving food regimes forward: Reflections on symposium essays," *Agriculture and Human Values*, vol 26, no 4, pp. 335–44.

Friedmann, H. and McMichael, P. (1989) "Agriculture and the state system: The rise and decline of national agricultures, 1870 to the present," *Sociologia Ruralis*, vol 29, no 2, pp. 93–117.

Fulton, M. (2006) "The Canadian Wheat Board in an open market: The impact of removing the single-desk selling powers," Saskatoon: Knowledge Impact in Society Project, University of Saskatchewan, available at kis.usask.ca/publications/pubcwbliterature.html, accessed February 3, 2012.

Furtan, W. H., Kraft, D. F., and Tyrchniewicz, E. W. (1999) "Can the Canadian Wheat Board extract monopoly rents? The case of the spring wheat market," *International Journal of the Economics of Business*, vol 6, p. 3.

Goodman, D. and Watts, M. (eds) (1997) *Globalising Food: Agrarian Questions and Global Restructuring*, Routledge, New York.

Hansen, D. (2007) "Federal Court of Canada Decision, docket T-1124–07," Federal Court Canada, decisions.fct-cf.gc.ca/en/2007/2007fc807/2007fc807.html, accessed April 26, 2013.

Harvey, D. (2005) *A Brief History of Neoliberalism*, Oxford University Press, New York.

Irwin, R. (2001) "Farmers and 'orderly marketing': The making of the Canadian Wheat Board," *Prairie Forum*, vol 26, no 1, pp. 85–106.

Kennett, J. (1997) "An examination of bread wheat quality and its effect on vertical co-ordination in the wheat supply chain," Master's thesis, Department of Agricultural Economics, University of Saskatchewan.

Knuttila, M. (2003) "Globalization, economic development and Canadian agricultural policy," Pp. 289–302 in H. Diaz, J. Jaffe, and R. Stirling (eds) *Farm Communities at the Crossroads: Challenge and Resistance*, Canadian Plans Research Centre, Regina.

Kraft, D. F., Furtan, W. H., and Tyrchniewicz, E. W. (1996) *Performance Evaluation of the Canadian Wheat Board*, Canadian Wheat Board, Winnipeg.

Magdoff, F. and Tokar B. (eds) (2010) *Agriculture and Food in Crisis: Conflict, Resistance and Renewal*, Monthly Review Press, New York.

Magnan, A. (2007) "Strange bedfellows: Contentious coalitions and the politics of GM wheat," *Canadian Review of Sociology and Anthropology*, vol 44, no 3, pp. 289–317.

Magnan, A. (2011a) "Bread in the economy of qualities: The creative reconstitution of the Canada-UK commodity chain for wheat," *Rural Sociology*, vol 76, no 2, pp. 197–228.

Magnan, A. (2011b) "The limits of farmer-control: Food sovereignty and conflicts over the Canadian Wheat Board," Pp. 114–33 in H. Wittman, A. A. Desmarais, and N. Wiebe (eds) *Food Sovereignty in Canada: Creating Just and Sustainable Food Systems*, Fernwood Books, Halifax.

McCann, B. (2011) "Canada's image for quality wheat bound to suffer," *Saskatoon Star Phoenix*, August 26.

McMichael, P. (2005) "Global development and the corporate food regime," Pp. 265–99 in Buttel and McMichael (eds) *New Directions in the Sociology of Global Development.*

McMichael, P. (2010) "The world food crisis in historical perspective," Pp. 51–67 in Magdoff and Tokar (eds) *Agriculture and Food in Crisis.*

McMichael, P. (2012) "The land grab and corporate food regime restructuring," *Journal of Peasant Studies*, vol 39, no 3–4, pp. 681–701.

McMichael, P. (ed.) (1994) *The Global Restructuring of Agro-food Systems*, Cornell University Press, Ithaca, NY.

Oleson, B. (1999) "The CWB in today's regulatory and trading environment: Adaptation to-date, building blocks for the future," *Canadian Journal of Agricultural Economics/Revue canadienne d'agroeconomie*, vol 47, no 4, pp. 509–18.
Painter, M. (2004) "Saskatchewan Wheat Pool," *International Food and Agribusiness Management Review*, vol 7, no 3, pp. 70–99.
Pelchaner, G. and Otero, G. (2008) "The third food regime: Neoliberal globalism and agricultural biotechnology in North America," *Sociologia Ruralis*, vol 48, no 4, pp. 351–71.
Pelchaner, G. and Otero, G. (2010) "The neoliberal food regime: Neoregulation and the new division of labor in North America," *Rural Sociology*, vol 75, no 2, pp. 179–208.
Pritchard, B. (2009) "The long hangover from the second food regime: A world-historical interpretation of the collapse of the WTO Doha Round," *Agriculture and Human Values*, vol 26, pp. 297–307.
Schmitz, A. and Furtan, H. (2000) *The Canadian Wheat Board: Marketing in the New Millennium*, Canadian Plains Research Centre, Regina.
Sinclair, S. and Grieshaber-Otto, J. (2009) *Threatened Harvest: Protecting Canada's World-Class Grain System*, Canadian Centre for Policy Alternatives, Ottawa, www.policyalternatives.ca/sites/default/files/uploads/publications/National_Office_Pubs/2009/Grain_Regulation_03_2009.pdf, accessed April 26, 2013.
Skogstad, G. (2005) "The dynamics of institutional transformation: The case of the Canadian Wheat Board," *Canadian Journal of Political Science*, vol 38, no 3, pp. 529–48.
Skogstad, G. (2008) *Internationalization and Canadian Agriculture: Policy and Governing Paradigms*, University of Toronto Press, Toronto.
Statistics Canada. (2006) "Census of agriculture, Highlights and analyses," Statistics Canada, www.statcan.gc.ca/ca-ra2006/analysis-analyses/sask-eng.htm, accessed April 26, 2013.
Steers, B. (1990) *Report of the Review Panel to the Canadian Wheat Board*, Winnipeg.
Stringer, C., Tamásy, C., Le Heron, R., and Gray, S. (2008) "Growing a global resource-based company from New Zealand: The case of dairy giant, Fonterra," Pp. 189–99 in Stringer and Le Heron (eds) *Agri-Food Commodity Chains and Globalising Networks*.
Waldie, P. (2012) "After wheat monopoly's demise, grain sellers see better profits," *Globe and Mail*, January 1, p. B1.
Weis, T. (2007) *The Global Food Economy: The Battle for the Future of Farming*, Zed Books, New York.
Weis, T. (2010) "The accelerating biophysical contradictions of industrial capitalist agriculture," *Journal of Agrarian Change*, vol 10, no 3, pp. 315–41.
Western Producer. (2011) "Big grain handlers should prosper," June 2, p. 15.
Wilkinson, J. (2010) "The globalization of agribusiness and developing world food systems," in Magdoff and Tokar (eds) *Agriculture and Food in Crisis*.
Winders, B. (2009) *The Politics of Food Supply: U.S. Agricultural Policy in the World Economy*, Yale University Press, New Haven, CT.

5

SITUATING NEOLIBERALIZATION

Unpacking the Construction of Racially Segregated Workplaces

Jill Lindsey Harrison

Introduction

In this chapter, I explore how neoliberalization works out on the ground in the context of the farm workplace. How, precisely, do neoliberal trends influence farm labor relations and the experiences of hired workers on farms? I pursue this question in the context of dairy farming in Wisconsin, where labor relations have become remarkably segregated by race and nativity in recent years. This chapter is part of a larger project in which I use original data from a survey of Wisconsin dairy farmers and workers, exploratory farmer focus groups, and in-depth interviews with employers and workers in order to describe and explain the emergence of agricultural workplace inequalities in the contemporary context of US agriculture (see Harrison and Lloyd, 2012, 2013).

I first briefly review the scholarship on the neoliberalization of agriculture and food; I note that few studies have investigated the ways in which neoliberal policies and ideology work out in the context of farm workplaces and describe how I will do so in this paper. I then describe recent changes in Wisconsin's dairy farm labor force, highlighting patterns of segregation by nativity and race in which the low-end dairy farm jobs are rapidly becoming "brown-collar" jobs conducted by Mexican and Central American immigrants. I then explain how these workplace inequalities came to be. As I will show, neoliberalization shapes the lives of farm workers in various ways and, in so doing, contributes to workplace inequalities along lines of nativity and race. I also show that we cannot understand the influence of neoliberalization on the farm workplace apart from its intersection with non-neoliberal structures, including longstanding Fordist policies that have fueled farm industrialization and contemporary immigration politics that discipline certain migrant workers to accept low-end jobs on the farm. I emphasize that contemporary immigration politics effectively strengthen the neoliberal project and thus render it more resilient. I conclude with a few suggestions about ways in which justice-oriented scholars can constructively apply these findings.

Research on the Neoliberalization of Agriculture and Food

Building off the broader literature on neoliberalism, many scholars in the interdisciplinary field of agri-food studies have worked to characterize the neoliberalization of the agriculture and food sector. Notwithstanding the important point that neoliberalization has unfolded unevenly in this sector (see chapter by Wolf in this volume, as well as Harrison, 2008) and that the trends and their consequences are often contested (see chapter by Bonanno in this volume), we can nonetheless recognize some important patterns in the ways that various actors have participated in neoliberalization. Notably, in the past thirty years, states have engaged in a significant amount of deregulation; at the same time that states have increasingly rolled back their entitlement programs, agri-environmental regulations, and protections for workers, they have embraced and endorsed market-based and voluntarist solutions to environmental and social problems. Such reforms have been promoted on a variety of grounds, including libertarian notions of justice, utilitarian arguments about the inefficiencies of Fordist policies and institutions relative to market-based solutions, and moral arguments about the "undeserving poor" that are usually based on gendered and racialized stereotypes (Harvey, 2005; Harrison, 2011).

Neoliberal deregulation has been accompanied (and in part compelled) by various forms of *re*regulation, notably through the ways in which international bodies like the World Trade Organization, World Bank, and International Monetary Fund mandate market liberalization and the privatization of state institutions, extend and enforce intellectual property rights, and restrict state-led protections of domestic producers, labor, and the environment (e.g., see Harvey, 2005 and chapter by Busch in this volume). These trends serve the interests of industry well, which in turn has developed an extraordinary array of private standards, labeling systems, and third party certification for regulating the flow of commodities and use of labor (Busch and Bain, 2004; Ransom, 2007; Busch, 2010; Bonanno and Cavalcanti, 2012; Quark, 2013). In some areas, states have started to reject neoliberal mandates by renationalizing key industries and decommodifying certain resources. Yet, as Kaup (2013) has shown, such efforts can be thwarted by the material legacies of neoliberal reforms.

Additionally, many of the practices that some refer to as "resistance to neoliberalism" are arguably better understood as embodying neoliberal principles. Many of the most vibrant forms of agri-food activism (organic, local food systems, fair trade, and diet reformism, as well as some forms of food charity and international agri-environmental treaties) abandon state-based solutions and/or embrace market-based, voluntarist, localist, entrepreneurial, and self-improvement-oriented approaches to addressing agri-food problems (e.g., Poppendieck, 2000;

Allen et al., 2003; Guthman and DuPuis, 2006; Brown and Getz, 2008; Guthman 2008a, 2008b, 2011; Harrison, 2008, 2011; Gareau, 2012). Moreover, although it does not employ the term neoliberalism, Szasz's (2007) recent work illustrates the neoliberalization of popular culture in terms of how the public reacts to environmental threats to bodily health.

Some scholars have started the important work of considering how these patterns of neoliberalization affect farm workers in the context of Latin America (e.g., Bonanno and Cavalcanti, 2012). However, little attention has been paid to how neoliberalization plays out in the agricultural workplace in the United States (though see Brown and Getz, 2008). In this chapter, I address this gap by describing how neoliberal politics work out on the ground in the context of dairy farm workplaces in Wisconsin. How do these trends influence farm labor relations and the experiences of hired workers on these farms?

In this paper, I showcase key findings from my case study of labor relations on Wisconsin dairy farms. These data come from a larger project in which, from 2007 to 2011, my research assistants and I conducted a statewide survey of dairy farmers (n = 83) and workers (n = 373), exploratory farmer focus groups (n = 50 farmer participants), and in-depth, follow-up interviews with a subset of the employers (n = 20) and immigrant workers (n = 12) who had participated in the survey. These methods were oriented toward describing Wisconsin's dairy farm labor force, identifying the primary factors shaping employers' hiring and management decisions, understanding immigrant workers' experiences living in Wisconsin, and identifying the primary factors influencing immigrant workers' time management decisions.

To understand how contemporary immigration politics work out on the ground in the farm workplace, I draw on recent scholarship on illegality (Coutin, 2000; De Genova, 2005; Ruhs and Anderson, 2007). Illegality scholars call for critically reflecting upon, rather than assuming, what it means to be illegal in any given context and identifying the implications of immigration policing practices for material outcomes such as workplace inequality. I therefore highlight in this paper how immigration politics has changed in recent years, how workers and employers alike perceive, experience, and respond to immigration politics, and the resulting implications for workplace relations. All quotations without citations come from my own surveys and interviews.

Changes in Work and Workers on Wisconsin Dairy Farms

Although Wisconsin dairy farms have long exemplified the pastoral model of small-scale "family" farms, the sector is nonetheless industrializing in ways similar to other agricultural and nonagricultural industries (Barham et al., 2005). In dairying, industrialization includes expanding the farm's herd size; regularly upgrading the barns, milking parlors, tractors, and other equipment with the latest technologies; confining the cows indoors (i.e., rather than managing them on pasture); milking the cows three times per day in machine-assisted milking parlors (compared to historical practice of twice per day, in tie-stall or stanchion barns), which enable cows to be milked more expediently and which reduce ergonomic strain for the worker than the older parlor designs; feeding scientifically formulated feed rations (i.e., not pasture) tailored to each animal's life stage; and, on some farms, using synthetic bovine growth hormone (Barham et al., 2005). Typically, these changes are motivated by a need or desire to increase productivity in order to maintain competitiveness. In many cases, expansion is the result of two existing farms joining together to increase efficiencies or to meet social obligations (i.e., so that children or other relatives can join and be supported by the business). In all cases, these investments are not scale-neutral; capitalizing on some of these technologies requires adopting some of the others and expanding the operation. These facets of industrialization unfold at various paces and on farms of a wide range of sizes; industrialization is thus best understood as a widespread (albeit uneven) process, which is obscured by popular but misleading narratives that characterize farms as being either industrialized/modern or nonindustrialized/traditional.

Most dairy farmers manage farm expansion by hiring employees. As of 2006, at least 23 percent of all Wisconsin dairy farms hired some nonfamily labor, a percentage that increases quickly as herd size grows (Lloyd et al., 2006). The number of employees varies according to farm size and family involvement. The highest number of employees recorded in our survey was sixty-two. However, most dairy farms are still quite small, and thus most farms that have any employees only hire one or two workers.

As dairy farmers expand their operations, they assign the vast majority of the workers they hire to one task: milking cows. Farmers emphasized that hiring employees to milk the cows frees them from this chore that they personally despise, given its monotonous and round-the-clock nature. Milkers work eight- to twelve-hour shifts in a milking parlor that typically runs nearly twenty-four hours a day. In some cases, milkers work split shifts, where they milk for approximately five hours in the morning and then again in the evening (or late at night). Although the milkers we interviewed expressed considerable appreciation for the fact that their jobs are (typically) full-time and year-round (unlike most other entry-level agricultural jobs, which are seasonal) and have been at their current place of employment for several years on average, they also lamented that milking tends to be monotonous, dirty, and physically arduous, and entails significant risks of ergonomic strain and injury from large animals. Milking jobs are largely deskilled, require working on the weekends, often require working night shifts, and receive the lowest wages on the farm (nine US dollars per hour on average, with few nonwage benefits) (Harrison and Lloyd, 2012).

To fill these new positions, dairy farmers have had to find and retain workers who would accept these terms. Overwhelmingly, the farmers we interviewed and those in our focus groups elaborated about the challenges in finding and retaining US-born workers to fill the milking positions on their farms (or white, local, or American workers, to use the employers' terms). The employers elaborated at length about how US-born workers would reject the milking positions and the weekend and late-night shifts. Accordingly, in just the past ten to fifteen years (since 2000, on average), Wisconsin dairy farmers started hiring immigrant workers for these milking positions (or, in the employers' terms, Mexican or Hispanic workers). Our 2008 survey of Wisconsin dairy farms found that nearly half of all hired workers on Wisconsin dairy farms are immigrants from Mexico and Central America, many of whom are unauthorized.[1] Most of those foreign-born workers are located in the milking positions (see Figure 5.1).

In contrast, most of the more desired, non-milking tasks are conducted by dairy farm owners and other US-born workers. Dairy farm owners do much or all of the non-milking work themselves: negotiating with milk processors and input suppliers; monitoring feed rations, breeding, and calf care for the herd; and managing cropland, feed purchases, and employees. Most dairy farms have between one and four owners; they are typically immediate kin whose social obligations have been brought into the business partnership. When farmers do need to hire employees to help conduct the non-milking tasks, they often hire US-born workers for those positions, as evidenced by the fact that most US-born dairy workers are located in non-milking positions (see Figure 5.1). Later in this chapter, I describe how farmers feel pressure to hire the US-born workers into the non-milking positions and

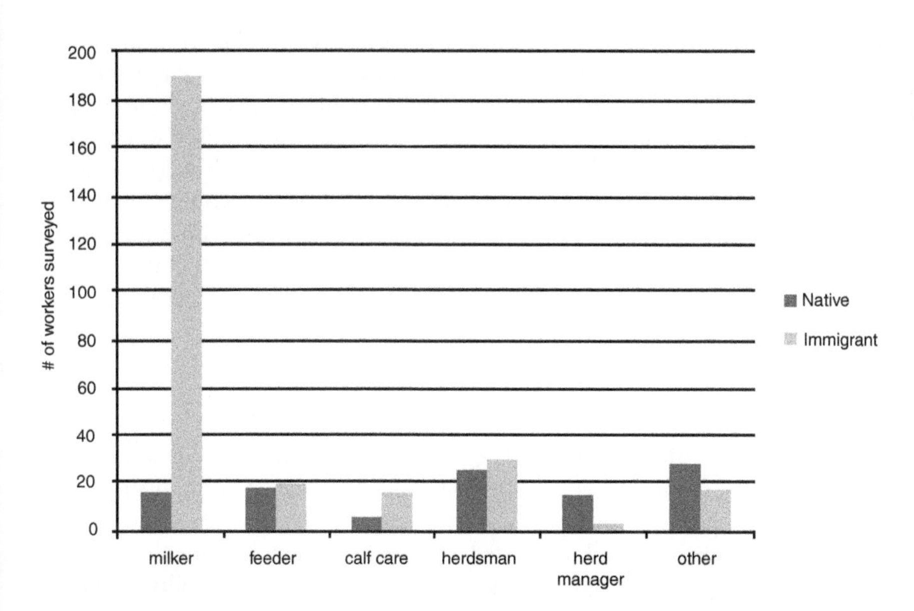

Figure 5.1 Distribution of surveyed hired dairy workers, by farm task and worker origin.

Note: This figure includes full-time and part-time hired employees; owners are not included.

to refrain from promoting their immigrant employees into the advanced positions (see also Harrison and Lloyd, 2013).

All US-born dairy workers in our survey were identified as white. The non-milking positions involve a greater variety of tasks, higher pay (eleven US dollars per hour on average), more decision-making authority, and more autonomy than the milking positions, as well as shifts that coincide with those of the traditional workday. Wisconsin's dairy farms thus demonstrate clear patterns of occupation segregation by both nativity and race, as the immigrant workers clustered in the low-end jobs are differentiated not only by their nonnative status but also their racially marked status as Mexican or Hispanic. Wisconsin dairy farms thus have come to reflect the broader pattern of occupational segregation by race and nativity that scholars have long documented in the United States and other countries in fruit and vegetable production, food processing, hotels, restaurants, landscaping, construction, and other industries.

Tracing the Roots of these Workplace Inequalities

In this section, I explain how dairy farm workplaces have become segregated by nativity and race. Through this narrative, I show that neoliberal politics play an important role in the resulting workplace inequalities but in ways that cannot be understood apart from non-neoliberal structures. Specifically, I first identify how Fordist agricultural and food policies and their attendant ideologies that emerged in the mid-twentieth century continue to compel dairy farmers to further expand and industrialize production, reorganize work, and hire tractable workers for the growing amount of low-end work on the farm. I then show how recent neoliberal policy reforms, although departing from Fordism in many ways, nonetheless fuel the trends of intensification of farm production and reliance on hired workers for low-end farm tasks. I then describe how transnational migrants have come to fill those low-end positions, explaining that neoliberal policies undermine the power of labor vis-a-vis capital and bolster these migration patterns, whose roots predate the neoliberal era. Finally, I describe how immigration politics in the United States influences these workers and employers in ways that channel immigrant workers into low-end jobs and keep them there.

Agricultural Intensification under Fordism

To a large extent, the occupational segregation that is just now emerging on Wisconsin dairy farms has its roots in the Fordist regime of accumulation and regulation in the twentieth century, whose emphasis on mass production and mass consumption included concerted state involvement in agriculture. Oriented toward maximizing agricultural production and the availability of cheap food, the federal government reduced farm commodity price uncertainties, enforced protectionist trade policies favoring domestic producers, absorbed surplus quantities of overproduced farm commodities through federally funded school lunch programs and foreign food aid programs, subsidized the development of manufactured farm inputs designed to boost farm productivity, established a public university research and extension system that fostered productivist agriculture, and helped to disseminate that productivist model internationally (Hightower, 1973; Marsden, 1992; Bonanno and Constance, 2001; Dowie, 2001; Winders, 2009). The state addressed the labor question in agriculture by helping to recruit low-wage foreign workers to maintain high productivity of cheap farm commodities, physically repressing workers during times of labor unrest, and exempting agricultural workers from many labor laws to make those workers affordable and docile when farmers need them (Calavita, 1992; De Genova, 2005). The widespread industrial ideal that dominated agricultural policy, practice, education, and broader discourse, the agrarian imaginary, and the widespread norm of consumerism have all ideologically legitimized such trends (Fitzgerald, 2003; Brown and Getz, 2008).

Although the Fordist regime is widely characterized as having peaked in the 1950s and 1960s and having been plagued by crises since the 1970s, its lingering agricultural and food policies and attendant ideologies nonetheless continue to fuel farm structural change, the industrialization of agricultural production, and the exploitation of hired labor. The commodity price support programs that constitute the overwhelming majority of today's agricultural policy allocations have long been scale-neutral, thereby failing to counteract well-capitalized farmers' advantages in the marketplace and the resulting farm concentration. Those policies also support unlimited production of just a few crops and thus encourage farms to specialize in those commodities and to expand and intensify production. The fact that those federal price supports have often been set below the cost of production further compels farmers to focus on increasing production through expansion, intensification, and extracting as much value as possible from labor, land, and animals. Farmers participating in programs that require setting aside degraded and ecologically sensitive lands have long responded by intensifying production and thus extracting greater value from their lands that remain in production. Today, many new federal programs subsidize the costs of farm expansion and modern technology adoption rather than prioritizing agroecological, non-industrial alternatives.[2]

All of these trends continue in the dairy sector as much as any other. As a result, dairy farms, like their counterparts in other commodity sectors, have continued to gradually expand and intensify production (Barham et al., 2005). To help run the growing farm operations, dairy farmers increasingly rely on hired workers, largely for the low-end jobs of milking cows. Working within tight profit margins, farmers seek out those workers willing to do that work for low wages and few nonwage benefits. As will become clear, neoliberalization and contemporary immigration politics have effectively stabilized farmers' labor issues—although not without generating a different set of instabilities.

Neoliberalization

In response to the crises of Fordism starting in the 1970s, the state's involvement in agriculture and other sectors shifted in significant ways. Under the growing popularity of neoliberal logic, the state's overarching responsibility has increasingly shifted toward enabling the free movement of capital and otherwise solving social problems through the market and appeals to individual responsibility. Although the influence of neoliberal ideology that spread throughout the global North starting in the 1980s has been less pronounced in agriculture than in other sectors, the sector has not been immune to neoliberalization. This is evident in multilateral trade policies that opened foreign markets to US farm commodities, the evisceration of environmental regulatory agency funding and authority to regulate industry, and the state's shift away from regulation and toward embracing voluntary solutions to agri-environmental problems (Harrison, 2008, 2011). Neoliberal reforms altered many of the Fordist farm policies. For example, the 1996 US Farm Bill reforms and recent rulings from the World Trade Organization shifted the US government's commodity price supports away from supply management and ostensibly start to dismantle the protectionist farm policies that flood global markets with cheap US farm products (Winders, 2009). Like other neoliberal reforms that have widely (albeit unevenly) transformed social and environmental policy, these agricultural policy reforms have been promoted in part as addressing the perceived inefficiencies and ineffectiveness of Fordist welfarist programs, reducing market distortions, and otherwise promoting trade.

Although such reforms mark a partial shift away from Fordist logic, the resulting policies and programs still effectively maintain, if not bolster, the productivist agenda in federal agricultural policy and encourage farmers to focus on expanding and intensifying production. Notably, while Fordist policies facilitated capitalist penetration into agriculture (Goodman et al., 1987), the power of off-farm agribusinesses has skyrocketed during the neoliberal era (Heffernan, 1998). Capitalizing on liberalized trade regimes by restructuring around global markets for cheap labor and inputs and aggressively creating new markets for its final products, off-farm agribusiness has grown so concentrated in recent years that only a few companies account for the majority of production in agricultural inputs (including seeds, machinery, fertilizers, pesticides, and genetic stock), processing, and retail (Heffernan, 1998). Under such circumstances, off-farm agribusinesses have been able to ratchet up the costs of farm production, decrease the prices of farm products, and effectively force farmers into disempowering contract relationships (especially in poultry and hogs). This combination of factors—faced by farmers across commodity sectors—shrinks farm profits, pressing farmers to leave the industry or find new sources of value. The dominant advice long given to farmers by university researchers and extension agents, dairy industry organizations, bankers, and policymakers is that the only sure way for farmers to survive this difficult "cost-price squeeze" is to maximize output and industrialize operations (Lobao and Meyer, 2001; Fitzgerald, 2003). The federal government has refused to regulate the near-monopolies in agricultural inputs, processing, and retail (Lynn, 2011), in part because of the Reagan Administration's neoliberal reinterpretations of antitrust legislation to focus strictly on cases where corporate concentration affects prices (i.e., rather than other forms of social welfare). These concentrations of corporate power under neoliberalization have further eroded the economic standing of most farmers and compelled farmers to squeeze more and more value from workers, animals, and the environment.

In Wisconsin's dairy sector, neoliberalization has thus fueled the patterns established under Fordism: farm expansion, industrialization of production, and exploitation of low-wage, hired workers to help run the increasingly large operations under tight profit margins. At the same time, neoliberal reforms have kept hired labor affordable and flexible. In the United States, the federal government has refused to increase the minimum wage in line with inflation, states have restricted the rights of workers to collectively organize and bargain (e.g., through "right to work" laws), and political institutions designed to empower farm labor (such as California's Agricultural Labor Relations Board) have been generally handed to appointees hostile to workers' rights (Majka and Majka, 2000; Harvey, 2005). As I explain below, neoliberal reforms in the global South contribute to these patterns as well.

Transnational Migration

At the same time that dairy farmers have industrialized and expanded their operations, transnational migrant workers, largely from Mexico and Central America, have increasingly sought work in Wisconsin and have been eager to accept low-wage, arduous, and dangerous jobs like those on dairy farms. Like farm industrialization, these migration patterns and demographic changes in rural Wisconsin stem from neoliberal policies as well as longstanding institutions that predate the neoliberal era. For example, the neoliberal austerity measures in structural adjustment policies imposed by the World Bank and International Monetary Fund on debtor countries in the global South have crippled those countries' domestic supports for workers, the poor, education, and health care—undermining the livelihoods of millions of the world's poor (Harvey, 2005). In Mexico, the North American Free Trade Agreement (NAFTA), which opened North American borders to the movement of capital, in turn dismantled economic protections for Mexican farmers and labor protections for workers. Under NAFTA, Mexico's agricultural sector was liberalized more than the agricultural sector in the United States, undermining the livelihoods of small-scale farmers in Mexico and fueling mass migration out of Mexico's agricultural communities (Delgado-Wise and Márquez Covarrubias, 2007; Nevins, 2007; Pechlaner and Otero, 2010; Otero, 2011). Such patterns parallel those that have been widely documented by scholars observing neoliberalization and migration in other geographic contexts (Sassen, 1998). Of course, such migration patterns have a much longer, pre-neoliberal history, one of largely unchecked direct recruitment of migrant workers by US agribusinesses since the late 1800s, state-sponsored direct recruitment efforts, such as the US Bracero Program (1942–64), and subsequent development of migrant social networks (Calavita, 1992; De Genova, 2005).

Nativist Immigration Politics

To understand the implications of unauthorized transnational migration between Mexico, Central America, and the United States for Wisconsin dairy farms, we must take into consideration recent escalations in the policing of migrants racially marked as Mexican and Hispanic and presumed to be illegal. Just as Higham (1955), Ngai (2004), and other historians have shown that past immigration policies were governed by and reinforced racist conceptions of national identity and belonging, contemporary immigration politics follow that pattern and demonstrate a resurgence of anti-immigrant fervor directed at certain racialized groups (Sanchez, 1997). As Varsanyi (2011) has done, I follow those scholars by defining contemporary immigration politics as *nativist* to draw attention to the fact that they are disproportionately enacted upon certain racialized groups (including Mexicans and Hispanics) seen as threatening an "idealized notion of the appropriate racial and cultural identity of 'real Americans'" (Varsanyi, 2011, pp. 298–99).

While migrant networks and drivers of migration have proliferated and accelerated, numerous factors have severely restricted the opportunities for migrating legally into the United States. US immigration policies of 1965 and subsequent policy reforms restricted the availability of temporary work visas and opportunities for permanent residency, especially for people from Mexico and elsewhere in the Western Hemisphere—a set of restrictions that have been supported by widespread and racially charged anti-immigrant sentiment (De Genova, 2005; Ngai, 2004). As a result, millions of people migrate without legal authorization, doing so because it "may be the best alternative from a very limited set of options available to improve their lives" or even simply to survive (Ruhs and Anderson, 2007, p. 4).

Yet, at the same time, migrating without legal authorization has become increasingly difficult and dangerous. Starting in the 1990s but gaining legitimacy and resources through the post-9/11 war on terror, the United States has invested tremendous resources into hardening its border with Mexico through massive increases in border walls, Border Patrol personnel and transport, and unmanned surveillance technologies; deputizing local police agencies to enforce federal immigration law; actively policing immigrants in the interior of the country through raids and apprehensions in workplaces, homes, and public spaces; and funding new immigration detention facilities to house those accused of violating immigration law. In the context of increased border enforcement, crossing has shifted to remote mountain and desert spaces, human smuggling operations have become more important and powerful, and border deaths and violence have skyrocketed (Andreas, 2001; Cornelius, 2001; Nevins, 2007). Fueled by ideological claims that Mexican immigrants contribute to crime, overuse social services, and pose a threat to national security, local municipalities across the United States have pressed to help enforce federal immigration law by passing anti-immigrant ordinances (Varsanyi, 2011; Walker and Leitner, 2011). Nativist immigration policing is conducted not only by sanctioned law enforcement but also by overzealous bureaucrats and everyday people intent on taking the enforcement of immigration law into their own hands (Romero, 2006). The court system has not received sufficient funds to keep pace with the growing number of detainees, thereby swelling the numbers of immigrant adults and children in detention and lengthening their detention times (for details, see Harrison and Lloyd, 2012).

These policing practices are not universally supported. Immigrant advocates and many law enforcement units contest these escalations in policing on grounds that they suppress immigrants' willingness to report crime and communicate with law enforcement (Decker et al., 2012). Additionally, industry actors often express concerns that by generating fear among immigrants, policing contributes to a labor shortage. This is especially the case in agriculture, given that farmers have long used cheap and docile immigrant workers to increase production and reduce costs. Local law enforcement officers who work under pressure to support industry interests often also contest immigration enforcement on similar grounds. Despite those contestations, immigrant workers still continue to bear the brunt of immigration politics, as they account for the overwhelming majority of immigration arrests (DHS, 2009). Moreover, industry actors who contest the heightened policing of immigrants instead call for guestworker programs that legalize workers temporarily but restrict their rights and mobility.

These recent escalations in nativist immigration politics unfold not just simultaneously with neoliberalization but in many ways interactively with it. For example, although NAFTA is widely regarded as a textbook neoliberal institution, it was partially justified through nativist rhetoric. Playing on xenophobic fears of an immigrant invasion, NAFTA's proponents argued that the multilateral trade agreement would create jobs abroad and thus reduce migration (Uchitelle, 2007). Additionally, Varsanyi (2011) argues that nativist municipal immigration policies are best understood as acts of "defensive nationalism" partially provoked by neoliberal policy reforms at the federal level and also that nativist local immigration laws in turn contribute to neoliberalization to the extent that localities wrest power from the federal government. As Hiemstra (2010) similarly points out, the devolution of federal authority to enforce immigration law to local authorities (i.e., through the Secure Communities initiative) is clearly consistent with broader neoliberal trends of rescaling governance (see also Roberts and Mahtani, 2010). Nativist immigration politics are not unique to the neoliberal era; as other scholars have shown, the state and industry have long used immigrant workers to weaken the power of labor and shift labor control away from labor markets and into the legal realm (through the policing of migrants) and the cultural realm (through xenophobic sentiment) (Bonanno and Cavalcanti, 2011, p. 5). That said, this strategy has become more acute and effective under neoliberalization, given the growing power of capital and the declining power of labor. The lack of meaningful federal immigration reform in the United States in recent years is partially attributable to these tensions, where industry has called for controlling immigration through guestworker programs that restrict migrants' rights, while organized labor has tended to oppose such programs in order to protect its own (declining) power.

Although Wisconsin does not constitute a major focus of the federal government's immigration enforcement efforts, policing happens there in various sanctioned and informal ways. Federal police have raided workplaces and homes; many employers have received "no-match" letters from federal officials advising them that an employee did not report a valid social security number; immigrants we interviewed report having been asked by neighbors and low-level bureaucrats if they are legal; and various municipalities have passed local anti-immigrant policies (i.e., English-only laws, or sanctions against landlords who rent to individuals without proof of legal residence; Walker and Leitner, 2011; Harrison and Lloyd, 2012). Indeed, in 2009, the Wisconsin Governor's Council on Migrant Labor initiated an investigation into the unsanctioned immigration policing practices conducted by Department of Transportation staff (Harrison, 2009). The immigrants we interviewed are acutely aware of such activity and regularly share relevant rumors and updates with each other.

Immigrant Workers' Responses

As I have elaborated elsewhere (Harrison and Lloyd, 2012), unauthorized immigrant dairy workers experience these escalations in nativist immigration politics as a palpable sense of deportability that makes them fearful of being in public and risking any interaction with law enforcement, intent on earning as much money as possible, compliant with work arrangements offered to them, and therefore highly exploitable by employers. Immigrant dairy workers emphasized that they live with memories of harrowing border crossings, under the weight of responsibilities to support their families and pay off debts (family medical expenses, or smugglers' fees), and under extraordinary fear of being apprehended and deported—so much so that they work long hours in the milking jobs (an average of fifty-seven hours per week), do not complain about their shifts or tasks, and minimize their use of public space. For example, several immigrant workers explained to us that they work very long hours and refuse to take any days off because they need to pay off their debts to their smugglers, who have been increasing their interest rates over time, and because the fear of coming into contact with law enforcement makes many feel that they have nothing "safe" to do besides work. One explained, "You feel the weight of the debt. I want to work more hours a day, two hours more. The more, the better. The debt you owe and the interest make you think." As another worker explained, "We paid $3,000 to get across the border. I went to work to repay the money that they lent us so we could come here, because it takes a lot of money to come here. The most important thing was we had work, and we needed work." One of the workers we interviewed, who works with a sibling on a small, geographically isolated dairy farm, similarly explained how the dangerous and uncertain nature of the border crossing, as well as their ailing mother's medical bills, have compelled them to work as many hours as possible:

> When we came here it was really difficult to leave the family. It is so sad. Believe me, it is so sad . . . And then the uncertainty when you leave them, because they know how we come here. And they are very worried, full of sadness, because people die trying to get here. So they don't know how we are going to cross, if they're going to get us, if they're going to assault us. They don't know. We leave them saying, "Well, don't worry about it, we're going to be okay." But so many things can happen, and there are people that die, they get assaulted, they get killed. They are so many dangers you have to confront when you come here. It is very difficult. That's why we have to stay a good amount of time here. We have to take advantage of the time because it costs us so much to come here.

Four of the twelve immigrant workers we interviewed told us that their fears of being apprehended by law enforcement have compelled them to leave home only to go to work and, twice per month, to buy groceries. Even legal immigrants live under the same "gaze of surveillance" (Stephen, 2004) as their unauthorized friends and family, are similarly racially targeted (for low-wage jobs and by law enforcement), and strategically comply with the role of the good immigrant (compliant at work and invisible otherwise) to protect their unauthorized friends and family.

Being workaholics stems also from the fact that heightened immigration enforcement has made returning home to visit with family simply too expensive because it has increased the need to hire smugglers. It has also increased the risks of apprehension and detention because travel increases a migrant's visibility to law enforcement. The immigrant workers we surveyed reported to have returned to their home country, on average, only once during their time in the United States, where the average time in the United States is 7.5 years. Sixty percent reported to have never returned to their home country since first arriving in the United States.

The workers we surveyed overwhelmingly want to learn new skills to advance within the workplace. However, few insist upon advancements (for fear of losing their jobs). Indeed, as the following interview excerpt demonstrates, some regard such opportunities as simply unavailable to them:

Immigrant Worker: Nine years. Same job. Same pay.
Interviewer: Is there any opportunity to get a different job there?
Immigrant Worker: No. Those jobs are not for immigrants.

Wacquant (2009) and others have argued that neoliberal welfare reforms' techniques of surveilling and punishing the poor help to manage the fallout of and in turn legitimize neoliberal cuts to the social safety net.[3] Nativist immigration politics function in this same way in the lives of Wisconsin's immigrant dairy workers. As the specter of immigration enforcement compels these migrants to work obsessively without complaint and dissuades them from going out in public or using medical services except for emergencies, nativist restrictions on migrants' rights and escalations in policing together effectively produce perfect neoliberal subjects: flexible, profitable workers who make few substantive demands on the state. These workers' combined flexibility and vulnerability, in turn, help to maintain the economic survival of marginal businesses like dairy farms that themselves have been rendered more vulnerable by neoliberal policy reforms. At the same time that nativist policies and neoliberalization together help to produce these workplace inequalities, immigrant workers' contemporary experiences of illegality in turn serve the neoliberal order by making the most vulnerable absorb its fallout. Hiemstra (2010) makes this argument nicely in her observations of immigration politics in Colorado: "Illegality disciplines immigrant labor in service of the neoliberal order, turns all residents into surveillers of immigrants' subordinant sociospatial position, and masks contradictions within neoliberalism that arise particularly at the local scale" (p. 74). In other words, in contrast to what neoliberal market theory would claim, labor in the neoliberal era is regulated not by the market but through cultural and legal means.

Employer Reactions

Immigrant workers' strategies for navigating the contours of the policing landscape coincide well with the ways in which employers have reorganized work in their expanded operations, and employers reap the benefits of this deportability. Dairy farmers laud immigrant workers for expressing two key characteristics that solved the problems they experienced with US-born workers: being committed to working long hours and being compliant with the tasks and shifts offered to them. Employers actively maintain this new workforce in several ways that turn milking jobs into "brown-collar jobs" and reproduce segregation on the farm: selecting the most subordinated workers for milking positions, recruiting new milkers through existing immigrant workers' networks, keeping the milking jobs as simplified as possible (which limits migrant workers' opportunities to develop skills that could help them attain better jobs), and generally refraining from promoting immigrant workers (see also Harrison and Lloyd, 2013).

Immigrant workers' sense of deportability and willingness to take the tasks, shifts, and hours handed to them clearly helps dairy farmers meet multiple goals. Hiring immigrant workers into the low-end milking jobs enables dairy farmers to fill the low-end jobs that are the most difficult to keep staffed, avoid the tasks they themselves resent, and take time off on weekends and evenings for family activities like sporting events and vacations. Additionally, not feeling pressure to promote their immigrant workers enables dairy farmers to meet their own social obligations to kin and peers, who insist on the better tasks and shifts. As I have elaborated elsewhere, the new organization of work and workers on dairy farms enables these employers to conform to the dominant middle-class, white, masculine norms of rural Wisconsin (Harrison and Lloyd, 2013).

Additionally, keeping immigrant workers in the low-end milking jobs helps employers manage the risk of immigration enforcement. For example, some farmers reported that they refrain from training and granting responsibilities to (presumed or actually unauthorized) immigrant workers, as they would lose that "investment" if the workers were arrested for immigration violations. Several employers explained that they were unwilling to promote immigrant workers into positions that require the use of tractors and other heavy machinery, as those jobs require paperwork that might trigger an immigration-related investigation (i.e., as insurance companies require copies of drivers' licenses and driving records for all workers operating such equipment). Others explained that restricting immigrant workers to milking positions keeps them in the barn—out of sight—and thus partially hides employers' hiring practices from scrutiny by nativist neighbors and law enforcement. In other words, keeping immigrant workers clustered in the low-end jobs enables these employers to meet their own commitments and aspirations and to delicately balance the benefits of capitalizing on immigrants' presumed or actual tractability with the need to minimize the likelihood and consequences of drawing the attention of law enforcement.

To rationalize the clustering of immigrants in the milking jobs, employers use racialized narratives of cultural and/or biological difference and ignore (and thus obscure) the structural contexts that produce immigrant worker vulnerability. As one employer stated, "I've never had anybody that seemed to want to work and just milk cows and be satisfied with that, like the Mexicans do. . . . They don't have ambition to drive tractors like American guys, who want to be in the tractor all the time." Another employer explained that immigrant workers are clustered in the milking positions because "it is a comfort zone for them. It is a job they can handle for their physical size." At the same time, dairy farmers insist that their hiring practices are "colorblind" and their workplaces meritocratic (as others have found; see Maldonado, 2009). As one farmer asserted, "We just want equal opportunity, give everyone the same chance no matter what." Claiming that everyone has an equal opportunity to earn promotions means that those in non-milking positions (including the employers) have earned their privilege fairly. Yet, because of how dairy farmers have reconfigured their operations, what they really need and value are workers who will comply with milking positions for years on end—not employees who expect to be able to work their way up. Race and nativity serve as markers for such compliance. Dairy farmers' claims that occupational segregation by nativity and race stem from biological or cultural difference and their claims to be "colorblind" in managing their workers, coincide well with broader neoliberal ideology. As Roberts and Mahtani (2010), Harvey (2005), and others have argued, neoliberal discourses of individualism and freedom obscure the role of racist nativism in producing inequalities experienced by certain immigrant groups. Common-sense neoliberal claims of meritocracy—that "modalities of difference, such as race, do not predetermine one's success as each individual is evaluated solely in terms of his or her contribution to society" (Roberts and Mahtani, 2010, p. 253)—silence and thus reinforce the racially differentiated experience of immigration policing and neoliberal policies alike, as well as the unequal ways they play out in the workplace.

These new labor relations on dairy farms contain their own tensions and contradictions. Many employers we interviewed expressed frustration with the legal and language barriers to promoting more of their immigrant workers, yet they also operate under considerable pressure to allocate those positions to their own kin and peers. Some dairy farmers publicly protest the escalating surveillance of their employees, yet they also do not want to draw attention to their own potentially illegal hiring practices. Limiting immigrant workers to the milking jobs effectively mitigates these tensions, though not without consequences.

Conclusions

In this paper, I set out to identify the ways in which neoliberalization works out on the ground in the context of agricultural workplaces in the United States. As I have shown through this study of shifting labor relations on Wisconsin dairy farms, understanding how neoliberal reforms work out in the agricultural workplace requires that we characterize those changes as they unfold together with longstanding, non-neoliberal policies and politics. Neoliberalization has contributed to occupational segregation by race and nativity by fueling farmers' needs to expand and intensify their operations and to exploit low-wage, hired farm workers—patterns established by Fordist policies decades before the neoliberal era began. At the same time, neoliberal reforms in the United States and abroad intersect with nativist immigration politics to keep hired labor abundant, affordable, flexible, and docile.

Dairy farmers, like employers in so many other industries around the world, capitalize on immigrant workers' structurally produced sense of deportability by hiring them into the low-end jobs and generally refraining from promoting them. As a result, dairy farmers have found a way to stay afloat economically amid increasingly precarious circumstances, meet their obligations to kin and peers, otherwise comply with dominant norms, and minimize their own risks associated with hiring workers who may lack legal status. Nonfarm agricultural capital (including input manufacturers, food processors, and retailers) and consumers all benefit as well, as these unequal labor relations enable farmers to produce cheap food commodities. Nativist and neoliberal policies also serve the interests of xenophobic residents of rural Wisconsin; the neoliberal reforms that undermined workers' livelihoods abroad, together with the nativist politics that make immigrants fearful and compliant in the workplace, make them reluctant to use social services and public space and intent on staying out of trouble.

The links between nativism and neoliberalism are worth emphasizing here, given that they have received less attention in the literature than have the connections between Fordism and neoliberalism. Contemporary immigration politics effectively strengthen neoliberalization and thus render it more resilient, in large part by disciplining those rendered most vulnerable by neoliberal policies—low-wage, transnational migrants—to play the role, at least in appearance, of perfect neoliberal subjects: flexible, compliant, profitable workers who make few substantive demands on the state. This structurally produced compliance, in turn, enables dairy farmers to manage their expanding farm operations in a context of increasing vulnerability vis-a-vis other industry actors in a neoliberalized global market. Additionally, by disciplining migrants in these ways, immigration politics subsidizes cheap milk and other dairy products, contributing to the resilience of the neoliberal state that has whittled away entitlements for the poor.

Of course, these and other transnational migrants are not fully disciplined. The insistence on migrating without authorization or overstaying a legal visa is a direct rejection of the neoliberal allocation of freedom to capital but not to labor. Indeed, unauthorized migrants' covert insistence on migrating and their visible public demonstrations and demands for pathways to citizenship, living wages, and rights to education and health care constitute potential moments of rupture to the fabric of neoliberalism, to the extent that they constitute demands that nation-states take responsibility for neoliberal fallout. Yet, this case demonstrates that nativist politics discipline those unruly bodies, forcing them (mostly) back into line and out of sight, and compelling them to behave in ways that serve the neoliberal agenda.

The case of Wisconsin dairy workers indicates that scholars interested in identifying ruptures to neoliberalism, contesting neoliberal fallout, and otherwise pursuing justice for labor in the neoliberal era will need to explicitly confront such non-neoliberal ideologies and institutions that effectively bolster the neoliberal agenda. Although these findings indicate that some of the roots of injustice are deeper, older, and more insidious than neoliberalism itself, they also indicate a few ways forward. For example, this case suggests that protecting farms and other firms from the pressures to expand and exploit could translate into improved conditions for workers; strengthened farm policies, antitrust actions, and cooperative farming arrangements could all serve such a function. Additionally, this case demonstrates that the neoliberal trade policies promoted by the United States have strengthened the rights of US-based capital, undermined the livelihoods of the poor and the status of labor, and fueled transnational migration. These findings thus suggest that the United States has an ethical obligation to open its borders much more widely than it currently does and to strengthen the rights of migrants in particular and labor in general. As scholars, we can continue to identify how neoliberal and other politics contribute to unjust outcomes and at the same time search for meaningful alternatives and contestations to neoliberalism.

Acknowledgments

I first presented these ideas at "The Neoliberal Regime in Agri-Food" mini-conference of the 2012 Rural Sociological Society annual meeting in Chicago. Thanks to Steven A. Wolf and Alessandro Bonanno for bringing these papers together into conversation with each other and providing extensive and constructive comments and to the participants in the mini-conference for their thought-provoking questions and helpful suggestions. Sarah Lloyd, Julia McReynolds, Trish O'Kane, and Brent Valentine helped with research design, data collection, and preliminary analysis. Data collection was supported by the Program on Agricultural Technology Studies (PATS) at UW–Madison, USDA Hatch Grant #WIS01272, and the Frederick H. Buttel Professorship funds.
Direct correspondence to Jill Lindsey Harrison, Department of Sociology UCB 327, University of Colorado–Boulder, Boulder CO 80309; E-mail: jill.harrison@colorado.edu.

Notes

1 Data suggest that the majority of Wisconsin's immigrant dairy workers lack legal authorization to be in the United States. Although we refrained from asking any of our research participants about individual workers' legal status (because we detected high levels of anxiety about immigration enforcement in the area at the time of our data collection and did not have time to establish significant rapport with the participants before meeting with them), other data provide insights into the legal status of these workers: eight of the twelve workers with whom we conducted in-depth, confidential interviews voluntarily divulged their lack of legal status to us; all of the twenty farmers we interviewed expressed many concerns about legal status issues, and most voluntarily reported knowingly having employed unauthorized workers; the farmers in a series of exploratory focus groups we conducted in 2007 as the first stage in this project consistently directed conversation to legal status issues on their own accord; the hired labor sessions at all major Wisconsin dairy industry meetings in the past several years have been dedicated to legal issues associated with hiring unauthorized workers; and other researchers consistently find that approximately half of immigrant agricultural workers in the United States are unauthorized (e.g., NAWS, 2005).

2 For example, substantial portions of the Environmental Quality Incentives Program subsidize the production of manure lagoons that large dairies use to prevent manure waste from contaminating water supplies (Martin, 2008; see also chapter by Wolf in this volume).

3 Thanks to Spencer Wood for pointing me to Wacquant.

References

Allen, P., FitzSimmons, M., Goodman, M., and Warner, K. (2003) "Shifting plates in the agri-food landscape: The tectonics of alternative agri-food initiatives in California," *Journal of Rural Studies*, vol 19, pp. 61–75.

Andreas, P. (2001) "The transformation of migrant smuggling across the US–Mexican border," Pp. 107–125 in D. Kyle and R. Koslowski (eds) *Global Human Smuggling: Comparative Perspectives*, Johns Hopkins, Baltimore, MD.

Barham, B. L., Foltz, J., and Aldana, U. (2005) "Expansion, modernization, and specialization in the Wisconsin dairy industry," Research Report No. 7, Program on Agricultural Technology Studies, University of Wisconsin, Madison.

Bonanno, A. and Cavalcanti, J. S. B. (2011) "Introduction," Pp. 1–31 in A. Bonanno and J. S. B. Cavalcanti (eds) *Globalization and the Time-Space Reorganization: Capital Mobility in Agriculture and Food in the Americas*, Emerald, Bingley, UK.

Bonanno, A. and Cavalcanti, J. S. B. (2012) "Globalization, food quality and labor: The case of grape production in north-eastern Brazil," *International Journal of Sociology of Agriculture and Food*, vol 19, no 1, pp. 37–55.

Bonanno, A. and Constance, D. (2001) "Globalization, Fordism, and Post-Fordism in agriculture and food: A critical review of the literature," *Culture and Agriculture*, vol 23, no 2, pp. 1–18.

Brown, S. and Getz, C. (2008) "Privatizing farm worker justice: Regulating labor through voluntary certification and labeling," *Geoforum*, vol 39, pp. 1184–96.

Busch, L. (2010) "Can fairy tales come true? The surprising story of neoliberalism and world agriculture," *Sociologia Ruralis*, vol 50, no 4, pp. 331–51.

Busch, L. and Bain, C. (2004) "New! Improved? The transformation of the global agri-food system," *Rural Sociology*, vol 69, pp. 321–46.

Calavita, K. (1992) *Inside the State: The Bracero Program, Immigration and the INS*, Routledge, New York.

Cornelius, W. (2001) "Death at the border: Efficacy and unintended consequences of US immigration control policy," *Population Development Review*, vol 27, no 4, pp. 661–85.

Coutin, S. B. (2000) *Legalizing Moves: Salvadoran Immigrants' Struggle for US Residency*, University of Michigan Press, Ann Arbor.

De Genova, N. (2005) *Working the Boundaries: Race, Space, and 'Illegality' in Mexican Chicago*, Duke University Press, Durham, NC.

Decker, S., Lewis, P., Provine, M., and Varsanyi, M. (2012) "Local policing, local communities, and immigration," *Law Enforcement Executive Forum*, vol 12, no 2, pp. 25–30.

Delgado-Wise, R. and Márquez Covarrubias, H. (2007) "The reshaping of Mexican labor exports under NAFTA: Paradoxes and challenges," *International Migration Review*, vol 41, pp. 656–79.

DHS. (2009) "Worksite enforcement overview," Department of Homeland Security, April 30.

Dowie, M. (2001) *American Foundations: An Investigative History*, MIT Press, Cambridge, MA.

Fitzgerald, D. (2003) *Every Farm a Factory: The Industrial Ideal in American Agriculture*, Yale University Press, New Haven, CT.

Gareau, B. J. (2012) "The limited influence of global civil society: International environmental non-governmental organisations and the Methyl Bromide controversy in the Montreal Protocol," *Environmental Politics*, vol 21, no 1, pp. 88–107.

Goodman, D., Sorj, B., and Wilkinson, J. (1987) *From Farming to Biotechnology: A Theory of Agro-Industrial Development*, Blackwell, Oxford.

Guthman, J. (2008a) "Neoliberalism and the making of food politics in California," *Geoforum*, vol 39, pp. 1171–83.

Guthman, J. (2008b) "Thinking inside the neoliberal box: The micro-politics of agro-food philanthropy," *Geoforum*, vol 39, pp. 1241–53.

Guthman, J. (2011) *Weighing In: Obesity, Food Justice, and the Limits of Capitalism*, University of California Press, Berkeley.

Guthman, J. and DuPuis, E. M. (2006) "Embodying neoliberalism: Economy, culture, and the politics of fat," *Environment and Planning D*, vol 24, no 3, pp. 427–48.

Harrison, J. (2008) "Abandoned bodies and spaces of sacrifice: Pesticide drift activism and the contestation of neoliberal environmental politics in California," *Geoforum*, vol 39, pp. 1197–214.

Harrison, J. L. (2009) Personal observations at the meeting of the Governor's Council on Migrant Labor, Madison, Wisconsin, January 28.

Harrison, J. L. (2011) *Pesticide Drift and the Pursuit of Environmental Justice*, MIT Press, Cambridge, MA.

Harrison, J. L. and Lloyd, S. E. (2012) "Illegality at work: Deportability and the productive new era of immigration enforcement," *Antipode*, vol 44, no 2, pp. 365–85.

Harrison, J. L. and Lloyd, S. E. (2013) "New jobs, new workers, and new inequalities: Explaining employers' roles in occupational segregation by nativity and race," *Social Problems*, vol 60, no 3, pp. 281–301.

Harvey, D. (2005) *A Brief History of Neoliberalism*, Oxford University Press, Oxford.

Heffernan, W. (1998) "Agriculture and monopoly capital," *Monthly Review*, vol 50, no 3, pp. 46–59.

Hiemstra, N. (2010) "Immigrant 'illegality' as neoliberal governmentality in Leadville, Colorado," *Antipode*, vol 42, no 1, pp. 74–102.

Higham, J. (1955) *Strangers in the Land: Patterns of American Nativism 1860–1925*, Rutgers University Press, New Brunswick.

Hightower, J. (1973) *Hard Tomatoes, Hard Times*, Schenkman, Cambridge, MA.

Kaup, B. Z. (2013) *Market Justice: Political Economic Struggle in Bolivia*, Cambridge University Press, Cambridge.

Lloyd, S., Bell, M. M., Stevenson, S., and Kriegl, T. (2006) "Life satisfaction and dairy farming study," Center for Integrated Agricultural Systems, University of Wisconsin–Madison, unpublished data.

Lobao, L. and Meyer, K. (2001) "The great agricultural transition: Crisis, change and social consequences of twentieth century US farming," *Annual Review of Sociology*, vol 27, pp. 103–24.

Lynn, B. (2011) *Cornered: The New Monopoly Capitalism and the Economics of Destruction*, John Wiley and Sons, Hoboken, NJ.

Majka, L. C. and Majka, T. J. (2000) "Organizing US farm workers: A continuous struggle," Pp. 161–74 in F. Magdoff, J. B. Foster, and F. H. Buttel (eds) *Hungry for Profit: The Agribusiness Threat to Farmers, Food, and the Environment*, Monthly Review Press, New York.

Maldonado, M. M. (2009) "'It is their nature to do menial labor': The racialization of 'Latino/a workers' by agricultural employers," *Ethnic and Racial Studies*, vol 32, no 6, pp. 1017–36.

Marsden, T. (1992) "Exploring a rural sociology for the Fordist transition: Incorporating social relations into economic restructuring," *Sociologia Ruralis*, vol 32, no 2–3, pp. 209–30.

Martin, A. (2008) "In the farm bill, a creature from the black lagoon?," *New York Times*, January 13.

NAWS. (2005) "Findings from the National Agricultural Workers Survey (NAWS) 2001–2002: A demographic and employment profile of United States farm workers," Research Report No. 9, United States Department of Labor.

Nevins, J. (2007) "Dying for a cup of coffee? Migrant deaths on the US-Mexico border region in a neoliberal age," *Geopolitics*, vol 12, pp. 228–47.

Ngai, M. M. (2004) *Impossible Subjects: Illegal Aliens and the Making of Modern America*, Princeton University Press, Princeton, NJ.

Otero, G. (2011) "Neoliberal globalization, NAFTA and migration: Mexico's loss of food and labor sovereignty," *Journal of Poverty*, vol 15, no 4, pp. 384–402.

Pechlaner, G. and Otero, G. (2010) "The neoliberal food regime: Neoregulation and the new division of labor in North America," *Rural Sociology*, vol 75, no 2, pp. 179–208.

Poppendieck, J. (2000) "Want amid plenty: From hunger to inequality," Pp. 189–202 in Magdoff, Foster, and Buttel, *Hungry for Profit*.

Quark, A. (2013) *Global Rivalries: Standards Wars and the Transnational Cotton Trade*, University of Chicago Press, Chicago, IL.

Ransom, E. (2007) "The rise of agricultural animal welfare standards as understood through a neo-institutional lens," *International Journal of Sociology of Agriculture and Food*, vol 15, no 3, pp. 26–44.

Roberts, D. J. and Mahtani, M. (2010) "Neoliberalizing race, racing neoliberalism: Placing 'race' in neoliberal discourses," *Antipode*, vol 42, no 2, pp. 248–57.

Romero, M. (2006) "Racial profiling and immigration law enforcement: Rounding up the usual suspects in the Latino community," *Critical Sociology*, vol 32, no 2–3, pp. 447–73.

Ruhs, M. and Anderson, B. (2007) "The origins and functions of illegality in migrant labor markets: An analysis of migrants, employers and the state in the UK," Center on Migration, Policy and Society (COMPAS), University of Oxford.

Sanchez, G. J. (1997) "Face the nation: Race, immigration, and the rise of nativism in late twentieth century America," *International Migration Review*, vol 31, no 4, pp. 1009–130.

Sassen, S. (1998) *Globalization and Its Discontents*, New Press, New York.

Stephen, L. (2004) "The gaze of surveillance in the lives of Mexican immigrant workers," *Development*, vol 47, no 1, pp. 97–102.

Szasz, A. (2007) *Shopping Our Way to Safety: How We Changed from Protecting the Environment to Protecting Ourselves*, University of Minnesota Press, Minneapolis.

Uchitelle, L. (2007) "NAFTA should have stopped illegal immigration, right?," *New York Times*, February 18.

Varsanyi, M. W. (2011) "Neoliberalism and nativism: Local anti-immigrant policy activism and an emerging politics of scale," *International Journal of Urban and Regional Research*, vol 35, no 2, pp. 295–311.

Wacquant, L. (2009) *Punishing the Poor: The Neoliberal Government of Social Insecurity*, Duke University Press, Durham, NC.

Walker, K. E. and Leitner, H. (2011) "The variegated landscape of local immigration policies in the United States," *Urban Geography*, vol 32, no 2, pp. 156–78.

Winders, B. (2009) *The Politics of Food Supply: US Agricultural Policy in the World Economy*, Yale University Press, New Haven, CT.

6

CREATING RUPTURE THROUGH POLICY

Considering the Importance of Ideas in Agri-food Change

Rebecca L. Som Castellano

Introduction

Some scholars have suggested that the current agri-food system is operating under a third food regime, which is characterized by increasing integration of transnational agri-food capital, global sourcing, and challenges to national regulation of agriculture by corporate-economic strategies (Pechlaner and Otero, 2010, p. 183).[1] As a result, some scholars assert that the hegemonic power of transnational corporations stymies the efforts of subordinate groups to improve social, ecological, and community well-being due in part to the lack of economic and political power of subordinate groups and their inability to mount a systematic critique of corporate agriculture and alter the distribution of costs and benefits in the neoliberal food regime (Magdoff et al., 2000, as quoted in Hassanein, 2003).[2] Others, however, argue that individuals, collectivities, and nation-states can create fissure in the neoliberal food regime, "mediating or perhaps even reshaping the globalization and global-value dynamic" and "temper[ing] its implementation" (Pechlaner and Otero, 2010, p. 181). Take, for example, the work of Wright and Middendorf (2008), who assert the importance of subordinate groups in challenging the global agri-food system. The potential power of subordinate groups is also noted by McMichael (2005, 2009), who identifies how the corporate food regime has engendered social movements, such as Vía Campesina, who promote small-scale agriculture as a way to improve social and environmental justice (La Vía Campesina, 2011).

The responses of producers, consumers, and activists to the neoliberal food regime, such as through farmers' market participation and the development of food policy councils, have been documented in the literature (Friedland et al., 2010), but limited attention has been paid to policy making as a potential mechanism for subordinate groups to resist and potentially transform the neoliberal food regime. While some scholars have argued that the neoliberal food regime operates in a way that cripples state autonomy, others have articulated the continued centrality of the nation-state as a sphere of struggle (Bonanno and Constance, 2006; Pechlaner and Otero, 2008, 2010). Pechlaner and Otero (2010, p. 179) assert that "social resistance at the level of the nation-state can produce an alternative trajectory to the neoliberal food regime" and that we need to look within nation-states to understand the potential of internal sociopolitical dynamics to "alter dominant trends in the world economy from the ground up." Bonanno and Constance (2006) demonstrate that while corporations influence state action, resistance from below occurs at local and state levels, and citizen groups can act as agents of change as "resistance from below prevents corporations from freely executing their strategies and allows the state to offer some forms of control of corporate behavior" (p. 79).

In this chapter, I further explore the question of whether subordinate groups can resist and potentially transform the neoliberal food regime via national policy making. To answer this question, I examine the recent case of the integration of farm-to-school (FTS) into the National School Lunch Program (NSLP). I employ a theoretical perspective developed by policy scholars, which asserts that policy entrepreneurs, promoting a new idea, can create policy change when a window of opportunity opens and three streams converge (political, policy, and problem) (Zahariadis, 2007; Cohen-Vogel and McLendon, 2009). In the next section, I explain how the FTS program may challenge the neoliberal food regime and follow that with a further elaboration of the analytical framework for understanding how a subordinate group might influence national policy making. I then use this framework to explore the case of FTS's integration into the NSLP, concluding with a discussion of what this case reveals about the power of subordinate groups to resist neoliberal trends in the agri-food system through policy mechanisms at the nation-state level.

Farm-to-School and the Neoliberal Food Regime

To begin to explore the question of whether and how subordinate groups might resist or transform the neoliberal food regime via national policy making, I provide some basic background on FTS, particularly in relation to the alternative agri-food movement and the movement's goals for agri-food system change. FTS "is broadly defined as any program that connects schools (K-12) and *local farms* with the objectives of serving healthy meals in school cafeterias, improving student nutrition, providing agriculture, health and nutrition education opportunities, and *supporting local and regional farmers*" (FTS, 2011 [emphasis added]). FTS activities can include local food procurement for school meals or classroom snacks, school gardens, farm tours, farmer in the classroom sessions, visits to farmers' markets, culinary education, food-based curriculum, or educational sessions for parents and community members (FTS, 2012). There has been a dramatic increase in FTS programs, growing from ten programs in 1998 to over 2,500 participating school districts in all fifty states in 2011 (Joshi et al., 2008; FTS, n.d.). The Community Food Security Coalition (CFSC) and the Farm to School Network (FSN) have been leading advocates of FTS development, and the CFSC acted as the primary FTS policy entrepreneur in the federal NSLP policy domain during the period of study. CFSC is not an organization that has been historically central to national agri-food politics and policy, and a national staff to promote the interests of FTS was not established until 2007 (FTS, n.d.).

FTS has become an increasingly popular strategy of food system development employed by the alternative agri-food movement, which generally aims to improve social, environmental, and community health and well-being by altering the conditions under which agriculture and food production, processing, distribution, and consumption occur (Allen, 2004). Proponents of alternative agri-food practices assert that activities such as farmers' markets, community supported agriculture (CSA), and fair trade can create long-term viability in local economies, build and support civic structure, promote democratic practice, and lead to more positive environmental outcomes (Kloppenburg et al., 1996; Feenstra, 1997; van der Ploeg, 2000). But there are questions regarding the degree to which alternative forms of agri-food production and consumption can oppose the neoliberal food regime, and some suggest that alternative agri-food movement practices do not necessarily oppose corporate agriculture (Allen, 2004). For example, some scholars assert that alternative agri-food movement participation can be motivated by pleasure rather than concern with altering social and economic relations in the agri-food system (Johnston and Szabo, 2011). In addition, alternative agri-food practices can be co-opted by corporations seeking to gain greater market share, which potentially delegitimizes the meaning of local within alternative agri-food movement discourse and limits the degree to which alternative agri-food movements oppose the neoliberal food regime (DeLind, 2010). Further, a number of scholars have challenged the idea that alternative agri-food movements are more equitable and can improve social justice in the food system. Scholars assert that racial and class hegemony as well as patriarchy can be maintained in alternative agri-food practices (Hinrichs and Kremer, 2002; Allen, 2004; Guthman, 2011).

Debate about the degree to which alternative agri-food movements can reverse negative trends in the neoliberal food regime has included FTS (Allen and Guthman, 2006; Kloppenburg and Hassanein, 2006). While Allen and Guthman (2006) have argued that the approaches and practices of FTS reinforce neoliberalism, Kloppenburg and Hassanein (2006) counter this argument by asserting that FTS advocates undertake "resistance and critical thinking and political action," and that they are "endeavoring to achieve equity, public funding, and state support for their proposed reforms" (Kloppenburg and Hassanein, 2006, p. 420). Considering the above definition of the neoliberal food regime, which is primarily characterized by the increasing integration of transnational agri-food capital, global sourcing, and challenges to national regulation of agriculture by corporate-economic strategies, I argue that FTS is indeed a form of resistance. While traditional school food programs and FTS have nearly identical goals, which are (1) to provide children access to healthy food, and (2) to provide markets for agricultural producers (National School Lunch Act, 1946; Allen and Guthman, 2006; CFSC, 2012), the emancipatory possibilities of FTS come via the ways in which these goals are accomplished. In short, FTS is not produced by the market system or by transnational corporations. Rather, the focus on support of small-scale, noncorporate producers, integration into a federal, welfare-state program, which provides meals at a free or reduced price to low-income students, and teaching food citizenship all represent a departure from the neoliberal food regime.

FTS is potentially even more powerful than other alternative agri-food movement practices, such as farmers' markets, given its public context, where participation is not limited by class, the frequency of exposure, nor the fact that FTS generally works within the context of a nationwide, welfare-state program.[3] The NSLP served more than 31.7 million school lunches each day in fiscal year 2010 and subsidizes the cost of meals for low-income students (USDA, 2012). Further, research suggests that FTS programs can lead to improved nutrition during the school day, as well as better classroom performance (Morris and Zidenberg-Cherr, 2002). An additional transformative aspect of FTS is that it works alongside or in conjunction with more industrial commodity chains, which can potentially lead to shifts in behavior from institutional buyers, undermining the historic monopoly of agri-food industries in the supply of food for the NSLP (Allen and Guthman, 2006). Further, the transformative potential of FTS increases as the program scales up to the federal level, allowing for greater regulation and increased access (Allen and Guthman, 2006; Kloppenburg and Hassanein, 2006).

As I discuss further below, the most recent reauthorization of the NSLP demonstrates the successful integration of FTS into the NSLP policy domain. But, how did this feature of the alternative agri-food movement achieve such success? To understand this, it is necessary to consider how policy change occurs and how subordinate groups can successfully promote policy innovation. In the next section, I describe an analytical framework that can help us discover how a subordinate group might successfully resist the winds of neoliberalism at the nation-state level.

Creating Change Through Policy

A theory of policy change can provide a framework for understanding how subordinate groups can resist the neoliberal food regime at the level of the nation-state. Public policy making is the outcome of a set of processes, including (1) agenda setting, (2) specifying alternatives for the agenda, (3) choosing among the alternatives, which occurs through legislative enactment or executive decision, and (4) implementation of the law (Roberts and King, 1991, p. 147). One of the hallmark theories of policy change is Kingdon's multiple streams theory, which asserts that policy innovation occurs during the agenda-setting and decision-making processes when three streams converge: the problem stream, the policy stream, and the politics stream (Kingdon, 1995; Zahariadis, 2007). The problem stream encompasses all of the conditions that policy makers define as problems (Zahariadis, 2007; Cohen-Vogel and McLendon, 2009). Statistical indicators, focusing events, such as a disaster or crisis, or feedback from existing programs can signal these conditions. The policy stream consists of the various solutions or policy alternatives available for a policy domain (Zahariadis, 2007; Cohen-Vogel and McLendon, 2009). Finally, the political stream involves macro-level political conditions, such as public opinion on a given issue or transition in administrative or legislative branches of government (Zahariadis, 2007; Cohen-Vogel and McLendon, 2009).

Kingdon's multiple stream theory asserts the importance of policy entrepreneurs, who use a new idea to gain traction during convergence of these three streams, which occurs when a window of opportunity opens (Zahariadis, 2007; Cohen-Vogel and McLendon, 2009). A policy entrepreneur is a "highly motivated individual or small team" advocating for specific and significant change to the "current ways of doing things in their area of interest" (Mintrom and Norman, 2009, p. 655). Scholars have identified a number of ways in which policy entrepreneurs can draw attention to new ideas and articulate them onto a policy agenda (Roberts and King, 1991; Mintrom, 1997; Mintrom and Norman, 2009). First, policy entrepreneurs can influence the problem stream. Effective policy entrepreneurs pay careful attention to problem definition, which is always a political act (Mintrom and Norman, 2009). Policy entrepreneurs must present evidence that suggests that there is a crisis at hand, and they must attempt to highlight the failure of the current policy domain. Policy entrepreneurs can also frame their ideas in relation to a perceived crisis or disaster to bring increasing attention to the issues they are attempting to incorporate into the policy-making process (Beland, 2005).

Policy entrepreneurs can also work to actively manipulate the policy stream. In order to develop viable solutions that will receive broad support, successful policy entrepreneurs must network and build coalitions (Mintrom and Norman, 2009), understanding "the ideas, motives and concerns of others" in the policy domain and responding "effectively" (Mintrom and Norman, 2009, p. 653). Effective networking can enable the development of alternatives that are appealing to other policy actors involved. This can be particularly important in policy domains with entrenched, institutionalized policy actors. This effective networking can lead to the important formation of coalitions. The size of a coalition can be vital; the larger the coalition, the more advocates there can be for a new idea. It is also important to have broad support for a proposal. At times, policy entrepreneurs will work to gain support from unlikely allies, as this can broaden their reach within the policy domain (Mintrom and Norman, 2009).

Policy entrepreneurs can also influence the policy stream by demonstrating the workability of a policy proposal. Models of a proposed change can provide information about how effective and practical a new programmatic idea is, which can help win credibility from other policy domain actors, and provide "momentum for change" (Mintrom and Norman, 2009, p. 653). Further, as a new programmatic idea appears increasingly viable, actors who oppose the new idea can look out of touch (Mintrom, 1997).

Policy entrepreneurs must also be attentive to the political stream. Effective policy entrepreneurs understand the broader political context and use this to their advantage. A shift in the broader political climate can lead to increased receptivity of the ideas put forth by policy entrepreneurs, resulting in dramatic turns in policy (Mintrom and Norman, 2009). The political stream is also shaped by the national mood and public opinion. By being attentive to the national mood related to their areas of interest, policy entrepreneurs can use public concern and opinion to bolster their own credibility on a subject (Burstein, 1998).

Timing is also important for successful policy entrepreneurship. Kingdon (1995) asserts that effective policy entrepreneurs take advantage of windows of opportunity. Windows of opportunity are essential in that they create moments where new ideas are needed to solve policy problems. Thus, policy makers are more open to innovative ways of understanding problems and identifying solutions, and actors are often forced to redefine their interests and policy goals during times of crisis or increased public awareness (Padamsee, 2009).

Finally, Kingdon's theory asserts the importance of ideas in policy change, and he is not alone in his claim that ideas can be a powerful tool in the policy-making process. Policy scholars have increasingly argued that policy change is more than economically and institutionally powerful actors asserting their will and that ideas, institutions, and interests all matter in achieving policy change (Campbell, 1998, 2002; Blyth, 2002; Lieberman, 2002; Beland, 2005; Streeck and Thelen, 2005; Andersen, 2008; Padamsee, 2009). The notion that economic and institutional power is important but not always sufficient in creating policy change parallels agri-food scholarship, which asserts that a revised political economy perspective can tell a more complete story about how agri-food system change can occur (Dixon, 1999; Goodman and DuPuis, 2002; Raynolds, 2002). A revised political economy perspective asserts that power does not primarily reside in the production process or in the hands of economically powerful, corporate actors, and that consumers play more than a passive, non-political role in the agri-food system. While they may not have the capacity to overturn corporations, subordinate actors, such as consumers, can shape the agri-food system (Goodman and DuPuis, 2002). The analytical approach employed here similarly argues that the established arrangement of actors within a policy domain and the informal rules and procedures that structure conduct, as well as the economic resources that actors bring to a policy domain, do matter. However, economically and historically powerful actors do not always get their way. By introducing a new idea during a window of opportunity, policy entrepreneurs representing the interests and new ideas of alternative agri-food movements can act as change agents by bringing increasing attention to policy problems, networking and building coalitions, promoting innovative policy solutions, and institutionalizing them within a policy domain by way of legislative action. To illustrate this point, I apply this framework to the recent integration of FTS into the NSLP policy domain.

Applying the Case of FTS Integration into the NSLP Policy Domain

In late 2008 and early 2009, data was gathered for a project examining the historical and contemporary actors engaged in constructing federal policy for the NSLP. Primary document sources were reviewed, such as policy publications, congressional testimony and reports, and United States Department of Agriculture (USDA) reports. In addition, semi-structured interviews with sixteen federal level NSLP policy makers were conducted. The interviews aimed to assess, among other things, the mission and goals of the organizations and their interests for NSLP policy change.[4] Table 6.1 identifies the category of organization that each interview participant represented.

Interviews were digitally recorded and transcribed, and thematic analysis was conducted, which is "a form of pattern recognition within the data, where emerging themes become categories for analysis" (Fereday and Muir-Cochrane, 2006, p. 82). This process revealed who had been involved in the NSLP policy domain over time, their interests, and how they aligned with subsequent legislative action. One of the most interesting findings from this research was that the theoretical assumptions originally guiding the research, mainly that those with the most economic and institutional power would consistently get their way in the policy-making process, did not always hold. Rather, subordinate groups repeatedly created significant programmatic and policy change in the NSLP, the most recent example being the integration of FTS into the NSLP (Som Castellano, 2009).

Table 6.1 List of interview subjects

# of Interview Participants	*Interview Category*
2	Legislative Representative
2	Industry
3	Nutrition and Health
4	Access
2	Education
2	Government Agency
1	Sustainable Agriculture

Source: R. Som Castellano (2009).

The Failure of FTS

Throughout the interview process, participants reported little to no support for FTS, demonstrating FTS advocates' lack of successful policy entrepreneurship. At the time of the interviews, respondents asserted that the policy alternatives promoted by FTS limited their integration into the NSLP policy domain, suggesting that they failed to effectively manipulate the problem stream and the policy stream. FTS policy entrepreneurs failed to convince historically and economically powerful NSLP policy domain actors that the conditions they defined as problems were worth considering. FTS promotes a new programmatic idea by encouraging procurement of local, fresh, and unprocessed foods directly from farmers and by suggesting that nutrition education should occur not just in the classroom but through activities such as farm visits or gardening. Interview respondents repeatedly asserted that these programmatic ideas diverged from their own priorities. For example, nutrition advocates viewed procuring food locally as less important than procuring food that met their nutritional ideals. One representative from a nutrition organization made this clear when he stated that "if that apple comes from California or if the apple comes from New York—that's not something we focus on." An education representative asserted that while local fresh fruits and vegetables are good, they are not cheap, and canned foods are less expensive. He further stated that programs that support local fresh fruit and vegetable consumption or school gardens are not realistic in every school. Other respondents similarly echoed concern with the feasibility of FTS. A representative from a nutrition organization asserted, "I don't think we—we're not against it. . . . but scaling up is a problem. Our goal is, obviously, to provide nutritious meals for kids." Another interview participant representing school food operators and the food industry asserted that FTS raised concerns about food safety, reliability of supply, and consistency of product. This same respondent focused on the financial constraints of the NSLP, stating that the NSLP is a "penny program," where saving a penny per meal can make a difference to those running school lunches. The interview participants made it clear that the alternative policy ideas promoted by FTS advocates deviated from the current policy stream and that FTS policy entrepreneurs did not have the support of powerful actors within the NSLP policy domain. The conditions FTS policy entrepreneurs were defining as problems were not convincing to other NSLP policy domain actors, and FTS advocates failed to present evidence that their policy innovation could feasibly address issues of public concern.

The failure of FTS policy entrepreneurs to frame their policy solutions in a way that appealed to other NSLP policy domain actors limited their ability to network or have their interests included on the agendas of the powerful NSLP coalitions. As stated above, in order to develop viable solutions that will receive broad support, policy entrepreneurs must understand the concerns, motives, and ideas of others in the policy domain and then respond effectively (Mintrom and Norman, 2009). Many of the above quotes demonstrate that FTS policy entrepreneurs had failed to achieve broad support for their policy solutions. This was in part due to their failure to incorporate the concerns, motives, and ideas of others in the policy domain. The importance of responding effectively to the interests of other policy domain actors was also made evident when interview participants spoke about the role of the food industry in the NSLP policy domain. The food industry was viewed as an important and powerful actor in the NSLP policy domain, and many interview participants asserted the importance of being aware of the food industry's interests and responding effectively. As one interview participant stated, "I mean you gotta have them—you gotta talk to these people . . . They're big time players. If you don't have them in the room or their viewpoint in the room and you haven't thought about it, you haven't dealt with it, you're gonna lose." This interview participant confirmed the importance of understanding the concerns and motives of other policy domain actors in order to achieve broad support for new policy ideas.

Networking and coalition building were also limited by the belief that FTS policy entrepreneurs didn't understand the history of the program or the ideas and motives they were promoting. A number of interview participants viewed FTS as a new actor that was questioning the practices of long-established NSLP policy makers and constituents and lacking "real" knowledge about the importance of the NSLP. An interview participant stated, "I think, sometimes, some folks who work on sustainability, local foods, sort of the food system piece, can lose sight of the value of the school meal programs. And I think sometimes the misperception around commodities comes in there."[5] The perception that FTS policy entrepreneurs were out of touch was also articulated by a representative of the hunger lobby, who stated the following:

> [The] bottom line is I'm an advocate for low-income people, and I don't see any of those people [alternative agri-food activists like Michael Pollan or Alice Waters] at 3 pm, outside the House Agriculture Community Room in the Longworth Building, lobbying for low-income people. All's I do is see them relaxing, having a nice glass of Chablis, doing another testimonial to something, and I find it offensive.

The belief that alternative agri-food movement advocates represented the interests of privileged consumers prevented FTS policy entrepreneurs from being favorably received by other NSLP policy domain actors.

FTS policy entrepreneurs were viewed as insignificant and lacking institutional power. As one former USDA official asserted, "the farm to school—kind of a mom to apple pie thing. . . . They have a small, very small voice." Another interview participant touched on the lack of institutional power of FTS when he stated this about the powerful coalitions working on Child Nutrition reauthorization: "it's kind of like when they're in the room talking to people, they don't mention it [FTS]. . . . It's too small and they have other priorities." This perception that FTS policy entrepreneurs were insignificant further limited their ability to network and build coalitions.

In addition to the problem stream and the policy stream, the importance of the political stream was also touched on by interview participants. For example, one participant spoke to the significance of having politicians promote a new idea in order to achieve policy change:

> FRAC [the Food Research and Action Council] and the School Nutrition Association have not done enough on Farm to School, and nor has NANA [The National Alliance on Nutrition and Activity], and it's just because it's little not big. Well—and it'll remain little until someone issues the four or the six most feared words in the English language. . . . Mr. Chairman, I have an amendment. Once those words are issued, everything changes. Everything changes.

The same interview participant continued to state that "it's [their] responsibility since they're small, it's their responsibility to find that Senator or that member of the House to issue those words. And if they do, they will be successful, and if they don't, they will not." This interview participant believed that FTS would receive more attention from established NSLP policy domain actors when they had the support of politicians. In addition to framing their interests in a way that would enable politicians to legitimize programmatic change to the public, changing public opinion or shifts in the broader political context could also influence the degree to which politicians would be interested in promoting the interests of FTS policy entrepreneurs.

In all, the data suggest that FTS policy entrepreneurs failed to get their interests represented on the policy-making agenda in part due to the disjointedness between their ideas and the ideas of other, institutionally and economically powerful NSLP policy domain actors. FTS policy entrepreneurs failed to propose solutions that were appealing for other policy domain actors, and the macro-level political conditions were not favorable at that time.

FTS Makes Inroads

In December of 2010, less than two years after these interviews, President Obama signed into law reauthorization of the NSLP via the Healthy, Hunger-Free Kids Act, which demonstrated the increasing power of FTS at the federal level. Section 243 of the legislation, titled *Access to Local Foods: Farm to School Program*, provided mandatory funding of forty million US dollars, distributed from 2012 to 2015 to administer technical assistance and competitive matching farm-to-school grants. These grants can be used for "training, supporting operations, planning, purchasing equipment, developing school gardens, developing partnerships and implementing farm to school activities" (USDA, 2011). This was a major step forward for FTS, representing a policy shift for the NSLP. How did FTS manage to achieve this legislative success? I argue that FTS policy entrepreneurs achieved stream convergence and asserted their new programmatic idea by taking advantage of two events that created a window of opportunity: the perceived crisis of childhood obesity and increasing concern with the changing agri-food system.

Problem Stream

In recent years, growing concern with the agri-food system has created public alarm, signaling a health crisis. The issue of childhood obesity in particular has garnered mounting public attention and concern (Lin, 2008), and the school food environment has been viewed as a culprit in the obesity epidemic and as an important location for change (Poppendieck, 2010).[6] At the same time, the American public has progressively become more engaged in alternative agri-food movement activities, such as shopping at farmers' markets (Mount, 2012). Participation in alternative agri-food system activities has not only increased among the everyday citizen but has also reached political elites, made evident by Michelle Obama's installation of an edible garden at the White House.

Shifting concerns about health influenced the problem stream for the NSLP policy domain, forcing NSLP actors to increasingly incorporate health and nutrition in their priorities for NSLP policy making. Every respondent I spoke with asserted concern about the nutritional quality of school meals, and many respondents suggested that childhood obesity was operating as a consensus frame. As one respondent stated, "I mean, I think that child obesity is really, the good side of the horror of it is kind of bringing everyone together to really think, rethinking how we eat and to try to address the problem. I mean, it's so alarming." In fact, no actors were immune from addressing increasing concerns with the changing agri-food system, including the agri-food industry. As one respondent stated, "there's acknowledgment from many food and beverage companies that nutrition standards will happen and should happen. And that the train's going." As concerns about childhood obesity created a window of opportunity, the problem stream for the NSLP shifted, and acceptability of innovative programmatic ideas brought forth by FTS policy entrepreneurs increased.

Policy Stream

As interest in alternative agri-food movement practices and concern with childhood health and nutrition increased, the alternative programmatic ideas promoted by FTS policy entrepreneurs became more acceptable and appealing to other NSLP policy domain actors. FTS policy entrepreneurs also engaged in active networking and increasingly framed their new programmatic ideas in ways that were appealing to other, more powerful NSLP actors. In particular, they made use of the consensus frame of childhood obesity. As one respondent asserted, "there are really no opponents to healthy school meals, just different interpretations of what is considered healthy, and how food quality is defined." FTS policy entrepreneurs began to explicitly frame their programmatic ideas as a solution to childhood obesity. For example, FTS advocates published a document in late 2009 asserting that "farm to school could help address childhood obesity, chronic disease and nutritionally inadequate food" (FTS, 2009). By focusing on childhood obesity, FTS advocates were able to effectively frame their new programmatic idea in a way that made it easier for other NSLP actors to support. The increasing number of models of FTS throughout the country also influenced the NSLP policy stream. These models demonstrated the feasibility of FTS's alternative to traditional NSLP programs and increased support at the local level for this new programmatic idea. Between 2006 and 2011, the number of farm-to-school programs more than doubled, growing from approximately 1,000 programs to over 2,500 (FTS, n.d.). FTS policy entrepreneurs, in turn, were able to get their ideas onto the agendas of the two most powerful coalitions: NANA and the Child Nutrition Forum (CNF). As the policy alternatives developed by FTS policy entrepreneurs became increasingly viable, their inclusion on these coalitions became an asset rather than a liability.

Political Stream

Increasing concern with the health of the agri-food system as well as increasing acceptance of and participation in alternative agri-food movements also altered the political stream, making the programmatic idea promoted by FTS policy entrepreneurs much more viable. Administrative change also shifted the political stream. A number of interview respondents touched on the importance of administrative change in relation to NSLP reauthorization. As one respondent stated, "You know, with the [forthcoming] democratic administration . . . my expectation is that we'll work pretty closely together." The election of President Obama created a more favorable policy-making environment for FTS policy entrepreneurs by bringing concern with childhood obesity and the feasibility of alternative agri-food movement alternatives front and center. For example, First Lady Obama specifically promoted FTS when she published an op-ed in the *Washington Post*, pushing for authorization of the Child Nutrition Act and connecting the act of gardening with alleviating childhood obesity (CSPI, 2010; Obama, 2010).

Stream Convergence

As the problem, policy, and political streams converged, dominant NSLP policy domain actors and political elites found advocating for the integration of FTS into the NSLP politically beneficial, and FTS policy entrepreneurs increased their institutional power (FRAC, 2008). In the end, three proposals promoting FTS were introduced into legislation, receiving sponsorship from prominent members of Congress, including Sen. Patrick Leahy (D-VT), Rep. Rush Holt (D-NJ), Sen. Barbara Boxer (D-CA), and Rep. Sam Farr (D-CA) (DeForest, 2010; FRAC, 2010; NANA, 2010). FTS policy entrepreneurs, armed with a new idea, took advantage of a window of opportunity created by the crisis of childhood obesity and the rise of alternative agri-food movements and achieved stream convergence, enabling them to create change in the NSLP policy domain.

Conclusion

For many decades, production and consumption processes in the agri-food system have been shaped by the neoliberal food regime. This dominance has brought with it a growing concern with the outcomes of the agri-food system, suggesting a potential legitimacy crisis for the neoliberal food regime. One potential and understudied tool for resisting the neoliberal food regime is the policy-making process, particularly at the level of the nation-state. The empirical case presented here suggests that national level policy making can be a mechanism linking subordinate groups with powerful decision makers and national policy change. During windows of opportunity, policy entrepreneurs can propose new solutions to policy problems previously outside the realm of possibility. FTS represents a departure from dominant trends in the neoliberal food regime and has successfully become integrated into federal level policy making, providing a small example of resistance to the neoliberal food regime via national policy making. The ability of FTS to achieve integration into federal level policy making hinged in large part on utilization of a window of opportunity and on actively working to define FTS programmatic ideas in an appealing way for other policy domain actors. FTS policy entrepreneurs also had to network, build coalitions, and pay attention to national mood and federal administrative shifts.

While FTS represents a small portion of the NSLP agenda and budget, programs such as FTS can be scaled up and out, further expanding the influence of national agri-food policy change. There are at least three ways in which the impact of this national policy change could extend beyond the fifty million US dollars set aside for the program, the increased institutional power of FTS policy entrepreneurs, and the program expansion, which this legislation enables. First, by achieving institutional and monetary support through national policy making, programs like FTS can gain greater exposure, adding to their ability to create greater change via national policy in the future. Second, one of the original goals of the NSLP was to influence consumption practices in the household (Levine, 2008), and this goal remains for FTS advocates. By exposing children to new ideas about what and how to eat, the influence of FTS's national policy achievement could be scaled out. Third, as stated above, FTS operates alongside or in conjunction with more industrial commodity chains, which can potentially lead to shifts in behavior from other institutional buyers. Thus, as alternative agri-food ideas and programs are integrated into national agri-food policy, the impact of this policy change can potentially lead to even greater rupture of the neoliberal food regime.

Some of the activities that helped FTS policy entrepreneurs achieve legislative success could undermine the potential for radical change via food policy and the agri-food system. For example, FTS's focus on obesity potentially draws attention away from larger concerns about social, environmental, and community health and well-being, weakening the potential of policy shifts to create rupture in the neoliberal food regime. Nevertheless, as the costs of the neoliberal food regime continue to rise, it appears that national policy making can be an important mechanism for change for those aiming to resist and transform the status quo of the agri-food system.

Acknowledgments

Support for this project was provided by the United States Department of Agriculture National Needs Graduate Fellowship Competitive Grant No. 2008–38420–18750 from the National Institute of Food and Agriculture and the Rural Sociological Society's Master's Thesis Research Award. I would also like to thank Jeff Sharp, Linda Lobao, Keiko Tanaka, and Patrick Mooney for their assistance at various stages of this project.

Notes

1 There has been debate about what characterizes a third food regime. In this paper, I utilize Pechlaner and Otero's (2010) conceptualization of the third food regime, which they label the neoliberal food regime.
2 In this chapter, I utilize the term "subordinate groups" to refer to actors who lack corporate wealth or the political clout commonly associated with large agri-food businesses that are assumed to hold sway over the policy-making process.
3 The NSLP, a federally assisted program administered by the Food and Nutrition Service (FNS) of the USDA, is a powerful institution, serving more than thirty-one million children school lunches each day and receiving more than $11.1 billion federal dollars in fiscal year 2011 (USDA, 2012). The NSLP acts as a welfare-state program in particular by providing school meals at free or reduced price based on the income of students.
4 All interview subjects signed an Internal Review Board (IRB) consent form, and all but two agreed to not be anonymous. I have opted not to utilize the names of interview subjects or the specific organizations they worked for in this analysis, but I do reveal the category of organization interviewed.
5 A number of respondents emphasized the decreased role of the commodities program (at 15 percent at that time), as well as the increased control school food operators have in determining what they purchase. See also Poppendieck (2010).
6 This claim has been contested by both NSLP advocates as well as by academics. See, for example, Gleason and Hedley Dodd (2009).

References

Allen, P. (2004) *Together at the Table: Sustainability and Sustenance in the American Agri-food System*, Pennsylvania State University Press, University Park, PA.
Allen, P. and Guthman, J. (2006) "From old school to farm-to-school: Neoliberalism from the ground up," *Agriculture and Human Values*, vol 23, pp. 401–15.

Andersen, E. (2008) "Experts, ideas and policy change: The Russell Sage Foundation and small loan reform, 1909–1941," *Theory and Society*, vol 37, pp. 271–310.

Beland, D. (2005) "Ideas and social policy: An institutionalist perspective," *Social Policy & Administration*, vol 39, pp. 1–18.

Blyth, M. (2002) *Great Transformations: Economic Ideas and Institutional Change in the Twentieth Century*, Cambridge University Press, Cambridge.

Bonanno, A. and Constance, D. H. (2006) "Corporations and the state in the global era: The case of seaboard farms in Texas," *Rural Sociology*, vol 71, no 1, pp. 59–84.

Burstein, P. (1998) "Bringing the public back in: Should sociologists consider the impact of public opinion on public policy?," *Social Forces*, vol 77, no 1, pp. 27–62.

Campbell, J. (1998) "Institutional analysis and the role of ideas in political economy," *Theory and Society*, vol 27, no 3, pp. 377–409.

Campbell, J. (2002) "Ideas, politics and public policy," *Annual Review of Sociology*, vol 28, pp. 21–38.

Center for Science in the Public Interest (CSPI). (2010) "Senate adopts historic improvements for school foods," Center for Science in the Public Interest, cspinet.org/new/201008051.html, accessed March 5, 2011.

Cohen-Vogel, L. and McLendon, M. (2009) "New approaches to understanding federal involvement in education," Pp. 735–48 in D. Plank, G. Sykes, and B. Schneider (eds) *Handbook of Education Policy Research. A Handbook for the American Educational Research Association*, Lawrence Erlbaum, Mahwah, NJ.

Community Food Security Coalition (CFSC). (2012) "National farm to school program," CFSC, http://www.farmtoschool.org/, accessed February 2, 2012.

DeForest, A. (2010) "Congressman Holt introduces farm to school improvement acts legislation," *Urban Farm Hub*, February 23.

DeLind, L. (2010) "Are local food and the local food movement taking us where we want to go? Or are we hitching our wagons to the wrong stars?," *Agriculture and Human Values*, vol 28, no 2, pp. 273–83.

Dixon, J. (1999) "A cultural model for studying food systems," *Agriculture and Human Values*, vol 16, pp. 151–60.

Farm to School (FTS). (n.d.) "Chronology of Farm to School," FTS, www.farmtoschool.org/chronology.php, accessed February 20, 2013.

Farm to School (FTS). (2009) "The farm to school collaborative," FTS, www.farmtoschool.org/files/LegislatorInfo-50milMandatoryF2S.pdf, accessed December 7, 2011.

Farm to School (FTS). (2011) "National farm to school month," FTS, www.farmtoschool.org/wp-content/uploads/2011/08/fact-sheet3.pdf, accessed October 10, 2011.

Farm to School (FTS). (2012) "Farm to school network, about," FTS, www.farmtoschool.org/aboutus.php, accessed December 10, 2012.

Feenstra, G. (1997) "Local food systems and sustainable communities," *American Journal of Alternative Agriculture*, vol 12, no 1, pp. 28–36.

Fereday, J. and Muir-Cochrane, G. (2006) "Demonstrating rigour using thematic analysis," *International Journal of Qualitative Methods*, vol 5, no 1, pp. 1–11.

Food Research and Action Center (FRAC). (2008) "Fresh from the farm," FRAC, frac.org/pubs/produceguide.pdf, accessed April 28, 2011.

Food Research and Action Center (FRAC). (2010) "Child nutrition reauthorization related legislation," FRAC, frac.org/pdf/cnr_matrix_apr10.pdf, accessed June 18, 2011.

Friedland, W., Ransom, E., and Wolf, S. (2010) "Agri-food alternatives and reflexivity in academic practice," *Rural Sociology*, vol 74, no 4, pp. 532–7.
Gleason, P. and Hedley Dodd, A. (2009) "School breakfast program but not school lunch program participation is associated with lower body mass index," *Journal of the American Dietetic Association*, vol 109, no 2, pp. 118–28.
Goodman, D. and DuPuis, M. (2002) "Knowing food and growing food: Beyond the production-consumption debate in the sociology of agriculture," *Sociologia Ruralis*, vol 42, no 1, pp. 5–22.
Guthman, J. (2011) " 'If they only knew': The unbearable whiteness of alternative food," Pp. 387–97 in A. Alkon and J. Agyeman (eds) *Cultivating Food Justice: Race, Class and Sustainability*, MIT, Boston, MA.
Hassanein, N. (2003) "Practicing food democracy: A pragmatic politics of transformation," *Journal of Rural Studies*, vol 19, pp. 77–86.
Hinrichs, C. C. and Kremer, K. (2002) "Social inclusion in a Midwest local food system project," *Journal of Poverty*, vol 6, no 1, pp. 65–90.
Johnston, J. and Szabo, M. (2011) "Reflexivity and the Whole Foods Market consumer: The lived experience of shopping for change," *Agriculture and Human Values*, vol 28, no 3, pp. 303–19.
Joshi, A., Misako Azuma, A., and Feenstra, G. (2008) "Do farm-to-school programs make a difference? Findings and future research needs," *Journal of Hunger and Environmental Nutrition*, vol 3, no 2–3, pp. 229–46.
Kingdon, J. W. (1995) *Agendas, Alternatives and Public Policies*, 2nd edn, Harper Collins, New York.
Kloppenburg, J. and Hassanein, N. (2006) "From old school to reform school?," *Agriculture and Human Values*, vol 23, pp. 417–21.
Kloppenburg, J., Hendrickson, J., and Stevenson, G. W. (1996) "Coming into the foodshed," *Agriculture and Human Values*, vol 13, no 3, pp. 33–42.
La Vía Campesina. (2011) "Organisation," La Vía Campesina, viacampesina.org/en/index.php/organisation-mainmenu-44, accessed March 3, 2013.
Levine, S. (2008) *School Lunch Politics: The Surprising History of America's Favorite Welfare Program*, Princeton University Press, Princeton, NJ.
Lieberman, R. (2002) "Ideas, institutions, and political order: Explaining political change," *American Political Science Review*, vol 96, pp. 697–712.
Lin, B. H. (2008) "Diet quality and food consumption: Food away from home," USDA Economic Research Service, available at www.ers.usda.gov/Briefing/Diet Quality/fafh.htm, accessed June 3, 2009.
Magdoff, F., Foster, J. B., and Buttel, F. H. (2000) *Hungry for Profit: The Agribusiness Threat to Farmers, Food, and the Environment*, Monthly Review Press, New York.
McMichael, P. (2005) *Development and Social Change: A Global Perspective*, 3rd edn, Pine Forge Press, Los Angeles, CA.
McMichael, P. (2009) "A food regime genealogy," *Journal of Peasant Studies*, vol 36, no 1, pp. 139–69.
Mintrom, M. (1997) "Policy entrepreneurs and the diffusion of innovation," *American Journal of Political Science*, vol 41, no 3, pp. 738–70.
Mintrom, M. and Norman, P. (2009) "Policy entrepreneurship and policy change," *Policy Studies Journal*, vol 37, no 4, pp. 649–67.
Morris J. L. and Zidenberg-Cherr, S. (2002) "Garden-enhanced nutrition curriculum improves fourth-grade school children's knowledge of nutrition and

preferences for some vegetables," *Journal of the American Dietetic Association*, vol 102, pp. 91–3.

Mount, P. (2012) "Growing local food: Scale and local food systems governance," *Agriculture and Human Values*, vol 29, pp. 107–21.

National Alliance for Nutrition and Activity (NANA). (2010) "Senate adopts historic improvements for school foods," NANA, www.cspinet.org/new/201008051.html, accessed February 2, 2011.

National School Lunch Act of 1946 § Pub. L. No. 396, § 60 Stat. 230 (1946).

Obama, M. (August 2, 2010) "A food bill we need," *Washington Post*, available at http://www.washingtonpost.com/wp-dyn/content/article/2010/08/01/AR2010080103291.html, accessed February 2, 2011.

Padamsee, T. (2009) "Culture in connection: Re-contextualizing ideational processes in the analysis of policy development," *Social Politics*, vol 16, no 4, pp. 413–45.

Pechlaner, G. and Otero, G. (2008) "The third food regime: Neoliberal globalism and agricultural biotechnology in North America," *Sociologia Ruralis*, vol 48, no 4, pp. 351–72.

Pechlaner, G. and Otero, G. (2010) "The neoliberal food regime: Neoregulation and the new division of labor in North America," *Rural Sociology*, vol 75, no 2, pp. 179–208.

Poppendieck, J. (2010) *Free for All: Fixing School Food in America*, University of California Press, Berkeley.

Raynolds, L. (2002) "Consumer/producer links in fair trade coffee networks," *Sociologia Ruralis*, vol 42, no 4, pp. 404–24.

Roberts, N. and King, P. (1991) "Policy entrepreneurs: Their activity structure and function in the policy process," *Journal of Public Administration Research and Theory: J-Part*, vol 1, no 2, pp. 147–75.

Som Castellano, R. (2009) "School lunch programs and the American diet: Exploring a contested food terrain," Master's thesis, University of Kentucky, Lexington.

Streeck, W. and Thelen, K. (2005) "Introduction: Institutional change in advanced political economies," Pp. 3–39 in S. Wolfgang and K. Thelen (eds) *Beyond Continuity: Institutional Change in Advanced Political Economies*, Oxford University Press, Oxford.

United States Department of Agriculture (USDA). (2011) "Farm to school FAQs procurement," USDA Food and Nutrition Service, www.fns.usda.gov/cnd/f2s/faqs_procurement.htm#6, accessed February 12, 2011.

United States Department of Agriculture (USDA). (2012) "National school lunch program," USDA Food and Nutrition Service, www.fns.usda.gov/cnd/lunch/AboutLunch/NSLPFactSheet.pdf, accessed September 10, 2012.

Van der Ploeg, J. D. (2000) "Revitalizing agriculture: Farming economically as starting ground for rural development," *Sociologia Ruralis*, vol 40, no 4, pp. 497–511.

Wright, W. and Middendorf, G. (eds) (2008) *The Fight over Food: Producers, Consumers and Activists Challenge the Global Food System*, Pennsylvania State University Press, University Park, PA.

Zahariadis, N. (2007) "Ambiguity and choice in European public policy," paper presented at the biannual meeting of the European Union Studies Association, Montreal, Canada, May 17–19, available at aei.pitt.edu/8031/1/zahariadis-n-10f.pdf, accessed February 20, 2013.

7

BEYOND FARMING

Cases of Revitalization of Rural Communities through Social Service Provision by Community Farming Enterprises

Haruhiko Iba and Kiyohiko Sakamoto

Introduction

For the last few decades, Japan has gone through drastic political, administrative, and fiscal reforms based on neoliberal ideologies toward a "small government," leaving a substantial impact on its rural societies. This study aims to elucidate struggles of some rural stakeholders in the political-economic and sociodemographic changes and assess the impact of the neoliberal reforms in rural Japan. More specifically, we will present two case studies of Community Farming Enterprises (CFEs) as social service providers in remote rural localities and discuss their significance, challenges, and potential to build socioeconomically sustainable rural communities. CFEs are groups consisting of members of agricultural communities collectively engaged in farming. By jointly owning equipment and arranging labor, CFEs have proven to be effective and efficient entities in building viable farming in even agroecologically and economically disadvantaged areas. Meanwhile, the neoliberal reforms led by the national government, including municipal mergers, have posed daunting challenges to local governments' ability to deliver public social services to aging and depopulated rural areas. As a result, the reforms have prompted some CFEs to embark on social service businesses, such as elderly care, in rural areas (Iba, 2012).

As noted by recent scholarships, in an era of neoliberalization, nongovernmental or civic entities, including farmers' organizations, have become increasingly significant actors enacting regulatory governance in a number of rural policy/administration settings, such as natural resource management, community development, or rural service provision (Alston, 2005; Lockie and Higgins, 2007; McKee, 2009). As we posit that the struggle of the CFEs engaged in social service provision signals a comparable trend happening in the neoliberalizing rural governance in Japan, gauging their capacity and challenges facing them is of timely importance towards a better comparative understanding of the neoliberal regime. Given this concern, to tackle this book's task of critically assessing the crisis of the neoliberal regime in the agri-food sector, we pose two research questions. First, what are strengths or advantages of the CFEs as social service providers in rural Japan undergoing the neoliberal reform? Underling this question is our "practical" interest in advocating actors, such as the CFEs, who strive to build viable economies and community lives under neoliberalizing rural sectors. And second, do the CFEs' struggles embody rural actors' resistance to, or rather, ironic enactment of neoliberalism? This question reflects our "critical" posture in gauging the complex effects of neoliberalism on agri-food sectors.

To address the research questions, we use two case studies of CFEs that provide social services in rural localities. We delineate characteristics and conditions enabling their businesses to supersede the public (i.e., governmental) sector and illuminate the consequences and challenges brought on by the neoliberal reforms. Foucault's "governmentality" concept will usefully inform our analytical framework to demonstrate that neoliberalism is embodied not simply as an exogenous force but through complex intertwining with endogenous processes of "*hybrid* assemblages" (Lockie and Higgins, 2007, p. 2, italic in original) by actors, including national and local governments, and individuals committed to mobilizing local resources. Our findings will demonstrate that CFEs' social service provision epitomizes the resilience of Japanese rural communities to weather major socioeconomic changes. However, we will conclude the chapter with an argument that the successful cases of CFEs would be cautiously evaluated vis-a-vis the multifaceted and contradictory nature of neoliberalism.

Neoliberalism and Neoliberalization in Japan

Our critical assessment of the neoliberal regime through the case studies of CFEs is not solely to focus on their farming endeavors but rather to examine social and political-economic conditions of rural Japan in which farming communities struggle for their sustainable development vis-a-vis proliferating neoliberal discourses and practices. Thus, in what follows, we will, first, problematize the backdrop of our study through configuring the concepts of neoliberalism and neoliberalization as its practices, and second, provide an outline of the neoliberal reforms that left significant impacts on rural societies of Japan.

Understanding Neoliberalism and Neoliberalization

Generally, neoliberalism is a political philosophy that exalts individual freedom, property rights, free market transaction, entrepreneurship, and minimal state intervention in individual activities. While the idea can be traced back to the classical liberalism of the eighteenth century with the limited scope of speaking specifically to human freedom in market (economic domain), neoliberalism emerging in the twentieth century has extended the scope to every aspect of human activity (Mirowski, 2009). Thus, "[n]eoliberalism seems to be everywhere" (Peck and Tickell, 2002, p. 380). Yet, neoliberalism is by no means a static ideal type representing a cohesive set of ideas; rather, it "actually exists" as a hybrid of multifaceted, even contradictory, plural "neoliberalisms," which are practiced or enacted in varying political contexts, and hence should be comprehended as "neoliberalization" (Brenner and Theodore, 2002; Peck and Tickell, 2002; Guthman, 2008a).

Reflecting such a "variegated" nature of neoliberalism/neoliberalization (Brenner et al., 2010), the current literature provides various strands of perspectives to understand it. Political economy literature inspired by Marxist traditions (e.g., Harvey, 2005) would grasp neoliberalism/neoliberalization as a comprehensive restructuring of regulations enabling the global spread of capitalism (Lockie and Higgins, 2007). From this perspective, the state accommodates institutions that allow capital accumulation, such as privatization, commodification, marketization, deregulation/reregulation, devolution, free market, and free trade. Harvey (2005) points to a contradictory nature of neoliberalism and neoliberalization—while, as with classical liberalism, the neoliberal ideal might extol personal freedom, the state is in favor of the corporate interest and thus limits this to the extent that the strong association of "free individuals" will not be formed to counter the hegemony, creating a tension or contradiction between individual freedom and collective social life.

In addition, recognizing changing modes of state intervention, Peck and Tickell (2002) have noted the transition from the "roll-back" stage during the 1980s where, as characterized by the Thatcher and Reagan administrations, the state dismantled the Fordist-Keynesian regime devastatingly, to the "roll-out" stage during the 1990s where the state employed conductive "metaregulation" or "reregulation" such that the state transformed itself into an "interventionist" form. "Roll-out" neoliberal agendas would include, among many others, regenerating communities to compensate for market failure, tapping into social-capital for development, public-private cooperations employing participatory schemes, and/or partnership with civic, nongovernmental sectors for program delivery for social welfare (Jessop, 2002; Peck and Tickell, 2002; McCarthy and Prudham, 2004; Harvey, 2005). An important observation by Peck and Tickell (2002, p. 391) points to neoliberalism's scalar shift, which entails "extensions of national state power, most notably in the steering and management of programs of devolution, localization, and interjurisdictional policy transfer." These insights, as detailed later, will helpfully guide us to situate Japan's neoliberal reforms within the global transformation of political economy, including a scalar shift in administrative and fiscal relationships between the state and local governments.

In the meantime, scholarship drawing on the neo-Foucauldian "governmentality" concept, while resonating somewhat with the observations of "roll-out" neoliberalization, provides distinctly important insights for our study. According to this strand of literature, neoliberalism deems human freedom not as naturally endowed (as the classical liberalism assumed) but as something to be constructed by fostering self-discipline, self-organization, self-actualization, self-help, self-responsibilization, and entrepreneurship (Gordon, 1991; Rose and Miller, [1992] 2010; Barry et al., 1996; Burchell, 1996; Guthman, 2008a; Mirowski, 2009; Dean, 2010). The neoliberal governmentality, fostering self-responsible and entrepreneurial individuals, is characterized as a "hybrid assemblage" of different rationalities, including economic efficiency, techniques, knowledge, expertise, or calculation (Higgins and Lockie, 2002). In Australia, for instance, different regulatory schemes combining standards and direct governmental involvement were employed as a governing technique to prompt farmers to become economically efficient entrepreneurs and simultaneously socially and environmentally responsible operators (Lockie and Higgins, 2007). Also, such hybrid assemblage of governing techniques would entail diffusive mechanisms "acting at a distance," such as devolution or the scalar-down of administrative authorities, or prompt partnerships between private, civic, and public sectors, which have been increasingly practiced in varying settings, such as agri-food and environmental regulation, social service provision, and community development (Alston, 2005; Lockie and Higgins, 2007; McKee, 2009).

Whereas the Marxist political economy tends to deem neoliberalization as a relatively cohesive mode of regulations, the Foucauldian scholarship highlights subtle and complex modes of neoliberal practices and their consequences embedded in, or path-dependent on, local political contexts (Brenner and Theodore, 2002; Brenner et al., 2010). For instance, in parallel with Harvey's observation above, Lockie and Higgins (2007) note that the hybrid assemblage of governing comprising different regulatory schemes in Australia has created a tension between the expectation of farmers as individual entrepreneurial actors pursuing economic rationality and the cultural expectation of them as responsible members of their collective community, whose interests are to be put before their own. Yet, such tensions, Lockie and Higgins go on to argue, continue to problematize the legitimacy of public intervention or participatory governance schemes, thereby creating spaces for alternative modes of actions as a "positive" or constitutive effect of neoliberalization. Importantly, moreover, the Foucauldian perspective draws critical and reflective attention to effects of discourses of a variety of actors, including scholars like ourselves, who could create—consciously or unconsciously—discursive influences through assessments of the neoliberal project. This reflection matters to us, as Guthman (2008a) has noted, that scholars lauding recent food politics movements to counter neoliberalism, such as consumer choice, localism, entrepreneurism, and self-improvement, ironically contribute to the neoliberal subject formulation. In essence, a progressive and critical assessment of, or even a counter-argument to, neoliberalism may end up with the self-enactment of neoliberalism (McCarthy and Prudham, 2004; Guthman, 2008b; Song, 2009).

In the meantime, the Foucauldian approach to neoliberalism has drawn critiques for its insensitiveness to neoliberalization as global-scale "pattern-making" processes and to constitutive roles of the state, which Marxist political economy approaches have aptly grasped (Barnett, 2005; Brenner et al., 2010). Thus, in the analysis of the case studies, we pay attention to both local specifics where "actually existing neoliberalisms" (Brenner and Theodore, 2002, p. 349) are realized, yet retain the analytical focus on neoliberalization as the global restructuring of political economy and roles of the state in it.

Neoliberal Reforms in Japan and their Impacts on Rural Societies

As mentioned, in recent years, Japan enacted massive reforms in political (including administrative, legal, and fiscal) and economic systems underlined by the neoliberal belief, although the contents were incoherent and stumbling (Hirashima, 2004; Itoh, 2005). Emulating the Reagan Administration in the United States, Japan instigated concrete neoliberal reforms under Yasuhiro Nakasone's Administration in the 1980s to curtail budgetary programs (except national defense expenditures), privatize major state-owned enterprises, and promote market-oriented economies through deregulation (Hirashima, 2004). To foster individuals with self-responsibility and free-choice, who would better adapt to market economies, Nakasone proposed to revise the education system (Inoue, 2013; Arai, 2013). While the cabinets after Nakasone also attempted neoliberal restructurings, it was Jun-ichiro Koizumi's administration (incumbent from 2001 to 2006) that embodied the neoliberal ideology in the most aggressive manner, leaving a significant impact on Japan's political economy and social lives. Speaking to the ideals of personal and individual responsibility, he eagerly cut down governmental welfare and education expenditures. To boost an entrepreneurial economy with competitiveness in global markets, he pushed further deregulations (e.g., facilitation of corporations to participate in farming) and privatization of state organizations (e.g., the national postal service company). He blamed the bureaucracy for its inability to promote private entrepreneurship and instead effectively employed neoliberal-rhetoric, such as "*Kan kara min e*" (from the bureaucracy to the private sector) or "*Shijô ni dekiru koto wa shijô ni*" (let the market do what the market can) to draw popular support to his drastic neoliberal reforms.

The above reforms seem to be in line with the tide of neoliberal restructuring of the political economy towards globalization since the 1980s. From the Marxist-inspired perspective, Nakasone's early bids could be comprehended as Japan's response to the crisis of the Fordist regime, which prompted the US and UK superpowers to aggressively seek global markets for its products (Inoue, 2013). Japan's stronger business sectors, including export-oriented manufacturers, were adamant supporters of deregulation and tax reduction (Hirashima, 2004). Global pressures—particularly from the United States—to open markets for foreign products were so intense that during the 1990s Japan accepted trade liberalization of long-protected agricultural products, such as beef, citrus, and rice, the staple crop of the nation.

However, Japan's neoliberalization in practice was more fragmented and contradictory than its rhetorical appearance, not least because the nation's economy and the state government were by no means a monolithic entity but because they were comprised of unevenly developed sectors with varying vested interests. While the stronger export-oriented businesses supported the reforms, small-scale manufactures, labor unions, and rural sectors were upset with advancing market economy, and policy makers, including bureaucrats, took the trouble to reconcile the conflicting parties' interests (Hiwatari, 1998; Hirashima, 2004). The role of the state bureaucracy, which had coordinated the postwar "developmental state" capitalism (Johnson, 1982), was ironically imperative in streamlining the role of the state government. Also, despite the call for "a small-government," the fiscal management since the 1990s was oftentimes operated as "Keynesian-like" mode with deficit policies to buoy the stagnant economy (Hirashima, 2004; Itoh, 2005). Historically in Japan, subsidies from the central treasury served not only to redistribute wealth to rural areas (thereby maintaining rural voters' allegiance to the leading party) but also to curb complaints of the farming sector pressed by accelerating market-economy and trade liberalization, which would in turn benefit the export-oriented industries. Indeed, after the 1993 rice-market opening as a condition to accept the GATT Uruguay Round agreements, six trillion yen (approximately sixty billion dollars) were earmarked to mitigate its impacts on farming communities for eight years by the Ministry of Agriculture, Forestry and Fishery (MAFF). Nonetheless, the effectiveness of rural subsidies, including this enormous budgetary program, has been doubted (Imamura, 2010) since it has left the uncompetitiveness of the heavily subsidized agricultural sectors intact.

Thus, Japan's political actors did not simply adopt the roll-back neoliberal ideology but rather opportunistically reconciled demands of conflicting parties to alleviate the direct impact of neoliberal globalization, especially in rural areas. In a sense, it could even be argued that Japan, by "bringing the state back in" (Lee, 2008, p. 2) (this "state back" concurs with the state "roll-out"), resisted rather than followed (US-led) neoliberal world order to continue on its state-led economic development policies. Our analysis of the CFEs, therefore, should not revolve solely around binary oppositions, such as the state/neoliberal world order, but be attuned to the apparently contradictory facades of neoliberalization and its embeddedness in and pathway dependency of neoliberalization on complex local politico-economic conditions. By doing so, we believe we can delineate the distinctive way through which neoliberal discourses and practices enabled and/or hindered actions of the CFEs as hybrid assemblages engaged in farming and social service endeavors.

Fiscal Reforms, Municipal Mergers, and Rural Governance

Throughout the series of neoliberal reforms since the 1980s, one of the key issues was decentralization or delegation of authorities of the central government to local administrative entities (Barrett, 2000; Tsukamoto, 2011). And the reform by Koizumi that is especially relevant to our research is the fiscal and administrative restructuring of the relationship between the national government and local administrative entities. To attain fiscal austerity and lessen the central influence over local governments, Koizumi implemented a series of fiscal restructurings called "*Sanmi-ittai no kaikaku*" (Integrated Triple Fiscal Reform). The "Triple Reform" consisted of (1) cutback in subsidies from the national government for local governments that were "tied" with specific policy-program purposes, (2) concession of fiscal sources (i.e., authorities to collect certain taxes) from the national government to local governments, and (3) reform, or reduction, of "untied" subsidies from the national government for local governments. Through the reform, while subsidies from the national government were reduced, local governments were supposed to obtain more fiscal freedom to control their tax revenues and allocations.

In addition, to cope with the fiscal austerities since the 1990s, which were accelerated by the "Triple Reform," the national government encouraged mergers of local municipalities, including cities, towns, and villages. The number of municipalities was reduced by nearly half from 3,234 in 1995 to 1,727 in 2010 (MIC, 2010). The Ministry of Internal Affairs and Communications (MIC), being responsible for local governance affairs, used "the carrot and stick" to prompt mergers, giving financial and legal privileges to willing municipalities while slashing fiscal supports for unwilling parties. Thus, by reducing the number of local entities, Koizumi attempted to build efficient and autonomous local fiscal governance.

However, the reform was self-contradictory since the fiscal freedom could hardly be beneficial when fiscal resources were diminished harshly or absent from the beginning. While local municipalities obtained a somewhat "free hand" to control their coffers, the decentralization reform increased the gap between local governments with stable financial revenues and local governments lacking such resources, as critics of decentralization had warned before Koizumi took office (Barrett, 2000). Conventionally in Japan, local governments with fiscal support (i.e., subsidies) from the national government have been undertaking social service provisions in rural areas where aging and depopulation have advanced. As the national fiscal supports were curtailed, however, some local governments, especially those in aging, depopulated, and economically frail, rural areas, came to face serious challenges in delivering public services. The municipal mergers exacerbated this trend because local governments were combined and streamlined, which resulted in a shortfall of human resources for social service delivery (MIC, 2010). This is ironic because the mergers were intended to make local governance more efficient and lower delivery costs of social services for remote areas. In this context of the neoliberal reform, CFEs as farming entities began their roles as providers of social services in rural areas.

CFEs as Social Service Providers

A CFE can be defined as an organization of members of a community who are collectively engaged, whether partially or fully, in agricultural production. CFEs, whose number has grown drastically since the 1980s, are innovative endeavors combining community members' self-help and mutual supports to tackle challenges for rural areas of Japan, as delineated below.

Since the early 1960s, rural and agricultural policies in Japan have aimed to establish economically autonomous, self-reliant farmers, thereby filling the income gap existing between the farm sector and other industries. In an attempt to modernize and consolidate small-scale farms into large-scale entities pursuing economies of scale, the government vigorously subsidized farmers to procure machinery and equipment or to launch land improvement projects. However, despite these efforts, called "agricultural structural reforms," the income gap between farms and other industries widened because the continuous economic growth driven by the export-oriented manufacturing was of little benefit to the farm sector. Because of the limited fluidity in transfer of farmland properties (that is, farmers were unwilling to sell their lands), consolidation of farmlands did not happen as expected, leaving Japan's uncompetitive mini-farm operation almost intact even when faced with the trade liberalization after the 1980s. The low profitability of farming never appeared attractive to heirs (farmers' children), while the thriving manufacturing industries absorbed the productive-age workforce from farming households, causing a dearth of young successors of full-time farm operations.

The shortage of farm laborers was accompanied by a few serious problems both at the individual farm-household level and the community level. At the individual level, to supplement the scarce workforce, many farmers became "over-equipped" with machinery for their small-scale operation. Costs of the equipment, including loan payment and maintenance expense, rendered the farmers "machinery poor," with the result being that the unprofitability and unsustainability of their farm economy were exacerbated. At the community level, the limited rural workforce made it difficult to maintain common farming resources indispensable for rice culture, such as irrigation canals, ponds, and farm roads, whose functions were sustained by community members' collective works. Thus, many farming communities came to struggle to arrange communal commitment to sustain the proper functions of the common farming resources, because this community problem could aggravate the unprofitability of the individual farming operation.

CFEs initially emerged as responses to the above described problems (decline of farm labor force, the state of being machinery poor, and the reorganization of workforce for communal commitment) and evolved as collective endeavors based on the community as the primary unit. Communal commitments are not necessarily new in Japanese rural areas. However, what makes the CFE distinctive from past collective rural activities is that it employs organizationally the principle of financial management to sustain business endeavors. For example, although in the past, rural community members would collectively engage in farming, monetary transactions (i.e., payment of wages for providing labor) would not be involved in such activities. Moreover, while CFEs can take different organizational formats or legal personalities, a substantial number of CFEs have incorporated to become collective, revenue-creating, farming entities. In essence, a rural community constitutes a CFE, and as an organization, it collectively engages in farming endeavors.

CFEs are considered as attempts to sustain farming as an autonomous economic sector within a rural area. Yet, the premise based on which CFEs can sustainably operate does not solely concern the economic rationality or profitability of their businesses. Rather, what enables communities to build a CFE is the motivation and commitment shared by community members to sustain the community per se. Thus, the business-oriented, revenue-creating activities of a CFE are the means to achieve its ultimate public purpose to sustain rural lives of the community.

From 2005 to 2011, the number of CFEs and commercial farmlands the CFEs cultivated increased by 50 percent; farmlands entrusted to CFEs for production increased by 30 percent, and the percentage of lands cultivated by CFEs increased from 10 to 14 percent of Japan's total farmlands (Table 7.1). Also, CFEs are engaged in diverse farm-businesses, such as contract farm operations collecting commissions, processing of value-added farm products, and farm restaurants. For the diversification of businesses, CFEs take on different organizational structures, whether as legal entities or voluntary groups. MAFF recognizes CFEs not only as farming entities but also as social vanguards undertaking important roles in governmental rural development programs, such as the Land, Water and the Environmental Conservation (LWEC) program launched in 2007. LWEC aimed to assist rural communities and through collective commitments for preservation of agricultural resources (e.g., soil and irrigation systems) that yield environmental and social goods, also known as multifunctionality (Iba, 2010, 2011). CFEs are, therefore, not only effective farming entities but also important contributors to values other than agricultural production.

Historically, Japan's small-scale rice cultivation made collective commitments essential to arrange intensive labor for production, such as planting, harvest, and distribution of water among community members. The collective engagement not only led community members to effectively allocate labor, but it also, through

Table 7.1 Extent of CFEs in Japan in 2005 and 2011

		Farmland areas (hectares)			
Year	*Number of CFEs*	*CFEs cultivated for their own commercial production (A)*	*Entrusted to CFEs for contract operations for commission (B)*	*Total areas cultivated by CFEs (A+B)*	*Percentage of lands CFEs cultivated in Japan's total farmlands*
2005	10,063	253,672	99,456	353,128	10
2011	14,643	372,490	128,969	510,459	14

Source: MAFF, "Survey of Community Farming Enterprises," http://www.maff.go.jp/j/tokei/kouhyou/einou/, accessed March 15, 2013.

day-to-day activities, strengthened solidarity. The living legacy of collective commitments and social ties built through farming makes CFEs vital actors to build active communities in rural societies. Especially in disadvantaged rural areas, CFEs are expected to be efficient farming entities and to deliver public and social goods to supplement the undermined capability of local governments.

To fulfill social service provisions, however, CFEs have to secure revenues in order to continue business as an economic activity. One of our aims is to provide practical insights from the case studies of CFEs to establish viable social service businesses. In the meantime, to critically assess consequences of the neoliberal reforms in Japan, we will examine social backdrops and discourses that enable and justify CFEs' commitments to social service.

Case Studies

The CFEs we investigated, Tsukinoya Healing and Green Work, are located in Shimane Prefecture in western Japan, which ranks third highest in average age and fourth lowest in population density, among forty-seven prefectures (NIPSSR, 2012). While the two CFEs are located in different municipalities, both are typical communities that suffer from underprivileged conditions for farming, aging, and depopulation in rural Japan.

Case 1: Tsukinoya Healing

Tsukinoya Healing (TH), a CFE established in 1998, is in the Tsukinoya area of Un-nan City. The area, a thirty-minute drive from the nearest urban center, consists of two smaller settlements, with a total population of ninety-two in thirty-one households. Its steep topography and substantial snowfall keep Tsukinoya's agriculture uncompetitive and accelerate the area's aging and depopulation process (average age is sixty-three years old; 47 percent of the population are sixty-five years or older; only three children are under seventh grade). The total commercial farmland is twenty hectares, nineteen of which are paddy fields. Out of the thirty-one households, only six are actively engaged in farming, while the rest entrust their properties to TH and are gaining income from nonfarming jobs. TH grows rice in eighteen hectares of paddies (ten hectares in Tsukinoya and eight hectares outside), manages entrusted farmlands for commission, and has greenhouses for some horticultural productions.

TH was born from deliberations in neighborhood organizations in the 1980s to revive the aging and depopulating community. The discussions resulted in two community-based organizations, including TH and a nonprofit organization (NPO), *Tsukinoya Shinkoukai* (TS).[1] While TS and TH are distinct entities with different purposes—TH as a CFE focuses on commercial farming, whereas TS, as a nonprofit organization, is committed to noncommercial community development affairs—nonetheless, they work in close collaboration, involving community members to synergistically achieve their missions of reinvigorating

Table 7.2 Human resources and sales of Tsukinoya Healing

		Sales (1,000 yen)			
Members	*Employees*	*Total sales*	*Farm production sales*	*Social service sales*	*Other subsidies*
20	1 (manager) 3 (employees)	28,000	24,000 (85%) Rice and vegetables Contract commission for farm operations	1,300* (5%) Support of elderly Contract park management, maintenance and cleanup of roads, snow removal, etc.	2,700 (10%)

*includes subsidies from the local municipal office.

Source: Interview by Iba.

Tsukinoya. Both entities divide roles and responsibilities in accordance with their characteristics and functions to perform their services effectively. This division of labor, as it were, allows them to engage in a variety of services and activities needed to reinvigorate the rural society. More specifically, depending on whether an activity is more private, revenue-creation-oriented, or public and social, volunteer-oriented, TS and TH divide and coordinate their roles and actions (Table 7.2).

For instance, oftentimes TH deploys its workers to remove snow or fallen trees on streets in the area, based on contract with TS. Since such services are considered public-oriented, serving for the community's welfare, TS takes an initiative to respond to the need; however, because TS does not have a sufficient workforce or equipment for physical labor, it cannot handle a task needing physical labor as volunteer service provision. Hence, TS entrusts the task to TH, which, as the provider of workforce and heavy machinery, handles the physical task and gains revenue based on the commission. TS plays principal roles in organizing other endeavors, such as rural-urban exchange (i.e., inviting urban residents to events held in the area), rural life improvement, and cultural activities, and TH complements them. Meanwhile, TH takes the responsibility to manage private, revenue-making business activities, a farm restaurant, and a farmers' market, which necessitates the economic rationality, and TS supports them.

TH employs two individuals year-round for both the farming operation and the social and public service, whereas TS is run by volunteers of the community. In addition, occasional demands for labor are oftentimes fulfilled by outsourcing to a Silver Human Resource Centre (SHRC),[2] which pools local seniors for temporary hire. This labor arrangement, while minimizing costs of the year-round employment, allows TH and TS to flexibly mobilize their workforce for diverse business endeavors, including a farm restaurant and processing of value-added products, as well as nonprofit public/social endeavors to respond to a variety of community needs.

Case 2: Green Work

Green Work (GW), a CFE founded in 2003, is in the Higashimura area, Izumo City of Shimane prefecture. Located thirty kilometers from the nearest urban center, Higashimura consists of five smaller hamlets, contains 100 households with 290 residents, and is undergoing aging and depopulation. While fifty-nine households are engaged in agriculture, only two are dedicated to full-time farming. The commercial farmlands total seventeen hectares, fourteen of which are paddy fields. GW manages 90 percent of the farmlands, including properties leased from its members and entrusted farmlands for commission.[3]

GW was founded as a result of deliberations by a committee of Higashimura residents as to how the community could sustain its viability and activity. Although initially two CFEs were founded, the growing problems eventually overwhelmed their capabilities. Hence, the Higashimura people unified those two and created GW to pursue the economy of scale for competitiveness and efficiency. In fact, the consolidation of farmlands and machinery and the resultant efficient farm operation have allowed GW to engage in agri-entrepreneur businesses, including environmentally sustainable production of rice and vegetables, its direct sale to consumers, grazing of sheep to weed farmlands, and wool products (from the weeding sheep). Still, the ultimate mission of the new GW is to build an active and viable community through diversification of rural businesses rather than the pursuit of economic efficiency per se. And, to embark on business activities beyond farming, GW became a limited liability corporation rather than a farm cooperative corporation (FCC), which is a specific type of farming company entitled to various advantages, including tax exemption, yet restricted to limited nonfarming businesses.

For its social service business, GW hires year-round employees, including newcomers to the area, who contribute to invigorating the community. For the aging community, it was an urgent issue to maintain social service programs for assisting the elderly population to accomplish errands. Currently, GW offers seniors assistance in remaining independent (e.g. assistance with transportation for shopping and other essential tasks). Meanwhile, the previous two CFEs had difficulty securing revenue sources for the winter off-season. GW's social service business solved this problem by maintaining seven employees during winter. Initial investments for GW were minimal as the business was inherited from for-profit companies and local governments.

Local governments in streamlining budgets and human resources would find it difficult to start or continue providing social welfare programs of their own for areas where aging and depopulation have advanced. And existing programs would be reduced as well. The social service business GW provides supplements the declining public services the local governments have been delivering in the aging

Table 7.3 Human resources and sales of Green Work

		Sales (1,000 yen)		
Members	*Employees*	*Total sales*	*Farm production sales*	*Social service sales*
32	7 Full-time	46,000	33,000 (72%) Rice and vegetables Contract commission for farm operations	13,000* (28%) Support of elderly Contract park management, maintenance, and cleanup of roads, snow removal, etc.

*includes subsidies from the local municipal office.

Source: Imai, 2012, p. 73.

and depopulated areas. The subsidy payment from the local government to GW not only serves for GW's financial stability but also functions as an effective and efficient policy measure to continue social services in rural areas.

In addition, as a locally rooted group, GW has been collaborating closely with resident volunteers and neighborhood organizations for diverse community activities, such as cleanup of parks and roads and "fun" social events. Since such "community-building" activities would not be profitable, the outside social service for-profit companies could not provide them. While GW is a self-sustaining business enterprise, it is also a community social-service group; oftentimes, the boundary between the two characteristics becomes indistinct.

Furthermore, combining small but diverse revenue-creating activities that span across the year allows GW to continue year-round employment (Table 7.3). This "small-but-diverse" business helps GW overcome the seasonality of employment inherent to farming and farming-related endeavors. Thus, GW and the local governments complement each other to maximize effectiveness and efficiency of their services by optimizing human and budgetary resources.

Discussion

The case studies above elucidate conditions that have led the CFEs to the social service businesses in rural areas. One important factor is the decline of farming and subsequent depopulation and aging of the communities. To achieve the ultimate goal of TH and GW to reinvigorate the local communities where agriculture has been vital in the economy, sustaining the farm sector is an essential and effective means. Pursuing the efficiency of farming and creating value-added products would improve competitiveness of the farm sector. For instance, by gathering resources of the local small-scale farms and incorporating the CFEs, production costs were lowered in the mountainous disadvantageous areas. However, as the

Japanese agriculture is continuously exposed to harsh global market competitions, such efforts might be insufficient to support the local economy. Once the viability of the farming endeavors of TH and GW is lost, the entire local societies they have supported would collapse. To hedge such a risk, it was crucial for both communities and CFEs to embark on a variety of revenue-creating activities, including both farm and nonfarm social-service businesses.

Critically important for the CFEs, however, has been the declining capability of local governments in rural areas that went through mergers as a part of Koizumi's fiscal reform. Kisugi Town to which Tsukinoya had belonged was merged with several municipalities and became Un-nan City in 2004; Sada Town to which Higashine had belonged was merged into Izumo City in 2005. As the independent towns were scrapped, the numbers of the municipal-council representatives for the communities were curtailed as well. The former town halls are now the new cities' branch offices. Such changes in the structure of administration made delivery of social services to remote areas with small populations difficult. Thus, understandably, the two CFEs committed to sustaining their home communities were called upon to complement and continue social service provisions. What enabled the CFEs to embark on the social service business was the private-public partnership, as encouraged by neoliberalism between efficient nongovernmental or private (including for-profit) actors and governmental entities. Despite the organizational streamlining of the merged municipalities, the public sectors maintain subsidies to social service providers, including the CFEs. Thus, availability of financial supports from governments can be a critical factor for a CFE to engage and continue social service endeavors.

Advantages and Strengths of the CFEs

Given the understanding of their roots in and strong ties with the communities, we argue in response to the first research question that the CFEs have an advantage in the local social service provision. Economically, for the CFEs providing social service business, it is essential to reduce costs and secure revenues that compensate costs. While the CFEs are striving to reduce expenditures, their close relationship and familiarity with the communities—namely, social capitals, help not only lower direct costs but also mobilize community volunteers for less profitable social services or social events. And, of course, the voluntary commitments by community members further strengthen their solidarity.

In addition, the fact that the CFEs are engaged in agriculture is also significant in prompting them to commit to social services. Many CFEs participating in LWEC have been committed to conserving local agroecological resources, including water, soil, and rural landscape. Since benefits from these resources are considered as agricultural multifunctionalities, or public goods coproduced by farming, the commitment by CFEs engaged in LWEC evidences their characteristic as public service providers beyond farming per se. Agricultural multifunctionalities are conducive not only for entrepreneurial farming-related businesses (Morgan et al.,

2010),[4] such as TH's farmers restaurant and GW's sheep grazing, but also, in our view, for commitment to provision of other public goods, such as the CFEs' social services.

In the meantime, following market models developed in Western nations, the fiscal crises that have occurred since the late 1990s pushed further deregulation and marketization in social services (especially nursery care) and encouraged private for-profit entities to enter the market (Goh, 1998; Kishida, 1998, 1999; Hirao, 2001; Rose and Miller, [1992] 2010). While the private sector was expected to offer more efficient services, there was a concern that profit-pursuing corporations would take on opportunistic behaviors, discriminating in places where sufficient returns could not be expected (Goh, 1998). However, the locally rooted CFEs have advantages over outside for-profit corporations in both cost efficiency and the missions shared with the communities.

Also, we have observed that the CFEs' human resource managements utilize both stable year-round employment for the continual social service and the ability to flexibly mobilize community members and the outside workforce (e.g., SHRC) to diverse social and community development activities. Although TH hires only two year-round full-time employees, their social service provision is made possible by flexibly outsourcing to the SHRC and/or collaborating with TS to mobilize volunteers from the community of Tsukinoya. GW hires seven year-round employees enabling stable social service provision. This means also significant job opportunities in the depopulated area.

From the finding above related to the managerial issues, we can elicit "practical" implications for other CFEs and/or local governments to begin a similar, private-government, partnership-based social service program. First, we point out that CFEs can have particular advantages if they have strong ties and missions shared with the communities, which can lower delivery costs. Still, to continue economically feasible businesses, CFEs would have to reduce operation costs, provide services with affordable prices for customers (i.e., community members), and develop new businesses to meet specific community needs, thereby expanding revenue sources. To balance costs and prices of sustainable social service provision, human resource management is also crucial. Both stable employment for continuous service provision and the capacity to flexibly mobilize additional laborers, including community volunteers, should be considered. It would be imperative to ensure human resources, depending on conditions of the workforce in the areas, including full-time employment from or outside of the community, availability of resident volunteers, or additional labor supply sources (e.g., SHRC). CFEs providing social services would have to handle seasonally fluctuating farm laborers and year-round social businesses in order to flexibly respond to ad hoc demands (e.g., the clean-up of fallen trees and festive events). In fact, many CFEs have had to handle human resources for their diversified farming businesses as well as for the LWEC program. Although it is challenging to stabilize employment in depopulated rural areas where workforce is limited, many CFEs engaged in diversified farm-businesses have demonstrated the potential to manage intricate workforces to respond to a variety of social and economic needs of communities.

Resistance to or Enactment of Neoliberalism?

Also, our case studies illustrate how the CFEs as emerging rural actors played multiple roles to sustain their communities in response to the neoliberal reforms in rural governance in Japan, when the state at least partially withdraws from social welfare support provision. Historically, in rural Japan, where the labor-demanding rice culture has prevailed, farming activities and rural social lives have been intricately intertwined; bucolic livelihoods are not solely about economic transactions but also reciprocal labor provisions for farming and mutual social supports. It may be tempting to argue that the CFEs with strong social ties and commitments to their communities are reviving the traditional social arrangement of reciprocity, or what Scott (1976) has documented as the moral economy in Asian rural communities. However, such an argument should be made cautiously so as not to romanticize rural traditions. The CFEs' engagement in social service should be understood not as revival of the good-old-day rural society but as a struggling response to neoliberalism/neoliberalization, which has multifaceted, or even contradictory and paradoxical, characteristics. Our examination of the CFEs and their roles responding to the reforms has to be situated in this paradoxical nature of neoliberalism/neoliberalization.

Recognizing the multifacetedness and apparent contradictions of neoliberalism/neoliberalization, we respond to the second research question on the resilience of rural communities. We posit that CFEs are becoming an important actor of "civil society," which stands between the individual family and the state, counters hegemonies of the state and the market economy (Ritzer, 2010), and plays vital roles in sustaining the communities they are rooted in by assembling hybrids that tap into resources and rationalities available for them. The CFE's flexible and stable internal labor arrangement, combined with the capacity to collaborate with the public sectors and the community members (volunteers) for nonrevenue-generating social activities, would build and reinforce social capital, which in turn could be remobilized for sustainable community development. With this ability of assembling, in response to the neoliberal fiscal austerity in the rural public sector, the CFEs could effectively and efficiently be substituted for the local governments in providing social services, including welfare provision. The CFEs we presented might signal the resilience of rural people's spontaneous, self-help, and entrepreneurial actions, which may be seen as resistance to the larger socioeconomic transformation imposed by neoliberalism.

However, the resilient CFEs "resisting" neoliberalism may also, paradoxically or ironically, be "neoliberalizing" their own rural milieu in Japan since their spontaneous, self-help, and entrepreneurial actions embody key values of neoliberalism. Discourses employed by the state government as a veiled promoter of neoliberalization would use exemplars, such as TH and GW, to justify advancing its neoliberal political reforms. Many exemplary CFEs are presented on MAFF's website (MAFF, 2012) with compliments (in Japanese) on their effectiveness and entrepreneurial achievements, such as "cost reduction," "higher productivity," "improved efficiency," "labour saving," "value-added production," "business diversification," "interactions with urban consumers," or "participation of village members in community activities," and so on. These compliments remind us of Lockie and Higgins's observation (2007) that neoliberalization is characterized as a hybrid of assemblage of different rationalities, including economic efficiency. The MAFF's praises are to encourage farmers and farming communities facing challenges imposed by neoliberalism to foster self-help entrepreneurial capabilities to survive competitive market or to revitalize aging depopulated communities; yet simultaneously and ironically, they embody neoliberal values.

Also, it is in this context that we should problematize discourses that critically assess neoliberalism/neoliberalization, including our own study. For even critical scholarship of neoliberalism could reinforce what it critically examines (Guthman, 2008a, 2008b; Song, 2009). Based on the experiences of the two successful CFEs, we have provided "practical" insights, which we hope help other CFEs tackle their own challenges. Nonetheless, detailing and assessing, if not extolling, the ingenuity of the exemplary CFEs may inadvertently mean enactment of the neoliberal project.

Given this, we should be reminded that capabilities of CFEs to embark on entrepreneurial farm-related and/or social service beyond farming vary across different political-economic and agroecological conditions. The exemplary cases do not guarantee success in other situations. We must remain cautious in political discourses that praise entrepreneurial CFEs, which may justify, without considering varying local contexts, exposing rural communities to global competition or curtailment of social welfare in disadvantaged areas under severe fiscal austerity.

Thus, recognizing paradoxical enactment of neoliberalization reminds us of the observation that neoliberalization engenders tension between the individual freedom and the collective social value based on free association (Harvey, 2005; Lockie and Higgins, 2007). The CFEs as actors of civil society may find themselves in multiple dimensions of tension or contradiction—they face the demand for efficient farm management while taking care of their communities' social needs by managing to arrange their limited resources without compromising the competitiveness. The CFEs also, though not deliberately, resist the hegemonic current of neoliberalization; yet, their resistance might be compromised and tamed by the neoliberal discourse that extols civil society as complementary for its own project (Ritzer, 2010; North, 2003) and thereby remains incapable to overturn it. Still, as noted earlier, the Foucauldian scholarship has found that tensions created by the hybrid assemblage of neoliberal governmentality could have the "positive" side where continuous problematization of tensions and contradictions creates new spaces for alternative actions (Lockie and Higgins, 2007). We hope that our analysis, exposing tensions, paradoxes, and contradictions of neoliberalization, plays a conductive role to support actors who are resiliently surviving the drastically changing rural milieu of Japan.

Conclusion

In this chapter, we intended to provide practical insights to advocate struggles of rural actors, as well as critical reflections of the effects of neoliberalism/neoliberalization in rural Japan, through depicting the way in which the country's neoliberal reforms have taken place, intertwining with the locally specific political contexts and the way the CFEs have responded to the consequences of the reforms. Drawing on the political economy and the neo-Foucauldian literature, our analysis sought to elucidate the local specific actions as hybrid assembling of varying rationalities while situating the political economic reforms in the global current of neoliberalization. However, critics have argued that it would not be a wise exercise to simply combine the two different perspectives on neoliberalism since they are focusing on different aspects of neoliberalization (Brenner et al., 2010). Still, in favor of the empirical value of insights from the case studies, we opted to eclectically employ the different perspectives rather than pursuing the theoretical consistency that would allow us to simultaneously explain variegations and the global pattern making of neoliberalization.

Finally, although our delineation was by no means to provide the whole picture of the transition of Japan's agri-food and rural sectors, our case studies, we believe, managed to convey a vivid illustration of contemporary rural social lives of Japan, which have gone through neoliberalization in unique manners. As works that address the Japanese rural society and economy are not made widely available in English, we hope that our case studies provide an opportunity for a wider audience to gain some insights about the complexity of political economy surrounding rural Japan and delve into whether the experiences of the CFEs are parallel with what is happening to other rural localities or a phenomenon that is unique to Japan.

Notes

1 *Shinkoukai* means "development committee."
2 The Silver Human Resource Centre is a public voluntary association of senior individuals/retirees, who provide contract or commission laborers for temporary works. The association is founded as per the Law for Stable Employment of Seniors. Approximately 1,300 centers exist across Japan.
3 The difference between the lease and the entrustment of farmlands is that products from leased land belong to the farm operator (in our cases, GW or TH), who pays rent to the landowner, whereas products from an entrusted land belong to the landowner, who pays the operator a commission for the labor.

4 The multifunctionality concept is in an ambiguous relationship with neoliberalism (Potter and Tilzey, 2007). While serving as a rationale created to subsidize uncompetitive farmers for producing environmental and/or social goods, the concept has prompted farmers to harness entrepreneurial—a neoliberal mantra—skills to tap into multifunctional values of agriculture (Morgan et al., 2010).

References

Alston, M. (2005) "Social exclusion in rural Australia," Pp. 157–70 in C. Cocklin and J. Dibden (eds) *Sustainability and Change in Rural Australia*, UNSW Press, Sydney.

Arai, A. (2013) "Notes to the heart new lessons in national sentiment and sacrifice from recessionary Japan," Pp. 174–96 in A. Anagnost, A. Arai, and H. Ren (eds) *Global Futures in East Asia: Youth, Nation, and the New Economy in Uncertain Times*, Stanford University Press, Stanford, CA.

Barnett, C. (2005) "The consolations of neoliberalism," *Geoforum*, vol 36, no 1, pp. 7–12.

Barrett, B. F. D. (2000) "Decentralization in Japan: Negotiating the transfer of authority," *Japanese Studies*, vol 20, no 1, pp. 33–48.

Barry, A., Osborne, T., and Rose, N. S. (1996) *Foucault and Political Reason: Liberalism, Neo-liberalism, and Rationalities of Government*, University of Chicago Press, Chicago, IL.

Brenner, N., Peck, J., and Theodore, N. (2010) "Variegated neoliberalization: Geographies, modalities, pathways," *Global Networks*, vol 10, no 2, pp. 182–222.

Brenner, N. and Theodore, N. (2002) "Cities and the geographies of 'actually existing neoliberalism,' " *Antipode*, vol 34, no 3, pp. 349–79.

Burchell, G. (1996) "Liberal government and techniques of the self," Pp. 19–36 in A. Barry, T. Osborne, and N.S. Rose (eds) *Foucault and Political Reason: Liberalism, Neo-liberalism, and Rationalities of Government*, University of Chicago Press, Chicago, IL.

Dean, M. (2010) *Governmentality: Power and Rule in Modern Society*, 2nd edn, Sage Publications, London.

Goh, K. (1998) "Introducing marketization of nursing care service business developed in Europe and America: Comparison and challenges for Japan (Japanese)," available at www.nli-research.co.jp/report/report/1997/02/9802-1.html, accessed March 15, 2013.

Gordon, C. (1991) "Governmental rationality: An introduction," Pp. 1–51 in G. Burchell, C. Gordon, and P. Miller (eds) *The Foucault Effect: Studies in Governmentality with Two Lectures by and an Interview with Michel Foucault*, University of Chicago Press, Chicago, IL.

Guthman, J. (2008a) "Neoliberalism and the making of food politics in California," *Geoforum*, vol 39, no 3, pp. 1171–83.

Guthman, J. (2008b) "The Polanyian way? Voluntary food labels as neoliberal governance," Pp. 64–85 in B. Mansfield (ed) *Privatization: Property and the Remaking of Nature-Society Relations*, Blackwell Publishing, Malden, MA.

Harvey, D. (2005) *A Brief History of Neoliberalism*, Oxford University Press, New York.

Higgins, V. and Lockie, S. (2002) "Re-discovering the social: Neo-liberalism and hybrid practices of governing in rural natural resource management," *Journal of Rural Studies*, vol 18, no 4, pp. 419–28.

Hirao, M. (2001) "Towards construction of value chains of nursing care service (Japanese)," available at www.nli-research.co.jp/report/report/2000/03/li0103a.html, accessed March 15, 2013.

Hirashima, K. (2004) "Regime shift in Japan? Two decades of neoliberal reforms," *Swiss Political Science Review*, vol 10, no 3, pp. 31–54.

Hiwatari, N. (1998) "Adjustment to stagflation and neoliberal reforms in Japan, the United Kingdom, and the United States: The Implications of the Japanese case for a comparative analysis of party competition," *Comparative Political Studies*, vol 31, no 5, pp. 602–32.

Iba, H. (2010) "Community supported conservation program in Japan," paper presented at the 73rd Annual Meeting of the Rural Sociological Society, August 13, Atlanta, GA.

Iba, H. (2011) "Enhancing social awareness of conserving agricultural resources: A case of farmland, water and environmental conservation program of Japan," paper presented at the 74th Annual Meeting of the Rural Sociological Society, July 29, Boise, ID.

Iba, H. (2012) "Developing social businesses of regional agricultural organizations (Japanese)," Pp. 185–92 in Y. Tsuya (ed) *Track and View in Farm Management Research*, Agriculture and Forestry Statistics Publishing Inc., Tokyo.

Imai, Y. (2012) "Social service businesses by community farming enterprises (Japanese)." *Agriculture and Economy*, vol 78, no 2, pp. 69–74.

Imamura, N. (2010) "My thoughts and proposal after reading 'Justice of opposing TPP' (Japanese)," available at www.jc-so-ken.or.jp/pdf/head/column168.pdf, accessed March 15, 2013.

Inoue, M. (2013) "Neoliberal speech acts the Equal Opportunity Law and projects of the self in a Japanese corporate office," Pp. 197–221 in A. Anagnost, A. Arai, and H. Ren (eds) *Global Futures in East Asia: Youth, Nation, and the New Economy in Uncertain Times*, Stanford University Press, Stanford, CA.

Itoh, M. (2005) "Assessing neoliberalism in Japan," Pp. 244–50 in A. Saad-Filho and D. Johnston (eds) *Neoliberalism: A Critical Reader*, Pluto Press, London.

Jessop, B. (2002) "Liberalism, neoliberalism, and urban governance: A state–theoretical perspective," *Antipode*, vol 34, no 3, pp. 452–72.

Johnson, C. (1982) *MITI and the Japanese Miracle: The Growth of Industrial Policy 1925–1975*, Stanford University Press, Stanford, CA.

Kishida, H. (1998) "Growing market of nursing care service (Japanese)," available at www.nli-research.co.jp/report/report/1998/05/li9805.html, accessed March 15, 2013.

Kishida, H. (1999) "Challenged values of nursing care businesses (Japanese)," available at www.nli-research.co.jp/report/report/1999/07/li9907a.html, accessed March 15, 2013.

Lee, Y. W. (2008) *The Japanese Challenge to the American Neoliberal World Order: Identity, Meaning, and Foreign Policy*, Stanford University Press, Stanford, CA.

Lockie, S. and Higgins, V. (2007) "Roll-out neoliberalism and hybrid practices of regulation in Australian agri-environmental governance," *Journal of Rural Studies*, vol 23, no 1, pp. 1–11.

MAFF (Ministry of Agriculture, Forestry and Fishery). (2012) "Examples of community farm enterprises in their activities and achievements (Japanese)," MAFF, accessed March 15, 2013.

McCarthy, J. and Prudham, S. (2004) "Neoliberal nature and the nature of neoliberalism," *Geoforum*, vol 35, no 3, pp. 275–83.

McKee, K. (2009) "Post-Foucauldian governmentality: What does it offer critical social policy analysis?," *Critical Social Policy*, vol 29, no 3, pp. 465–86.

MIC (Ministry of Internal Affairs and Communications). (2010) "On the Heisei grand municipal merger (Japanese)," MIC, www.soumu.go.jp/gapei/pdf/100311_1.pdf, accessed March 15, 2013.

Mirowski, P. (2009) "Postface: Defining neoliberalism," Pp. 417–55 in P. Mirowski and D. Plehwe (eds) *The Road from Mont Pèlerin: The Making of the Neoliberal Thought Collective*, Harvard University Press, Cambridge, MA.

Morgan, S. L., Marsden, T., Miele, M., and Morley, A. (2010) "Agricultural multifunctionality and farmers' entrepreneurial skills: A study of Tuscan and Welsh farmers," *Journal of Rural Studies*, vol 26, no 2, pp. 116–29.

NIPSSR (National Institute of Population and Social Security Research). (2012) "Archives of population statistics 2012." available at www.ipss.go.jp/syoushika/tohkei/Popular/Popular2012.asp?chap=0, accessed March 15, 2013.

North, L. (2003) "Rural progress or rural decay? An overview of the issues and the case studies," Pp. 1–22 in L. North and J. D. Cameron (eds) *Rural Progress, Rural Decay: Neoliberal Adjustment Policies and Local Initiatives*, Kumarian Press, Bloomfield, CT.

Peck, J. and Tickell, A. (2002) "Neoliberalizing space," *Antipode*, vol 34, no 3, pp. 380–404.

Potter, C. and Tilzey, M. (2007) "Agricultural multifunctionality, environmental sustainability and the WTO: Resistance or accommodation to the neoliberal project for agriculture?," *Geoforum*, vol 38, no 6, pp. 1290–1303.

Ritzer, G. (2010) *Globalization: A Basic Text*, Wiley-Blackwell, Malden, MA.

Rose, N. and Miller, D. ([1992] 2010) "Political power beyond the state: Problematics of government," *British Journal of Sociology*, vol 61, no 1, pp. 271–303.

Scott, J. C. (1976) *The Moral Economy of the Peasant: Rebellion and Subsistence in Southeast Asia*, Yale University Press, New Haven, CT.

Song, J. (2009) *South Koreans in the Debt Crisis: The Creation of a Neoliberal Welfare Society*, Duke University Press, Durham, NC.

Tsukamoto, T. (2011) "Devolution, new regionalism and economic revitalization in Japan: Emerging urban political economy and politics of scale in Osaka-Kansai," *Cities*, vol 28, pp. 281–9.

Part 3

RESEARCH OPPORTUNITIES

8

TO BT OR NOT TO BT?

State, Civil Society, and Firms Debate Transgenic Seeds in Democratic India

Devparna Roy

Introduction: Neoliberal Globalization Project in Crisis

Unlike the development project, the successor globalization project sought to undermine the postcolonial state's policies of welfarism and developmentalism, depicting such statist intervention as eroding market efficiencies and economic growth (McMichael, 2013).[1] The globalization project combined a neoliberal blueprint for continuing development through private means with a project of crisis management. The project of neoliberal globalization is currently facing a crisis, and I argue that it may even be unraveling, as the crisis of confidence in this project has provoked much opposition in the global South and is now infecting the North.

What are the responses of the nonindustrialized countries such as India to the profound crisis of the neoliberal globalization project? As McMichael (2013) argues, India has long been part of a countertrend of sorts to the globalization project, as the Indian state never completely accepted neoliberal principles because of compelling social and ecological reasons. This nationalist turn can be credited to factors such as a six-decade-old Indian experiment with democracy, huge agrarian population, and recurring rural unrest. In response to the neoliberal globalization project, the Indian state has initiated many welfare measures for rural citizens. This Indian form of twenty-first-century development finds resonance with recent Latin American initiatives to bring markets under social control. McMichael (2013) states that given a stagnant global economy and mounting socioecological crisis, consolidation of forms of the developmental state is likely.

The objective of this chapter is to understand the crisis of neoliberal globalization through the lens of transformations in the seed sector of India. I will attempt to understand the possible unraveling of neoliberalism in India and the simultaneous consolidation of the Indian developmental state through a comparative analysis of two case studies: the successful introduction of Bt cotton seeds and the stalled introduction of Bt brinjal[2] seeds.

Bt seeds are types of transgenic, or genetically engineered, seeds. A mélange of actors—drawn from the fields of government, judicial, parliament, civil society, business, and agriculture, among others—have jousted with each other for at least two decades regarding the introduction of transgenic seeds in India, which is the world's largest democracy as well as the nonindustrialized world's most vociferous site for public and policy debates over transgenic seeds. In this chapter, I will take into account the roles of state actors (both at the central and regional levels),[3] civil society actors (including anti-biotech civil society organizations, or CSOs), and business actors (including pro-biotech transnational corporations, or TNCs) in weaving together my narratives of the introduction of Bt cotton and Bt brinjal.

The thesis of my chapter is: (1) in the era of neoliberal globalization, there is a three-way contest going on between state actors, anti-biotech "seed sovereignty" interests, and pro-biotech corporate interests regarding the transformation of the Indian seed sector; (2) in the case of the introduction of Bt cotton, the pro-biotech corporate actors won the battle against the other two interest groups because of the support of certain farmers' groups, which were impressed with the efficacy of unauthorized Bt cotton seeds against pests in western India; and (3) in the case of the introduction of Bt brinjal, the anti-biotech groups have so far won the contest, but so have the state actors that support the ideology of "democratic developmentalism." Neoliberalism may be unraveling in the Indian seed sector, but it is far from clear as to what will ultimately replace a neoliberalizing India: a return to developmentalism or the rise of autonomous communities.

After more than a decade of acrimonious debates, the Indian government commercialized Bt cotton in 2002, making it the first transgenic crop that could be legally cultivated in India. Disagreements on whether Bt cotton has proved to be a success in India continue to this day (Herring and Rao, 2012; Stone, 2012). Efforts have been made to launch another transgenic crop, Bt brinjal, but events led to an indefinite moratorium being imposed by the same government on the commercialization of that crop in 2010. There are no signs that the moratorium will be lifted soon. In this context, the phrase "to Bt or not to Bt" is not just a pun on the oft-quoted words "to be or not to be, that is the question" of Prince Hamlet in the famous Shakespearean play. The phrase depicts the reality of the Indian debates over transgenic crops.

During the 1990s, state actors, civil society groups, and business actors first wrestled with the vexing question over whether to introduce Bt cotton or not. The question was settled by the discovery of unauthorized cultivation of Bt cotton in western India, a story that I will recount later in this chapter. Bt cotton was legally introduced in 2002 and proved to be popular with Indian farmers. Despite this popularity—it has been estimated that 92 percent of the cotton grown in India was Bt in 2012—there is no national-level consensus emerging on the introduction of a second transgenic crop, Bt brinjal. In other words, more than ten years after the commercialization of the first Bt crop, Indian society is still grappling with a question that may be stated as: to Bt or not to Bt?

This chapter is divided into six sections. Following the introductory section, I will examine the relationship between international development and technology transfer. Next, I will present a historical perspective on the different phases of "democratic developmentalism" in India. After that, I will offer a typology of different seed markets. I will link various types of seed markets with the three phases of democratic developmentalism that India has so far witnessed. Next, I will analyze the factors that enabled the commercialization of Bt cotton in 2002. I will highlight the events that led to the 2010 moratorium on the commercialization of Bt brinjal. I will compare the two cases, examining the factors that led to the successful introduction of Bt cotton and the stalled introduction of Bt brinjal. In the concluding section, I will reflect on the roles played by the state actors, civil society actors, and business actors in the possible unraveling of neoliberalism and the possible ascent of a form of developmentalism in the Indian seed sector.

International Development and (Bio)Technology Transfer

The era of "development" is widely acknowledged to have begun when President Harry S. Truman announced in 1949 that the United States would make effort to mobilize "our store of *technical knowledge* [emphasis added] in order to help [the peoples of underdeveloped nations] realize their aspirations for a better life" (Escobar, 1995, p. 3). Since 1949, many technologies, such as the Green Revolution technologies, were exported from the West to India. While being instrumental in making India self-sufficient in food grains, Green Revolution technologies (which were imported in the 1960s and 1970s) have created many sociopolitical and environmental problems and taught Indians that the success of a given technology depends on the sociocultural, political, and ecological contexts in which it is introduced.

It was in this context of the bittersweet experiences with imported Green Revolution technologies and stagnating crop yields in the 1990s that debates began in the public and private circles in India on the import of agricultural biotechnologies—which are designed to launch a Second Green Revolution. The interlocutors in India have been joined by many opinion makers from Western countries, giving the Indian debates on the introduction of transgenic crops an international flavor. Let us briefly examine the main arguments of the two opposing sides of the debate over agricultural biotechnology in India.

Advocates of transgenic crops have linked agricultural biotechnology to the issue of development through the storyline of "biotechnology for the poor."[4] They point to the environmental and health benefits of transgenic crops. Further, neo-Malthusians have argued that population growth in developing countries will overwhelm the availability of food grains, and hence yield-enhancing new technologies (e.g., transgenic seeds) are necessary. The pro-biotech groups acknowledge that transgenic crops currently available in the market may not enhance yields. They argue that it is important to give biotechnology a chance, as newer generations of transgenic crops will have traits (such as drought-tolerance and nitrogen-fixing capability) that will enhance yields and help the poor directly.

Just like the critics of the Green Revolution, anti-biotech groups have hit back with counterclaims that biotechnology should be equated with hunger. They have termed transgenic crops as a failure and linked these crops to suicides by agriculturists in India.[5] Many anti-biotechnology activists are also antiglobalization activists who advocate the concepts of "food sovereignty" and "seed sovereignty" as alternatives to the concepts of food security and corporate seeds. Further, anti-biotech civil society groups argue that food insecurity is the direct result of the industrial model of agriculture, which is characterized by large-scale monocultures of transgenic crops; many of them advocate that agriculturists should abandon industrial agriculture and instead adopt agroecological methods.

Return of the "Democratic Developmentalist" State in the Indian Seed Sector

Given this global-level debate on the utility and desirability of transgenic seeds in developing countries, where does the Indian state stand on this issue? In order to understand the responses of the Indian state actors to transgenic seeds, it is useful to trace the history of the Indian state from the perspective of the ideology of "democratic developmentalism."

In the post-World War II era, newly sovereign postcolonial countries such as India faced an important question: can a developing country afford to simultaneously pursue the goals of development and democracy?[6] Unlike South Korea (which chose the route of "authoritarian developmentalism"), the Indian state chose to simultaneously pursue the goals of development and democracy since January 1950. The Indian state chose to construct a democratic civil society while simultaneously pursuing state-led industrialization. In simplified terms, drawing upon Kumar (2008), it may be argued that the postcolonial Indian state has journeyed through three phases of "democratic developmentalism" between 1947 and the present: the Nehruvian socialist era (1947–64), the populist era (1967–84), and the neoliberal reforms era (1985 onwards).

While there is scholarly consensus that the capacity of the democratic developmental state was strong during the Nehruvian socialist era and was fairly strong during the populist era, it is a matter of debate as to whether or not the Indian state continues to adhere to the ideology of democratic developmentalism in the post-1985 phase. When it comes to the agriculture sector, has the post-1985 Indian state turned into a neoliberal state that regulates the market so as to favor TNCs? Based on my reading of extant literature and my fieldwork in India, I will contend that the neoliberal globalization project has been localized in India, and neoliberal economic policies have been applied to the Indian agricultural sector, but this application has not followed international trends. For example, there is no significant focus on export agriculture at the cost of the domestic market (Vasavi, 2012). Neoliberalism has been implemented gradually in the rural and agrarian contexts. Policies have been varied in different domains of the agricultural sector.

Even this partial, gradual, and cautious implementation of neoliberal policies by the Indian state has resulted in a severe agrarian crisis since the 1990s, one symptom of which is the wave of suicides by Indian agriculturists. It has been estimated that a quarter of a million Indian farmers have committed suicide since 1995. The noted development economist K. Nagaraj (2010) has stated that the wave of suicides is largely a consequence of a number of neoliberal policy measures that the governments—both at the center and in the states—have instituted as part of economic reforms. Economists have sought to explain the welfare declines of Indian agriculturists in sharply polarized terms. One group (e.g., Gulati and Narayanan, 2003) has termed the pace of agricultural liberalization as "slow" and held this slow pace responsible for the severe agrarian crisis. They have advocated an increased role for markets. The other group (e.g., Vakulabharanam and Motiram, 2011) has sought to lay the blame for the agrarian crisis on the state's withdrawal from its support for agriculture and also on the integration of Indian agriculture with global markets. The solution lies in an increased role for the state in agriculture, according to this group.

Many actors (including certain farmers' groups, economists, policy makers, politicians, and TNCs) are actively seeking to further liberalize the agriculture sector, including the subsector of seeds, while others are seeking to roll back neoliberal economic reforms. Before I turn to a discussion of what actually transpired in the cases of the successful introduction of Bt cotton and the stalled introduction of Bt brinjal, I will delineate the types of seed markets that exist in India.[7]

A Typology of Seed Markets in India

Currently, the domestic seed market in India is one of the largest in the world. It is worth about US$1.3 billion and has been growing at a compounded annual growth rate of about 15 percent, according to a report in the leading Indian financial daily, *The Economic Times* (2012). The Indian seed market can become enormously profitable, since the vast majority of the agriculturists in India do not currently purchase seeds. Most Indian agriculturists use seeds saved from earlier harvests and exchange such seeds informally with others.

This was especially true during the Nehruvian era, when Indian agriculturists participated in what I would term as the "traditional seed markets" (farmers saved and exchanged local varieties of seeds; farmers selected and improved their varieties by relying on their own experience).[8] Importing technologies of modern seeds was not a priority for the Nehruvian state. In Nehruvian India, the "traditional seed markets" led to local or community-level control of the seed sector and the agricultural economy. Traditional seed markets continue to exist in many parts of India today. In contemporary India, the concept of autarkic (i.e., not relying on imports) "traditional seed markets" is supported by some of the antiglobalization CSOs that also advocate for "seed sovereignty." Autarkic traditional seed markets are no longer supported by Indian state actors or the private sector.

The Green Revolution in India took place during the populist era. High yielding varieties (HYVs) of seeds in food grains such as wheat were introduced. These seeds were based on imported technology that Indian public-sector scientists adapted to the local conditions. Further, these seeds formed the basis of a technological package that included other inputs, such as mechanization (through tractors for plowing), chemical fertilizers and pesticides, and irrigation (often through tubewells). Thus, HYVs replaced the traditional Indian methods of farming with modern methods. This modern method of cultivation is capital-intensive and leads to high dependence on credit. The modern agriculturist in India often relies upon labor for sowing, weeding, fertilizer and pesticide application, and harvesting. The modern method emphasizes yields and is premised on state support in the provision of inputs, marketing, infrastructure, credit, and extension.

During the populist era, the developmental state was quite strong; though private seed firms existed, seed markets were dominated by the public sector firms. Indian agriculturists participated in what I would call "state-dominated seed markets" during the populist era. "State-dominated seed markets" led to a state-controlled seed sector and agricultural economy. Traditional seed markets and state-dominated seed markets coexisted during the populist era. This "populist" model is still supported by those interests that continue to support state intervention in the national economy.

Since the neoliberal economic reforms era began around 1985,[9] growth rates of some sectors of the Indian economy (e.g., information technology industry) have looked up. A vibrant civil society continues to flourish in democratic India. In terms of private sector growth in agriculture during the neoliberal reforms era, two points are worthy of note. First, there was an exponential increase in the number of private seed firms in India since the late 1980s. Second, transnational seed firms have entered India since the late 1980s, with 100 percent foreign equity allowed in seed industry since 1991. In terms of technology, note that the first transgenic crop (Bt cotton) was introduced in 2002.

Indian agriculturists participate in what I would term as "private-sector dominated seed markets" during the neoliberal era. Both public sector and private sector seed firms coexist, but the private sector's role continues to grow during the neoliberal era. The "private-sector dominated seed market" leads to increasingly TNC control of the agricultural economy, and this situation is supported by those interests that support globalization and favor the neoliberal ideology of the regulatory state acting in ways to facilitate market capture by the TNCs.

There are three types of seed markets currently operating in India: the majority of Indian agriculturists still continue to save and exchange seeds and participate in traditional seed markets; some Indian agriculturists grow crops (e.g., rice) in which the public sector firms are the main suppliers in modern seed markets, while other Indian agriculturists grow crops (e.g., cotton) in which private sector firms are the main seed suppliers in the markets. When it comes to agriculture, there are three types of interest groups currently existing in India. The first type of interest groups are against industrial agriculture (including transgenic seeds) and instead support agroecology and autarkic seed markets; the ideology driving these groups is that of "seed sovereignty" and examples of such groups are the anti-biotech CSOs. The second type of interest groups support industrial agriculture (and transgenic seeds) and state-dominated seed markets; the ideology motivating such groups is that of "democratic developmentalism," and examples of such groups are certain state actors. The third type of interest groups support industrial agriculture (including transgenic seeds) and private-sector dominated seed markets; the ideology propelling these groups is that of "neoliberal globalization," and examples of such groups are pro-biotech corporate actors.

While the media and many scholars strategically focus on only two of the groups of actors (the anti-biotech activists and the corporate actors) and posit the debate over agri-biotechnology as a struggle between pro-biotech corporate forces and anti-biotech CSOs, I emphasize that we need to focus on the interactions of all three groups of actors, paying special attention to the role of the state actors. Based on my reading of the events and my fieldwork, I contend that the Indian state is not merely the adjudicator of the raging battle between the pro-biotech forces and the anti-biotech forces regarding the transgenicization of Indian agriculture, but it also has its own interests in furthering its own control over the biotechnology sector and the seed markets. The Indian state has invested in the biotechnology sector since the 1980s. India became the first developing country to set up a Department of Biotechnology (under the Ministry of Science and Technology in New Delhi) in 1986. There is a strong biotechnology public sector in India in terms of sheer numbers and expertise, and it is too soon to write off the Indian state as a group of actors with their own set of interests.

In my formulation, the Indian state is a "hybrid" state—exhibiting neoliberal impulses in some sectors of the economy (e.g., information technology) but strong developmental impulses in others (e.g., the seed sector). The Indian developmental state prioritizes the role of seed markets in meeting certain social and developmental needs (e.g., addressing the livelihood concerns of small and marginal agriculturists) rather than just generating profits.

The Indian seed market is potentially worth billions of dollars, should most farmers be somehow persuaded to buy seeds from the market. The possible transgenicization of Indian agriculture (especially food crops) also has implications for the food security of the country as well as the sovereignty status of the Indian nation-state. Food, after all, can be used as a political weapon. For all these reasons, the debate over the second round of modernization of Indian agriculture (i.e., transgenicization) has assumed great political significance in the global arena.

Two Case Studies: Bt Cotton and Bt Brinjal

Indian agriculturists have cultivated cotton landraces for centuries. Today, Indian agriculturists sow three types of cotton seeds: varieties, hybrids, and transgenic seeds. Indian agriculturists have been used to growing traditional and modern seeds of cotton for many decades; the same farmer may grow both varieties and hybrids to suit her or his specific needs.

The first move to introduce Bt cotton dates back to 1990, the year during which Monsanto initiated talks with the Government of India for transferring the Bt gene (Ramanna, 2006). Monsanto's offer was refused in 1993 because according to the Indian government, the technology transfer fees demanded by Monsanto were too high. In response, Monsanto began a process of alliance building with domestic actors. In 1995, the Government of India granted permission for a domestic seed company, MAHYCO (Maharashtra Hybrid Seed Company), to import Bt cotton seeds from Monsanto. Imported Bt seeds were used by MAHYCO for backcrossing into Indian cultivars. In 1996–8, MAHYCO was granted permission by the Indian government to conduct field trials on these Bt cotton hybrids. In 1998, Monsanto acquired a 26 percent share of MAHYCO. In 2002, a joint venture, MAHYCO Monsanto Biotech Limited (MMB), was set up. A long period of testing the Bt technology took place following the approval of MAHYCO to import Bt cotton seeds. Despite elaborate tests, MMB was refused permission by the central government to commercialize Bt cotton in June 2001 and was asked to conduct further tests. Thus, in mid-2001, it was not at all clear as to how long it would take before the Indian government approved the commercialization of Bt cotton. However, all this was to change due to an unforeseen event in the fall of 2001: the discovery of unauthorized Bt cotton growing on hundreds of hectares in India.

In fall 2001, there was a severe bollworm pest attack in Gujarat (a state in western India), which devastated all cotton varieties, except for one, which was known as Navbharat 151 (NB 151). Subsequently, it was discovered that NB 151 was an unlicensed Bt cotton hybrid carrying the *Cry1Ac* gene. NB 151 was found growing on hundreds of hectares in Gujarat. Although the NB 151 seed contained the same Bt toxin gene as the MMB Bt cotton, it was crossed with a different parent.[10] Agriculture is decided at the regional state government level (whereas biotechnology is decided at the central government level). The central government's Genetic Engineering Approval Committee (GEAC) ordered the state government to burn the illegal plantations. The Gujarat government questioned this policy. Considering that some cotton had already been marketed, the GEAC changed its order and asked for recovery of the unpicked cotton to the extent possible and the destruction of the crop residue (Herring, 2005).

In March–April 2002, the Indian government approved the commercial release of MMB Bt cotton. According to Ramanna (2006), the reasons for the decision moving from *de facto* to *de jure* acceptance of transgenic crops must be understood in terms of a powerful story line of "GM as farmers' choice," which emerged following the events in Gujarat and which posed a challenge to the discourse of the anti-biotech CSOs. News of farmers growing Bt cotton in Gujarat and other states prior to the central government's approval led to a shift in the way transgenic crops were portrayed in the media and policy-making circles. The rationale was then put forward that if farmers want the technology, what right does the central government have to deny them transgenic crops? The anti-biotech CSOs could not rebut this powerful logic. If the anti-biotech CSOs opposed Bt cotton, it made them appear to be indifferent to the real interest of the farmers they were supposed to represent. The pro-biotech lobby presented Indian farmers as decision makers and voters for transgenic technology, which trumped the portrayal of Indian farmers as hapless victims of globalization (a picture which had been put forth by the anti-biotech lobby and which had garnered widespread attention because of farmers' suicides). The pro-biotech lobby's strategy following the Gujarat incident was not to stress the intellectual property rights violation but rather to emphasize the issue of farmer's choice (Ramanna, 2006). Pro-biotech farmers' groups demanded the approval of Bt cotton. The discourse on farmer's choice essentially blurred the distinction between NB 151 and MMB Bt cotton, when it came to the Bt hybrid's success.[11] Note that the cultivation of NB 151 continues to be illegal in India (though there are many legal Bt hybrids being cultivated in India). Hundreds of Gujarati farmers have created what has been informally called the "greatest participatory plant breeding experiment in human history" as they have crossed the parent line of NB 151 that contained the transgene with different local cultivars to create many new Bt cotton hybrids. These new underground Bt cotton hybrids—called "loose seeds" in local parlance (Roy et al., 2007)—have led to a situation lauded as "vigorous rural anarcho-capitalism" by some, while others have decried the regulatory failure of the state when it comes to Bt cotton hybrids. The Indian state continues to turn a blind eye to the cultivation of illegal Bt cotton hybrids, which continues to this day in Gujarat and perhaps other states. There are no official statistics regarding how many hectares in India are under illegal Bt cotton, though journalists (e.g., Bhattacharya, 2012) have reported that illegal Bt hybrids account for 25 to 30 percent of the Gujarat seed market.

When it comes to brinjal, Indian farmers cultivate both hybrids and open pollinated varieties (OPVs). Hybrids and varieties cover about 40 and 60 percent of the brinjal crop area, respectively. This market segmentation facilitated collaboration between the public sector and the private sector, beginning in 2003 (Herring and Shotkoski, 2011). MAHYCO shared its biotechnology (which it had developed in collaboration with Monsanto) with Indian public institutions for development of Bt brinjal. This collaboration was assisted by an international partnership: The Agricultural Biotechnology Support Project II (ABSPII). To create regionally appropriate cultivars of Bt brinjal, ABSPII worked with three public-sector partners in India. These public institutions developed locally popular varieties with the IR trait donated by MAHYCO, while MAHYCO itself continued to concentrate on Bt hybrids, assuming that many farmers would eventually favor them for their yield advantage. In contrast to Bt cotton, more brinjal OPVs from the public sector than hybrids from the private sector were planned for release. This would give Indian agriculturists a choice between two types of IR cultivars: the lower-cost and savable seeds of varieties, and the higher-yielding, more expensive hybrid seeds.

The GEAC's Expert Committee concluded in October 2009 that the IR trait in brinjal, for both hybrid and OPVs, was effective in controlling target pests, safe to the environment and humans, and had the potential to benefit farmers. However, the Minister of Environment and Forests Jairam Ramesh announced that he would not accept the GEAC recommendation for commercial release of Bt brinjal but would instead open public consultations in a tour of seven Indian cities. After the consultations, in February 2010, Minister Ramesh placed an indefinite moratorium on the commercialization of Bt brinjal. Critiques from Indian and foreign scientists pressured the minister to reconsider his decision. In 2010, he asked six of India's leading science academies to assess Bt brinjal. Their report confirmed the original conclusions of the GEAC, but the moratorium still holds.

If 2010 was the year that began the opening up of debates regarding Indian agriculture—largely due to Minister Ramesh's public consultations regarding the commercialization of Bt brinjal—then I would argue that 2012 may go down as the year that marked the decisive return of the democratic developmental state in India, at least as far as the seed sector is concerned. First, in August 2012, the government of Maharashtra cancelled the license of MAHYCO (the Indian partner of Monsanto) from selling Bt cotton seeds in that state. That same month, a 359-page Parliamentary Standing Committee (PSC) Report was issued. The PSC was chaired by Basudeb Acharia, a member of Parliament representing the Communist Party of India (Marxist), and the PSC had members from other political parties. The PSC report was highly critical of the 2009 GEAC's decision to approve the commercialization of Bt brinjal. In October 2012, the Supreme Court of India-appointed Technical Expert Committee (TEC) issued a twenty-four-page report, which recommended a ten-year moratorium on field trials of all Bt transgenics of food crops. The central government has raised objections to what it calls a "scientifically flawed" report. However, about one hundred scientists, many of them well known, have written an open letter to the Supreme Court asking it to comply with the TEC's recommendations. Other scientists have disagreed with this open letter. Thus, the Indian scientific community stands divided on the issue of the TEC's report.

The three factors that led to the 2002 Government of India decision to approve cultivation of Bt cotton are: first, due to entry of unapproved seeds, the introduction of Bt cotton in Indian farmers' fields was *fait accompli*. There was clear demonstration to the farmers' groups that Bt cotton (NB 151) works against pests, can decrease the usage of pesticides, and can also increase yields. These cotton farmers' groups were politically strong, and they lobbied with their respective state governments. Second, there was more or less consensus among

India's scientists that Bt cotton was a "public good." Third, cotton is a nonfood crop important to the national economy as it is a major industrial input and earns export revenues. Despite strong opposition by anti-biotech "seed sovereignty" CSOs, the chain of events set in motion by the silent entry of unauthorized Bt cotton seeds won the day. Given the uproar created by some farmers' groups, who demanded that Bt cotton seeds be made legally available to Indian agriculturists, the anti-biotech CSOs fell relatively silent and were sidelined during fall 2001 and spring 2002.

It may be speculated that if the discovery of unauthorized Bt cotton in Gujarat fields had not been made in fall 2001, the Indian government may have further delayed the release of MMB Bt cotton hybrids as it was not in the interest of state actors (who support the ideology of "democratic developmentalism") to permit the release of private-sector transgenic seeds before the public sector could offer such seeds. In 2001–2, the Indian public sector was conducting research on transgenic cotton seeds. The Indian public sector was to release its own Bt cotton seeds in 2009.

In the case of Bt brinjal, the five factors that led to the 2010 moratorium are: first, there is either no widespread release of such illegal seeds, or there is no devastating attack by pests over an extensive area. Consequently, there is no demonstration effect of the insect-resistance capacity of transgenic brinjal in actual farmers' fields. Second, even if unauthorized Bt brinjal seeds are released quietly and a demonstration effect of its efficacy takes place (as in the case of Bt cotton in Gujarat), it would be difficult to mobilize brinjal producers in the way that cotton producers could be mobilized to support transgenic cotton. There are only 1.4 million brinjal farmers, controlling about 550,000 hectares of land (Herring and Shotkoski, 2011). Most of these farmers grow multiple crops. There is no cohesive brinjal producers' lobby. In contrast, cotton farmers number 6.3 million, on 9.4 million hectares, and many of them are politically organized, largely due to the long-standing cotton cooperative movements in western India.

The third factor is the role played by anti-biotech CSOs; they have undergone a steep learning curve since the Bt cotton episode, and they have effectively lobbied with many state governments and expressed their concerns regarding Bt brinjal. The fourth factor is the opposition by state governments to Bt brinjal. Many state governments have expressed their anxieties about adoption of transgenic food crops; these concerns range from consequent contributions to food security to the interests of small and marginal agriculturists. Fifth, there was a lack of scientific consensus on the issue of Bt brinjal, with respected scientists such as M.S. Swaminathan advising further discussion on the issue. All these factors resulted in Minister Ramesh declaring a moratorium on the release of Bt brinjal until independent scientific studies established the safety of the product in terms of its long-range impact on human health and the environment (including the rich biodiversity existing in brinjal in India) to both the public and the professionals.

Conclusion: Unraveling of the Neoliberal Globalization Project in the Indian Seed Sector?

These two cases demonstrate that even though a neoliberal food regime (Otero, 2012) exists in India as in many other countries, state actors can play a role in contesting the power of TNCs provided the state faces sufficient pressure from CSOs (for a similar case of the Guatemalan state keeping TNCs and their transgenic technologies at bay because of pressure from civil society, see Klepek, 2012). This is all the more true in the case of countries like India that have a potent public sector in seed biotechnology as well as a reinvigorated democratic developmental state. In the case of Bt cotton, the pro-biotech corporate interests were able to outmaneuver the anti-biotech CSOs and the state actors (at least some of whom support the public sector and the ideology of "democratic developmentalism") due to the alignment of powerful cotton farmers' interests with the TNCs in the wake of the discovery of unauthorized Bt cotton and its efficacy over pests. In the case of Bt brinjal, at the moment, the anti-biotech CSOs seem to have won the battle against the biotech TNCs. During this process, the anti-biotech CSOs have strengthened the hands of the developmental Indian state, which seeks to control the seed markets and will not allow the TNCs to capture these markets.

I contend that the debate over the introduction of transgenic crops in India is intimately linked to the struggle over agricultural models (industrial or agroecological?), development (authoritarian or democratic?), and capitalism (state-led or corporate-led?). Let me explain what I mean by this statement. Many advocates of transgenic crops tend to adopt a "There is No Alternative" (TINA) approach to modern/industrial agriculture (with its large-scale monocultures and corporate-owned seeds) if we are to feed the whole of India, but the reality is that India currently produces enough food to feed all citizens. Many citizens lack monetary resources to buy food from the market. It is debatable whether increasing yields via deployment of transgenic food crops can alone provide a magic bullet solution to the problems of endemic hunger and malnutrition in India. A more rational approach may be to let the poor grow their own food, wherever possible, through use of agroecological methods and noncorporate (traditional) seeds. Agroecological and/or organic methods generally eschew the use of transgenic seeds, though the use of Bt cotton seeds by some self-identified organic farmers in India has been documented (Roy, 2010; Roy, 2012).

If the goal of a community of agriculturists (well-informed about the various alternatives available to them and also democratically literate) is to achieve and maintain individual and community autonomy over seeds at any cost, then agroecological methods (with its small-scale farming, polycropping, and use of farmers' varieties) may be the right solution for them. However, individual agriculturists and farmers' groups may democratically decide to purchase transgenic seeds from the market, and in this case, the state, CSOs, and corporations should make all efforts to inform and educate farmers about the advantages and disadvantages of different types of seeds. The pursuit of agroecological farming may not exclude transgenic seeds, but this is for the informed agriculturists to decide. Agriculturists would have to rely on either the public sector or TNCs for transgenic seeds.

Perhaps we can make a normative statement that Indian agriculturists, either as individuals or as communities, should not be coerced (directly or indirectly) by those who have power over them into buying a particular kind of seed. If the state is authoritarian, it can force the agriculturists to buy transgenic seeds instead of limiting its role to making such seeds available and also educating the farmers about the pros and cons of modern farming and transgenic seeds. Thus, the debate over transgenic crops is also a struggle over the nature of development models: authoritarian or democratic? If the Indian state actors, civil society actors, and business actors abide by the principles of democracy, then the empowerment of citizens so that they can make informed choices is paramount. This necessitates free and fair public debates in the media regarding the merits and demerits of different types of seed markets and seed technologies, truthful advertising regarding the advantages and disadvantages of transgenic crops (especially for small agriculturists), and transparency in the government's policy making and regulation of transgenic technology.

The debate over transgenic crops is ultimately a contention over the kind of capitalist structures that Indian citizens wish to live with in the foreseeable future: state-led capitalism or corporate-controlled capitalism (neoliberal globalization), or neither. Two questions follow: to what extent should modern seed markets replace traditional seed markets in India, and what roles should state-owned enterprises, private domestic firms, and TNCs play when it comes to the seed markets? The search for answers to these normative questions (should we adopt agroecological methods or industrial agriculture, democratic developmentalism or authoritarian developmentalism, state-led capitalism or corporate-controlled capitalism?) will continue to engage Indian citizens as they participate in formal democracy (e.g., vote in elections) and more substantive forms of democracy (e.g., participate in rallies either in support of or against transgenic seeds).

Despite the onset of the crisis of the neoliberal globalization project in the late 1990s, the private sector (including TNCs) and their civil society supporters continue to be influential actors in the Indian polity and society, as evidenced by the successful introduction of corporate Bt cotton seeds in 2002 in face of the opposition by anti-biotech CSOs. The democratic tradition of India as well as the rural and agrarian unrest engendered by citizens' experiences with neoliberalism led to a situation where anti-biotech CSOs were able to successfully challenge the introduction of Bt brinjal seeds and convince state actors to impose an indefinite moratorium on Bt brinjal. Though Bt cotton has proved to be immensely popular with Indian agriculturists, the state hesitates to give the green signal to Bt brinjal, possibly because of the health and environmental ramifications of transgenic food crops as well as its own developmental priorities.

The indefinite moratorium on the commercialization of Bt brinjal marks the crisis of neoliberalism in the Indian seed sector and may even signal the unraveling of neoliberalism as the power of TNCs in the seed market may be limited by the joint action of the developmental state and anti-biotech CSOs. The moratorium raises a number of questions for the future. Will the influence of TNCs and their transgenic technologies erode in favor of the developmental state and its transgenic technologies, or will the TNCs stage a comeback with the help of some pro-biotech farmers' groups? Will the developmental state's public sector disband its R&D work on transgenic technologies and opt for disseminating inputs for agroecological farming? Will the anti-biotech CSOs somehow persuade the state and agriculturists to adopt agroecology instead of industrial agriculture?

The two case studies related to the introduction of two different transgenic crops in India have hopefully demonstrated that the consolidation of the neoliberal globalization project and the market domination by TNCs are being openly challenged by anti-biotech civil society actors and somewhat secretly by developmental state actors. Using the opportunity provided by the moratorium on Bt brinjal, anti-biotech "seed sovereignty" CSOs are seeking to consolidate support among agriculturists for autarkic traditional seed markets and landraces. Whether it was intentional or not, the actions of the anti-biotech CSOs have strengthened the hands of the democratic developmental state and weakened those of corporate actors.

What are my predictions for the future? I will argue that in the short term, the Indian state will play a more important role than the anti-biotech CSOs in deciding the future of neoliberalism in India. The extent to which the Indian state stays true to its original mission (through continuing to fulfill its twin goals of nurturing democracy by actively listening to CSOs and encouraging development by bringing seed markets under social control) will decide the future of transgenic food crops, TNCs, and neoliberalism in India. In the long term, it is unclear as to what will eventually replace the neoliberal project in India; it is too early to say whether the future belongs to autonomous local communities and traditional seed markets or to a national community and state-controlled seed markets (with public sector transgenic seeds perhaps). Since India is arguably the global South's most influential site for debates over transgenic seeds and an important site for the neoliberal globalization project, future developments in India are likely to have spillover effects in other nonindustrialized countries.

Notes

The author gratefully acknowledges the helpful review and comments on an earlier version by the editors and participants at the Rural Sociological Society conference held in Chicago, July 2012. All errors and omissions are the responsibility of the author alone.

1 McMichael (2000, p. 259) defines the globalization project as "an emerging vision of the world and its resources as a globally organized and managed free trade/free enterprise economy pursued by a largely unaccountable political and economic elite." For a brief breakdown of its components, see McMichael (2000, p. 187).

2 Bt crops are created by introducing gene(s) from the soil bacterium *Bacillus thuringiensis* into the crop genome. The Bt gene(s) confer the capacity to the

plant to resist certain pests. Scientists from the transnational corporation Monsanto introduced genes from *B. thuringiensis* into the cotton genome to create Bt cotton, which is resistant to certain lepidopteran pests. In this chapter, I have avoided the use of the term "genetically modified" (GM) plants in the interests of scientific accuracy. Almost all plants being cultivated by humans today are genetically modified—that is, generations of peasants have carried out on-farm seed selection to create the varieties that we have today. The term "transgenic" refers to those plants that are the result of scientists' manipulation of genomes. Scientists can introduce genes from other kingdoms into plants (e.g., put a bacterial gene into the plant genome). The vegetable "brinjal" (*Solanum melongena*) is also known as "eggplant."

3 India began its journey as a sovereign nation-state in August 1947 after winning political independence from British colonial rule. Over the decades, India has metamorphosed from a quasi-federal structure with a strong center and relatively weak regional states to a polity where the central government is often challenged by strong regional governments.

4 The storyline of "biotechnology of the poor" argues that transgenic crops are scale-neutral technologies and thus can benefit the small farmers as much as the big farmers. This is important for India, which has millions of small and marginal agriculturists.

5 Following Vasavi (2012), I use the term "agriculturist" to encapsulate and represent all cultivators who either engage in production for subsistence (or only for their own and family's needs) and/or for the market and who occupy a range of positions from that of cultivators-cum-laborers to that of landlords. The term "farmer" in this chapter refers to all types of cultivators.

6 Here, I define "development" as the pursuit of both economic growth and certain social objectives or public goods (such as alleviating poverty, hunger, and malnutrition, providing meaningful work to all able-bodied adult citizens, and protection of the environment).

7 Since the term "market" can represent any collection of trades that takes place under a certain set of rules and circumstances, markets can range from seasonal exchanges of seeds between neighbors to more codified and anonymous exchange of seed for money (Lipper et al., 2010).

8 This "selection" by agriculturists was the origin of landraces and other varieties that scientists subsequently used to begin systematic improvement of crops.

9 Some scholars date the beginning of neoliberal economic reforms to 1980, which marked the beginning of the second government of Indira Gandhi, while others opt for 1991. In this chapter, I have chosen 1985 as the year marking the entry of the neoliberal moment in Indian political economy as I consider the Rajiv Gandhi era (1985–9) as signaling a definite break in the economic thinking of the political class of India.

10 It is unclear how the MAHYCO gene construct got into the Navbharat 151 Bt cotton hybrid. It has been speculated that "gene transfer" from Bt cotton hybrid plants (in field trials) to neighboring non-Bt cotton hybrids may have occurred, resulting in the formation of new Bt cotton hybrid plants.

11 See Roy (2006) and Roy et al. (2007) for a discussion of how Navbharat 151 performed better than MMB's Bt cotton hybrids for many farmers interviewed in Gujarat during the years 2002–4.

References

Bhattacharya, P. (2012) "As cotton fields thrive, so do concerns," LiveMint, October 2, http://www.livemint.com/Politics/LoGoMERwbaK4uc7pw5kogP/As-cotton-fields-thrive-so-do-concerns.html, accessed December 10, 2012.

The Economic Times. (2012) "India's seed industry to grow by 53 % by 2015: Assocham," http://articles.economictimes.indiatimes.com/2012-12-09/news/35705464_1_hybrid-seeds-quality-seeds-high-yielding-varieties, accessed September 1, 2013.

Escobar, A. (1995) *Encountering Development: The Making and Unmaking of the Third World*, Princeton University Press, Princeton, NJ.

Gulati, A. and Narayanan, S. (2003) *The Subsidy Syndrome in Indian Agriculture*, Oxford University Press, New Delhi.

Herring, R. J. (2005) "Miracle seeds, suicide seeds and the poor: GMOs, NGOs, farmers and the state," Pp. 203–232 in M. F. Katzenstein and R. Ray (eds) *Social Movements in India: Poverty, Power and Politics*, Rowman & Littlefield, Lanham, MD.

Herring, R. J. and Rao, N. C. (2012) "On the 'failure of Bt cotton': Analyzing a decade of experience," *Economic and Political Weekly*, vol 47, no 18, pp. 45–53.

Herring, R. J. and Shotkoski, F. (2011) "Eggplant surprise: The puzzle of India's first transgenic vegetable," *Scientific American Worldview*, http://www.saworldview.com/wv/archive/2011/eggplant-surprise/accessed September 1, 2013.

Klepek, J. (2012) "Against the grain: Knowledge alliances and resistance to biotechnology in Guatemala," *Canadian Journal of Development Studies*, vol 33, no 3, pp. 310–25.

Kumar, N. (2008) "India: A failed democratic developmental state?," Pp. 148–70 in V. Kukreja and M. P. Singh (eds) *Democracy, Development and Discontent in South Asia*, Sage Publications, New Delhi.

Lipper, L., Anderson, C. L., and Dalton T. J. (2010) *Seed Trade in Rural Markets: Implications for Crop Diversity and Agricultural Development*, FAO, London.

McMichael, P. (2000) *Development and Social Change: A Global Perspective*, 3rd edn, Sage Publications, Thousand Oaks, CA.

McMichael, P. (2013) "Globalization: A project in crisis," Pp. 75–87 in R. Palan (ed.) *Global Political Economy: Contemporary Theories*, 2nd edn, Routledge.

Nagaraj, K. (2010) "Neoliberal deaths," *Himal*, www.himalmag.com/component/content/article/349.html, accessed December 10, 2012.

Otero, G. (2012) "The neoliberal food regime in Latin America: State, agribusiness transnational corporations and biotechnology," *Canadian Journal of Development Studies*, vol 33, no 3, pp. 282–94.

Ramanna, A. (2006) *India's Policy on Genetically Modified Crops*, Asia Research Center Working Paper 15, Asia Research Center (ARC), London School of Economics & Political Science, London.

Roy, D. (2006) "Farming 'white gold': Early experiences with genetically engineered cotton production in Gujarat, India," PhD thesis, Cornell University, Ithaca, NY.

Roy, D. (2010) "Of choices and dilemmas: Bt cotton and self-identified organic cotton farmers in Gujarat," *Asian Biotechnology and Development Review*, vol 12, no 1, pp. 51–79.

Roy, D. (2012) "Cultivating Bt cotton in Gujarat (India): Self-identified organic cotton farmers revisited," *Asian Biotechnology and Development Review*, vol 14, no 2, pp. 67–92.

Roy, D., Herring, R. J., and Geisler, C. C. (2007) "Naturalizing transgenics: Loose seeds, official seeds, and risk in the decision matrix of Gujarati cotton farmers," *Journal of Development Studies*, vol 43, no 1, pp. 158–76.

Stone, G. D. (2012) "Constructing facts: Bt cotton narratives in India," *Economic and Political Weekly*, vol 47, no 38, pp. 62–70.

Vakulabharanam, V. and Motiram, S. (2011) "Political economy of agrarian distress in India since the 1990s," Pp. 101–26 in S. Ruparelia, S. Reddy, J. Harriss, and S. Corbridge (eds) *Understanding India's New Political Economy: A Great Transformation?*, Routledge, New York.

Vasavi, A. R. (2012) *Shadow Space: Suicides and the Predicament of Rural India*, Three Essays Collective, Gurgaon, India.

9

TURNING OF THE TIDE

Rising Discontent over Transgenic Crops in Brazil

Karine Peschard

Introduction

In this chapter, I explore the relation between neoliberalism and agricultural biotechnology through the lens of the controversy over transgenic crops in Brazil. The development of agricultural biotechnology in the 1980s and 1990s coincides historically with the height of neoliberalism, and the agbiotech industry has been driven and shaped by neoliberal policies and institutions (McAfee, 2003; Kloppenburg, [1988] 2004; Otero, 2008). In the neoliberal era, biotech firms have been able to capitalize on technological developments by shaping legal regimes to suit their commercial interests. The resulting neoregulations—in the form of strengthened intellectual property rights regimes for plant varieties, patents on genetically engineered organisms and seed contracts—have been controversial and are increasingly being questioned. Brazil is a case in point. Since 2009, farmers have been expressing growing discontent at the new legal regimes governing seeds and have challenged their legality through several class action lawsuits.

Transgenic crops have proven highly controversial worldwide, and many efforts have been devoted to exploring and understanding the resistance they generate, including in Brazil.[1] The original contribution of this chapter is to explore the resistance among hitherto supportive sectors of Brazilian society rather than those who have long been opposed to transgenic crops. The fact that soy farmers who are well integrated into capital-intensive farming are now becoming increasingly critical of transgenic crops (and their legal framework) points to a crisis in the neoliberal food regime. These developments are especially significant in Brazil, an agricultural giant and the second-largest soybean producer and exporter after the United States. Brazil is also a regional power and in recent years has played a leadership role among countries from the global South.

In the first section of this chapter, I examine how the development of the agbiotech industry is intimately linked to neoliberal policies and institutions and how the new legal regimes governing transgenic seeds have been generating resistance among farmers. I then offer a brief overview of the controversy surrounding transgenic crops in Brazil and how farmers' organizations, government ministries and agencies, nongovernmental organizations (NGOs), and the private sector have positioned themselves in the debate. I proceed to show how cracks have been appearing in the protransgenic coalition since 2009, as more and more soy farmers show signs of dissatisfaction with the legal framework regulating Roundup Ready (RR) soybeans. In the following three sections, I examine in more detail each of the three issues of discontent: royalties, overall profitability, and lack of access to conventional varieties. Lastly, I discuss how these more recent expressions of discontent differ from earlier ones, and I analyze their broader political significance. In the conclusion, I reflect on the possible outcome in the coming years of the growing rift in Brazilian agri-food politics between industry and soy growers.

Transgenics and the Paradoxes of Neoliberalism

Contemporary biotechnologies, such as crop genetic engineering, must be understood in relation to the economic markets within which they emerge. As Rajan (2006) points out, biotechnology might more accurately be called technoscientific capitalism. Indeed, the development of agricultural biotechnology took place at the height of neoliberal capitalism, characterized by liberalization, global trade regimes, and the financialization of the economy. Its most salient aspects—extensive patenting, the concept of substantial equivalence or trade disputes before the World Trade Organization (WTO)—are intimately linked to neoliberal policies and institutions.

The favoring of private corporations under neoliberalism has influenced the direction taken by research and development in crop biotechnology, and this has resulted most significantly in the almost exclusive genetic engineering of plant varieties. It is important to keep in mind that genetic engineering is only one dimension of crop biotechnology; there are nontransgene approaches to incorporating molecular-biological knowledge into crop improvement.[2] However, genetic engineering—the use of recombinant DNA methods to genetically modify a living organism—has been by far the most prominent. For the industry, it offers a decisive advantage: the possibility to patent its products. By successfully arguing that genetically engineered organisms are human inventions rather than products of nature, industry opened the door to the patenting of life forms and to an entirely new source of profit. While the analogy between the Green Revolution and Gene Revolution is appealing, it glosses over a key difference between the two regarding the roles of the state and the public/private spheres; while the Green Revolution was largely conducted in the public domain, the Gene Revolution is first and foremost a private enterprise.[3]

Agricultural biotechnology illustrates the paradoxes of neoliberalism. One such paradox is that while the latter advocates a minimalist role for the state, states are in fact playing an active role in the implementation of neoliberal policies. As Otero notes, "in spite of the free trade rhetoric, the US government has worked hard to facilitate the development of its biotechnology industry" (2008, p. 14). This support can be found at various levels, from the ease with which the products of public research can be appropriated by the private sector, to the willingness of the United States Patent and Trademark Office (USPTO) to grant patents on living organisms. Industry has lobbied hard on the one hand for a stronger intellectual property rights regime and on the other for a lax regulatory framework for the regulation of transgenic crops. The role played by the US industry in formulating the WTO agreement reveals the way in which neoliberal interests have been successful in using the state to promote capital-friendly policies, both nationally and globally (Oh, 2000). Of course, the state is not a monolith. In Brazil, the federal government has been internally divided over the issue of transgenic crops. But even these divisions reflect the class dimension of the neoliberal project: those government bodies that have been supportive of transgenic crops, such as the Ministry of Science and Technology and the Ministry of Agriculture, are closer to capital and wield more influence than those government bodies that have been critical, such as the Ministries of Health, of the Environment, or of Agrarian Development.

A second and related paradox is that while neoliberalism advocates deregulation and the free play of the market, the implementation of neoliberal policies in fact involves a complex and dense set of regulations. At the international level, efforts at the global regulation of genetically modified organisms have led to a host of overlapping, and sometimes conflicting, agreements regulating the different dimensions of their production and circulation—primarily the UN Cartagena Protocol on Biosafety and a host of World Trade Organization agreements.[4] As Randeria suggests (2007), we are in a world of reregulation rather than deregulation.

Nowhere is this more evident than in plant breeding. In the space of two decades, the legislation governing intellectual property rights in plants has evolved from a flexible legislation to one that confers on plant breeders extensive patent rights. The WTO TRIPS Agreement (1994) established the obligation to provide patents for transgenic microorganisms and a form of intellectual property protection for plant varieties. To further strengthen its control over seed technologies and to prevent the practice of saving seeds, Monsanto introduced legally binding Technology Stewardship Agreements in 1996. These private contracts, signed by farmers upon the purchase of patented seeds, further restrict their access and rights over seeds.

As Harvey (2003) and others have pointed out, the flip side of the extension of intellectual property rights is that people become dispossessed of their means of subsistence. So far, dispossession has been felt most acutely by small farmers who (re)produce their own seeds and are losing the right to do so under new intellectual property rights regimes covering transgenic seeds (Peschard, 2010). For those farmers who were already dependent on the market for seeds, buying patented seeds was not perceived as such a threat. My contention, which I will explore in the remainder of this chapter, is that this may be changing as Brazilian farmers discover the longer-term implications of the biotech seed economy.

Blurring the Lines

The state government of Rio Grande do Sul was for a brief period (1998–2003) the epicenter of resistance to transgenics in Brazil. After the Workers Party lost the 2003 state elections, most of the political momentum in the opposition to transgenic varieties shifted to the state of Paraná, under the leadership of governor Requião (2003–10). On the federal scene, the Lula government (2003–11) rapidly betrayed hopes that it would adopt a precautionary approach and pursued the protransgenics policy of the government of its predecessor, Fernando Henrique Cardoso.

The first genetically engineered variety was legalized in 2005 after a long and protracted battle. Since 2008, however, the authorization of transgenic plant varieties has picked up, bringing the total to thirty-six varieties of soybean, cotton, corn, and bean.[5] These crops covered a total of 36.6 million hectares in 2012, making Brazil the second largest producer of transgenic crops worldwide (James, 2012).[6] There are no indications that President Dilma Rousseff, who took office in January 2011, will change the government's transgenics policy.

Over the past fifteen years, the lines of the transgenics controversy in Brazil have been drawn clearly (see Figure 9.1). Proponents of transgenics include the Ministry of Agriculture (*Ministério da Agricultura, Pecuária e Abastecimento*, MAPA) and the Ministry of Science and Technology (*Ministério da Ciência e Tecnologia*, MCT); the National Technical Commission on Biosafety (*Comissão Técnica Nacional em Biossegurança*, CTNBio), responsible for regulating all recombinant DNA technology-related products and activities, and the National Biosafety Council (*Conselho Nacional de Biossegurança*, CNBS), the highest authority on biosafety. Transgenic crops are also promoted by industry groups, mainly the Brazilian Seed Producers Association (ABRASEM) and the Brazilian Plant Breeders Association (*Associação Brasileira dos Obtentores Vegetais*, BRASPOV), and by large farmers' unions, such as the National Agricultural Confederation (*Confederação Nacional da Agricultura e Pecuária do Brasil*, CNA) and the Agricultural Federation of Rio Grande do Sul (*Federação da Agricultura do Estado do Rio Grande do Sul*, FARSUL). Brazil's influential public research corporation, EMBRAPA, has been supportive of genetic engineering in agriculture, entering into technical cooperation agreements with Monsanto for the development of transgenic soybeans adapted to Brazilian soils and investing in its own research on transgenic crops. At the same time, it is concerned with maintaining germplasm resources and conventional varieties (Sousa, 2012).

Among those who are critical of transgenics are the Ministry of the Environment (*Ministério do Meio Ambiente*, MMA), the Ministry of Health (*Ministério da Saúde*, MS), and the Ministry of Agrarian Development (*Ministério do Desenvolvimento Agrário*, MDA). Founded in November 1999 by Brazilian NGOs, the

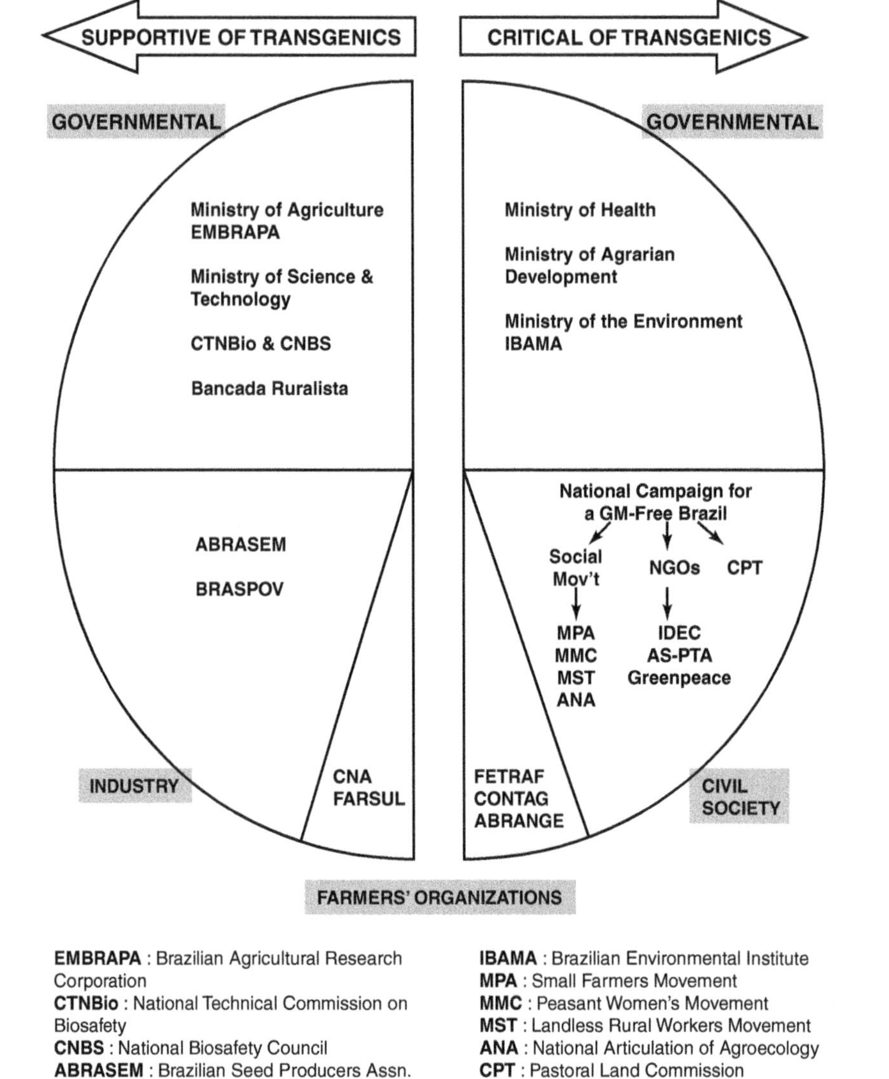

EMBRAPA : Brazilian Agricultural Research Corporation
CTNBio : National Technical Commission on Biosafety
CNBS : National Biosafety Council
ABRASEM : Brazilian Seed Producers Assn.
BRASPOV : Brazilian Plant Breeders Assn.
CNA : National Agricultural Confederation
FARSUL : Agricutural Federation of Rio Grande do Sul

IBAMA : Brazilian Environmental Institute
MPA : Small Farmers Movement
MMC : Peasant Women's Movement
MST : Landless Rural Workers Movement
ANA : National Articulation of Agroecology
CPT : Pastoral Land Commission
IDEC : Institute for Consumer Protection
AS-PTA : Consultancy and Services for Projects in Alternative Agriculture
FETRAF : Family Farming Workers Fed.
CONTAG : National Confederation of Agricultural Workers
ABRANGE : Brazilian Assn. of Non-GM Grain Producers

Figure 9.1 Structural map of key actors in the Brazilian transgenics debate (1996–2009).

National Campaign for a GM-Free Brazil (*Campanha Nacional por um Brasil Livre de Transgênicos*) plays the role of watchdog over government approval of transgenic varieties and promotes a broad debate over transgenics within Brazilian society. Finally, family farming unions, such as the Family Farming Workers Federation (*Federação dos Trabalhadores na Agricultura Familiar*, FRETAF) and the National Confederation of Agricultural Workers (*Confederação Nacional dos Trabalhadores na Agricultura*, CONTAG), also tend to be critical of transgenics. In 2008, the Brazilian Association of Non-Genetically Modified Grain Producers (*Associação Brasileira de Produtores de Grãos Não Geneticamente Modificados*, ABRANGE) was created to defend the interests of conventional grain producers.

In short, government ministries with close links to the agricultural industry, seed producers, plant breeders, and mainstream farmers' organizations have been strong supporters of transgenic crops. By contrast, government ministries concerned with health and the environment, as well as farmers who produce within an alternative, agroecological paradigm, have been more critical of transgenic crops. These alliances held true from the beginning of the controversy, in the mid-1990s, up until recently. Since 2009, however, there have been dissenting voices within the protransgenic coalition. The object of discontent is Roundup Ready (RR) soy, genetically engineered to tolerate Monsanto's herbicide Roundup (generic name glyphosate).[7] RR soy was introduced illegally in Brazil in the late 1990s and was the first transgenic variety to be authorized in 2005. It is the most widely grown variety today, with approximately three in four hectares (or eighteen million hectares) sowed in the 2010–11 harvest (Inácio, 2011). The first evidence of discontent regarding RR soy can be traced back to a public hearing held in 2009.

The 2009 Public Hearing: A Turning Point?

In September 2009, the Agricultural Commission of the Brazilian Chamber of Deputies held a public hearing entitled "Debating the demands of soy producers in Rio Grande do Sul who want to protect the right to save the product of transgenic soy harvests for replanting in their fields and the right to sell this production without paying royalties to Monsanto." Present at the public hearing that day were the Ministry of Agriculture (MAPA), the Brazilian Seed Producers Association (*Associação Brasileira de Sementes e Mudas*, ABRASEM), soy producers (*Associação de Produtores de Soja do Rio Grande do Sul*, APROSOJA-RS), a Monsanto lawyer, and federal deputies.

What prompted the public hearing was a class action filed against Monsanto by soy growers in the southernmost state of Rio Grande do Sul in 2009 (Athayde, 2009; Canal Rural, 2009). Rural unions were demanding that soy growers be given the right to (1) save the product of soy cultivars for replanting on their properties, (2) sell their harvest as food or raw material, and (3) give or exchange seeds without having to pay royalties, indemnities, or fees. These rights, until recently taken for granted, are being eroded under the new legal regimes governing transgenic seeds. (I will discuss this at greater length later.)

An interesting exchange took place at the public hearing that day. The representative for soy producers noted that payments for access to proprietary seeds (or royalties) now represent approximately 14 percent of farmers' profits. Indeed, the cost of royalties has soared since 2005. For example, in August 2009, Monsanto announced a 26 percent increase in royalties on RR soybeans (from R$0.35 to R$0.44 per kilo of seeds). To get a sense of the numbers involved, this increase represented an additional twenty million reais (ten million US dollars) in profits for Monsanto in the state of Mato Grosso alone (Agência Estadual de Notícias, 2009; Zanatta, 2009). For the 2009–10 harvest, revenues in royalties totaled six hundred million reais in Mato Grosso (three hundred million US dollars) and one billion reais (five hundred million US dollars) for Brazil as a whole (Diário de Cuiabá, 2011).

Soy producers went on to make comments such as "we're being exploited"; "the production cost of soybean in 2000 was 13.9 bags per hectares, while in 2009, the cost is 34.2 bags per hectare"; "we thought that the cost of production would decrease, but it has increased"; "Monsanto lures us with glyphosate and soybean, and doubles its earnings"; "we don't want Monsanto . . . to be a financial partner in our harvest"; "when comes the time to sow, if we don't have seeds at home, there is no harvest"; and "RR technology does not increase anyone's productivity" (Reis, 2009, my translation).

At first sight, the above statements may not seem noteworthy, but for anyone who has followed the Brazilian transgenic debate, they are mind-boggling. Indeed, such a public demonstration of tension and disagreement between Monsanto and soy growers on the issue of genetically engineered organisms and intellectual property was unheard of. What is significant here is that these actors are traditionally embedded in the dominant capital-intensive agricultural paradigm; they welcomed the advent of transgenic crops and have been its staunch supporters for over a decade. The fact that natural allies of the agbiotech industry are becoming increasingly critical has important implications for the neoliberal food regime. But first, let me discuss the three issues that are the focus of discontent: royalties, overall profitability, and lack of access to conventional varieties.

Controversy over Royalties

The first and foremost issue is also the one most directly related to the crisis of neoliberalism since it takes aim at the new legal regimes governing seeds, the so-called neoregulations that form an integral part of the current food regime.

Unlike other countries, such as Canada and the United States, where royalties on transgenic seeds were introduced without much ado, in Brazil, the issue of royalties on transgenic varieties has been controversial all along. Monsanto came up with a "dual remuneration scheme" that was unique worldwide at the time it was implemented (2005). It combines royalties on the sale of seeds with a technological fee on the sale of grains. Here is how the system works in practice. When a farmer goes to the grain processor to sell his grain, he is asked if it is transgenic. If he can show a receipt for the payment of royalties on the purchase of seeds, he is exempted from additional payment. If the farmer declares that he used RR seeds but did not pay royalties on seeds, he has to pay Monsanto 2 percent of the value of his crop (Monsanto also refers to this technological fee as an "indemnity for the unauthorized use of a patented technology"). Finally, if the farmer declares that he does not use RR soybean, his grain is tested; if the test is positive, the farmer has to pay a fine in addition to the 2 percent technological fee. It must be noted that it is 2 percent of total gross production, which represents approximately 13 percent of net revenues (Diário de Cuiabá, 2011). Since contamination is widespread, many farmers who grow conventional soy declare their crop to be transgenic in order to avoid being fined. The dual remuneration scheme is a shrewd system: The farmer who does not pay royalties when he purchases the seeds is forced to pay them when he sells his crops. This system is dependent on the seed companies' collaboration, which Monsanto secured by giving them a share of the royalties (Peschard, 2010).

This royalties collection system is controversial. As a rule, royalties are collected on seeds only, and there is some legal debate as to whether intellectual property rights extend to farmers' production (Heimes, 2010). According to the patent exhaustion doctrine, a patent holder exhausts his rights upon selling the patented good and cannot restrict its use any further. However, patents on transgenic seeds have been reinforced with license agreements (also known as Technology Use Agreements or, informally, bag-tag agreements), which farmers must sign upon acquiring seeds. The seed industry argues that it is not selling seeds but granting a limited license to use the seeds purchased (for an example of a license agreement, see Monsanto, 2008). The question yet to be settled is whether private contracts like license agreements can override the patent exhaustion doctrine (Heimes, 2010). This question is essential for the following reason: If intellectual property rights extend to production, companies can prohibit farmers from saving seeds.

In Brazil, seed saving is allowed under the 1997 Plant Variety Protection Act (PVPA). Moreover, the PVPA specifies that it is the "sole form of protection in the country for plant varieties" (Government of Brazil, 1997). Seed companies' prohibition of seed saving through license agreements has been brought before the courts on several occasions since 2005. In their decisions, judges have tended to recognize the legitimacy of the royalties collection system, suspending the right to save seeds when it comes to transgenic varieties (Reis, 2005, n.d.).

However, in April 2012, a first instance decision suspended the collection of royalties on RR soybeans in all of Brazil. This decision was the result of a class action lawsuit filed against Monsanto in 2009 by three rural unions of Rio Grande do Sul to question the collection of royalties on commercialized grain. The Soybean Producers Association (APROSOJA-RS), the Federation of Agricultural

Workers (FETAG-RS), and over 350 rural unions of Rio Grande do Sul are now involved in the class action. In addition to suspending the collection of royalties, the judge ruled that Monsanto pay back the royalties collected since the 2003–4 harvest, bearing interest at the legal rate and adjusted to the cost of living. This decision also stops Monsanto from prohibiting the exchange of seeds among farmers. Monsanto immediately petitioned for and obtained the suspension of the first instance ruling until a decision is rendered by a higher court (Ruralbr e Canal Rural, 2012). There is no doubt that Monsanto will appeal the decision; if it is upheld, five million soy producers will be entitled to fifteen billion reais (USD 7.5 billion) (Valor Econômico, 2012). In the meantime, the Superior Court of Justice ruled on two related issues. It rejected Monsanto's claim that rural unions could not file a class action. It also ruled against Monsanto—who wanted any decision to be restricted to the plaintiffs—that any future decision in the case will apply to the whole of Brazil (Última Instância, 2012; Valor Online, 2012).

It will be interesting to see how Brazilian courts rule on the matter of royalties after the European Court of Justice's decision in the case opposing Monsanto to Argentina (European Court of Justice, 2010). The European Court determined that Monsanto could not use its European patent on RR soybean to block Argentinean soymeal imports (Correa, 2007). According to the decision, the biotechnology patent legislation does not extend to the product when the latter no longer exercises the function for which it was patented. In other words, even if the soymeal contains the patented RR gene, Monsanto cannot exercise its patent rights since the herbicide tolerance function is irrelevant to soymeal (ASPTA, 2010). This decision creates a precedent in international law and could have important implications for the Brazilian royalties collection system. Indeed, under the current system, a farmer is forced to pay royalties on commercialized grain if the RR gene is detected, regardless of the degree to which it is present. There are already signs that the legitimacy of this practice may be reconsidered: In the legal case discussed above, the judge asked for laboratory analyses on transgenic soybean samples from the 2009–10 harvest (Correio do Povo, 2010). The objective is to determine the extent to which the transgenic trait is present in the grain. Farmers hope to demonstrate that in cases where the presence of the transgenic gene is only marginal, royalties should not be charged since the patented trait no longer exercises the function for which it is patented—that is, tolerance to Roundup herbicide (if such soy was sprayed with Roundup, it would be destroyed).

Legal challenges over royalties have multiplied in recent years, and there are a number of other cases before the courts. In September 2013, APROSOJA, the Mato Grosso Agriculture and Livestock Federation (*Federação de Agricultura e Pecuária de Mato Grosso*, FAMATO), and forty-seven rural unions filed a class action lawsuit against Monsanto demanding the suspension of royalties and indemnity on Bollgard I (Bt cotton) and Roundup Ready technologies. The organizations base their case on a technical and legal opinion according to which the intellectual property rights on these technologies expired in September

2010.[8] They argue that they are therefore in the public domain and that Monsanto has been collecting royalties illegally for years. In a first instance decision, the judge suspended the payment of royalties and determined the judicial deposit of royalties until the case comes to a final resolution (Ascom Aprosoja and Ascom Famato, 2012).

In January 2013, the National Agricultural Confederation (CNA) and eleven state agricultural federations signed a Declaration of Principles in which they reaffirmed their commitment to "work together to facilitate approvals of technologies that can be applied in Brazil and result in Brazilian export to international markets" and to "reinforce the recognition of intellectual property rights on agricultural technologies and improve, by mutual agreement, the royalties collection system." In return, Monsanto committed to "permanently and irrevocably suspending the patent on RR1 soybeans for those producers who agree to individually sign on to the licencing agreement for the use of Intacta RR2 technology" (RR2 is the second generation of Roundup Ready soybeans) (Farsul, 2013, my translation). FAMATO, APROSOJA, and rural unions unanimously rejected the agreement and advised their members not to sign it (O Estado de São Paulo, 2013). According to them, farmers would give up their rights and commit to pay royalties for RR2 in exchange for the suspension of a patent (RR1) which is no longer valid (Aprosoja, 2013).

On February 21, 2013, the Superior Court of Justice denied the extension application filed by Monsanto on its patent on the first generation of Roundup Ready soybean (RR1), thus confirming that the company does not have the right to collect royalties on this variety (O Estado de São Paulo, 2013). Monsanto appealed the decision but in a reversal of its earlier position, announced that it would defer the collection of royalties on RR1 soybeans in all of Brazil until the case comes to a final resolution (Monsanto, 2013).

In these legal challenges, producers contend that Monsanto is abusing its intellectual property rights but do not question the legitimacy of intellectual property rights over plant varieties. As FAMATO, APROSOJA, and rural unions state in a joint press release: "Because we recognize and value the important research and technological contributions made by both public and private corporations, we want to clarify that we support payments for intellectual property (royalties). Moreover, we defend the fair collection of royalties in accord with Brazilian patent legislation" (Aprosoja/BrasilWorks, 2013).

In this tense context, three bills aimed at amending the Plant Variety Protection Act hold potential to stir further conflict between Monsanto and rural producers. Two bills are sponsored by members of Congress, while the third, broader in scope, comes from the executive branch. These bills vary in their particulars but share the same underlying goal: to amend the existing legislation so that patent protection is extended to products and seed saving prohibited (with the exception of small farmers) (Santilli, 2009). The explicit aim is to bring Brazil's plant protection legislation into line with the 1991 Convention of the International Union for the Protection of New Varieties of Plants (UPOV 1991). Brazil is currently a member of UPOV 1978. The main difference between the two conventions is that UPOV 1991 extends protection to harvested materials. This means that farmers are not allowed to save seed unless individual governments, with the consent of the breeder, allow limited exceptions. UPOV 1991 also introduces patents alongside breeders' rights, which means UPOV members can patent plant varieties. This means that a farmer would need the breeder's authorization to sell his crop, effectively losing control over his harvest. The seed industry is pressuring hard for these amendments; if passed, they will no doubt fuel greater discontent among farmers.

Diminishing Profitability

The second reason explaining the growing dissatisfaction with transgenic varieties is their decreasing profitability. Experience suggests that transgenic varieties' initial superiority vanishes after a few years. While in 2005–6, transgenic soy was 10.2 percent more profitable than conventional soy, by 2009–10, it was 13.6 percent *less* profitable (Salgado, 2009). Another study by a private seed research foundation comparing forty transgenic soybean varieties and twenty conventional ones in Rio Grande do Sul concludes that the transgenic varieties produced on average 9 percent less than the conventional ones, with equivalent production costs (ASPTA, 2009a).

In addition, exposure to a single herbicide (Roundup or glyphosate) accelerates the development of weed resistance, thus forcing farmers to use greater quantities of Roundup in combination with other herbicides. In Brazil, at least five different weeds have developed resistance to glyphosate (Reuters/Brasil Online, 2010). According to the Brazilian Institute of the Environment and Renewable Natural Resources (*Instituto Brasileiro do Meio Ambiente e Recursos Naturais Renováveis*, IBAMA), between 2000 and 2004, a period of rapid expansion for RR soybeans, the use of glyphosate in the state of Rio Grande do Sul, the main producer in Brazil at the time, increased by 162 percent (95 percent for the country as a whole) (Agência Estadual de Notícias, 2007). Sales of agrochemicals have reached a record high in recent years, and since 2008, Brazil has been competing with the United States as the first consumer of agrochemicals worldwide (ASPTA, 2009b; Valor Econômico, 2011b).

As a result, the cost of production is now generally higher for transgenic soy than for conventional soy. Various studies conducted in Mato Grosso by the Brazilian Agricultural Research Corporation (*Empresa Brasileira de Pesquisa Agropecuária*, EMBRAPA), the Mato Grosso Institute for Agricultural Economics (*Instituto Mato-grossense de Economia Agropecuária*, IMEA), and the National Food Supply Company (*Companhia Nacional de Abastecimento*, CONAB) show that production costs are lower for conventional soy than for transgenic soy (Faria, 2010; Folha de São Paulo, 2010; Revista Ótima S/A, 2010; Só Notícias, 2010). According to the Centre for Advanced Studies in Applied Economics (*Centro de Estudos Avançados em Economia Aplicada*, CEPEA) of the University of São Paulo, in Sorriso (Mato Grosso), the leading soy producing municipality in Brazil, the cost of production for the 2009–10 harvest was 6 percent higher for transgenic soy than for conventional soy. When the production cost of conventional soy is higher, the premium paid by European importers for conventional soy makes up for the difference. Some cooperatives offer a premium of two to four reais (one to two US dollars) per sixty kilogram bag of conventional grain, up to 80 percent of which goes to the farmer (Gomes, 2010; Marques, 2010). In the 2011 harvest, farmers who planted conventional soy in Mato Grosso earned one hundred reais (fifty US dollars) per hectare more than those who planted transgenic soy (Folha de São Paulo, 2011).

A consensus is now emerging that RR soy is neither more productive nor more profitable than conventional soy. As one soy grower states: "Transgenic soy is not economically attractive to us. It spread here because it simplifies management on the farm, not because it reduces costs" (Zanatta, 2009, my translation). If this is the case, then why aren't more farmers reverting to conventional soy? Part of the answer lies in the rapid disappearance of conventional soybean varieties.

Access to Conventional Seeds

The latest expression of discontent involves Monsanto's corporate practices and its domination of the Brazilian seed market. In May 2010, the Soybean Producers Association (APROSOJA) and the Brazilian Association of Non-Genetically Modified Grain Producers (ABRANGE) announced that they were considering launching an appeal against Monsanto before the Administrative Council for Economic Defence (*Conselho Administrativo de Defesa Econômica*, CADE), affiliated to the Department of Justice. The two organizations are considering launching a formal complaint for abuse of economic power and lack of transparency in the collection of royalties. They contend that Monsanto is deliberately restricting farmers' access to conventional soy. According to them, Monsanto, through its 70 percent share of the market, is imposing a quota of 85 percent transgenic seeds and 15 percent conventional seeds (Agência Brasil, 2010). The problem is that this quota does not correspond to the demand; in other words, farmers who would like to buy conventional seeds are being forced to buy genetically engineered varieties (Macedo, 2010; Zanatta, 2010).

This is an interesting turn of events given that Monsanto's main line of response to critics of genetic engineering is that it is up to farmers to decide and that they should be given options. In fact, it seems as if more and more farmers are forced to grow transgenic varieties for lack of access to conventional ones.

Research is one dimension of the problem. According to one agronomist, "Companies have been focusing their research on transgenic soy varieties more than on conventional ones. So in ten years, we could have 100 per cent of the area planted with transgenic soy not because this was the farmers' choice but because development of new conventional varieties is getting scarce" (Riveras, 2009). The director of a processing company specializing in conventional soy concurs: according to him, the bottleneck is not in the field or in the market, but in research (Gomes, 2010). Some producers, however, accuse Monsanto of deliberately imposing transgenic varieties by fixing limits on the production of conventional seeds in its contracts with seed producers (Maschio, 2009). Some farmers also report being obliged, in the same transaction, to buy transgenic seeds in order to obtain conventional seeds; this practice is called, in Portuguese, *compra casada* (Folha de São Paulo, 2010).

In July 2010, the Mato Grosso Seed Producers' Association (APROSMAT) announced it was leaving the National Seed Producers Association (ABRASEM) over the latter's stand in the conflict with Monsanto (Monsanto is also a member of ABRASEM). ABRASEM had declined to act on APROSMAT's complaint regarding Monsanto's role in restricting access to conventional seeds. According to APROSMAT's director, Monsanto offers an artificially low ratio of conventional to transgenic seeds to farmers responsible for multiplying seeds for the next harvest (Inácio and Barros, 2010). In its press release, APROSMAT states: "unfortunately, ABRASEM is no longer defending the legitimate interests of seed producers, who are concerned with maintaining themselves and their independence in the national context" (APROSMAT, 2010, my translation).

In response to the problem of access to conventional seeds, EMBRAPA, APROSOJA, and ABRANGE launched the Free Soy Program (*Programa Soja Livre*) in November 2010 (Folha de São Paulo, 2011; Gazeta do Povo, 2010; Jornal Hora do Povo, 2010). The objective is to strengthen conventional soybean production in Mato Grosso and provide farmers with better access to conventional varieties. According to the participants, the increasing dependence of soy growers on a limited number of transgenic varieties is a source of concern, and expanding the conventional soy area has become an issue of national security and sovereignty (Romeu, 2011).

Opposition from Within: The Neoliberal Food System in Crisis

Over the past decade, the National Campaign for a GM-Free Brazil has spearheaded the opposition to transgenics. By monitoring the CTNBio, taking legal actions, and lobbying and organizing public demonstrations and direct action, it was pivotal in raising essential questions regarding farmers' autonomy, agricultural biodiversity, the ethics of patenting, and food sovereignty. Its most important contribution has been to force a redefinition of the terms of the debate over transgenics away from a narrow focus on health and environmental risks toward broader issues of social and environmental justice. While it has obtained a number of successes, it has not been able to counter the powerful interests in favor of transgenics.[9]

Interestingly, some of the arguments long advanced by the Campaign for a GM-Free Brazil—regarding seed industry concentration or the development of weed resistance, for example—are now being made by farmers' organizations, which until recently supported transgenics. The authorization of transgenic varieties was fought for in the name of farmers' freedom of choice. In an ironic reversal, farmers are now defending better access to conventional varieties in the name of the very same freedom of choice.

Brazilian farmers have never been unanimous in their support of transgenic crops, and a number of small and large cooperatives have sought alternatives in conventional and organic varieties early on in the controversy (Jepson et al., 2008; Hisano and Altoé, 2008). However, the events described here denote a qualitative and quantitative change in the nature of the opposition.

First, the sheer number of farmers involved in contesting the new regimes governing seeds has greatly increased. As of 2013, hundreds of rural unions are engaged in a number of class action lawsuits. For example, the number of rural unions implicated in the main lawsuit in Rio Grande do Sul grew from three to over 350. It is already considered to be a precedent-setting case because of the financial stakes involved and because it has been determined that the ruling will apply to the whole of Brazil.

Second, for the first time, the opposition involves large soy producers in two leading soy-producing states—Mato Grosso and Rio Grande do Sul—in a country where rural interests are well-entrenched in the political system.[10] Indeed, the *bancada ruralista* is the most influent interest group in Congress. In the 2007–11 legislature, it had the support of approximately 20 percent of deputies and 15 percent of senators (Congresso em Foco and DIAP, 2007). Formed during the 1988 Constituent Assembly to block any constitutional initiative at agrarian reform, it defends the interests of large landowners and agribusiness and is actively pushing for the swift approval of transgenic crops. The growing discontent among farmers now puts *ruralistas* deputies (known for their links to agribusiness) in an awkward position, caught between their allegiance to Monsanto and their allegiance to big farmers. Interestingly, at the public hearing, they remained silent and did not speak up in defense of royalties, leaving it up to the Ministry of Agriculture to do so.

Third, the controversy surrounding royalties has evolved from a largely private matter being discussed between Monsanto and farmers' organizations to a national issue, as witness the Superior Court of Justice decision that any future decision in the case will apply to the whole of Brazil. Soy producers are also resorting (or threatening to resort) to the highest court and public authorities, such as the Administrative Council for Economic Defence. In a context in which corporations effectively use the state to promote their interests, farmers are forced to resort to the courts, leading to a judicialization of politics.

Finally, the nature of complaints has evolved. Until recently, soy growers' complaints pertained only to the increase in royalties—that is, to negotiating the terms of the seed contracts. They are now increasingly contesting key issues, such as the validity of Monsanto's patents on RR varieties and its corporate practices. These include restricting access to conventional seeds, collecting royalties not only on seeds but also on production, and, most importantly, restricting the right to save seeds. As a matter of fact, it is questioned whether restricting the right to save seeds is permitted under Brazilian legislation.

For all of these reasons, this movement holds the potential for bringing changes to the neoliberal food regime in Brazil. The nature of these changes is open to question. But even if soy growers' complaints have broadened, they fall short of questioning the basis of the neoliberal food regime. Their discontent is primarily aimed at patents and royalties, which they consider abusive, but they accept the legitimacy of collecting royalties on seed biotechnology. Soy growers contest Monsanto's domination and control of the Brazilian seed market and the increasingly unfavorable terms imposed upon them in the new seed economy. However, their demand for better access to traditional varieties is consistent with the freedom of choice ethos of neoliberalism. In sum, their critiques are aimed at one specific dimension of the neoliberal food regime: the key role of large agribusiness transnational corporations, who have become the crucial economic actors in global capitalism (Otero, 2012).[11]

In May 2012, the Agricultural Commission of the Senate held a public hearing on royalties. During the public hearing, seed and soybean producers denounced transgenic corporations' monopoly over the seed market and asked for the reassessment of patents and more transparent rules for the collection of royalties. According to Blairo Maggi, an influential politician who is widely considered the largest individual soybean producer worldwide, "this is where the danger lies: the capacity of this technology to prevent seed producers from reproducing conventional seeds. This is a monopoly, a crime against the country that must be denounced" (Altafin, 2012).

This statement reflects the growing "dissatisfaction that has emerged from the unchecked behavior of corporations under neoliberalism" (see Bonanno, this volume). Interestingly, in their conflict with Monsanto over patents and seed contracts, soy growers now put forward the national interest. Issues of national security and sovereignty with regards to seeds point to the continued relevance of the "national" in our globalized world. Despite all the talk about globalization, the state is never far away when there is a crisis, and crises have been a defining characteristic of neoliberalism.

Conclusion

The events described here clearly reveal a crisis in Brazilian agri-food politics. After RR soybean enjoyed unwavering support from large soy producers, the tide now seems to be turning. Whereas opposition to transgenics was long restricted to vocal, though politically marginal, sectors of Brazilian society, since 2009, it has spread to mainstream groups in Brazilian politics—namely, large soy producers' organizations. Assuming they have the political will, these actors have the necessary political muscle to force, for example, a reexamination of the current royalty collection system. The court decision in the case opposing Monsanto to rural unions (pending in 2013) will be decisive in this regard as it could set a precedent for other countries to question Monsanto's corporate practices.

While the Brazilian Campaign for a GM-Free Brazil represents a radical critique of transgenic crops from peripheral sectors of society, the more recent discontent among soy growers represents a moderate critique from politically and economically influential sectors. It is important to keep in mind that soy growers are primarily contesting the role of large agribusiness corporations in the neoliberal food regime. Given this fact, and capitalism's historical resilience, the most likely outcome is that the neoliberal food regime will adapt to accommodate large soy growers' grievances and interests without altering its essence. It is too early to know where this movement is heading, but it should not be dismissed lightly.

Acknowledgment

This research was supported by a graduate scholarship and a postdoctoral fellowship of the Social Sciences and Humanities Research Council of Canada.

Notes

1 For an overview of the literature on resistance to transgenic crops, see Stone (2010, pp. 386–7). On opposition to transgenic crops in Brazil more specifically, see Bauer (2006), Pelaez and Da Silva (2008), and Scoones (2008).
2 For example, marker-assisted selection combines conventional plant breeding with genetic and molecular biology; it uses DNA analysis and trait-linked molecular markers to select traits of interests, which are then incorporated into plant genomes using traditional breeding techniques.
3 For a detailed comparison of the Green and Gene Revolutions, see Parayil (2003).
4 The Cartagena Protocol on Biosafety is a supplement to the Convention on Biological Diversity that deals with transboundary movements of living modified organisms. The relevant WTO agreements are (1) the Agreement on Trade-Related aspects of Intellectual Property Rights (TRIPS), (2) the Agreement on the Application of Sanitary and Phytosanitary Measures (SPS), and (3) the Agreement on Technical Barriers to Trade (TBT).
5 The list of transgenic plant varieties commercially authorized in Brazil is available on the following website: http://cib.org.br/biotecnologia/regulamentacao/ctnbio/eventos-aprovados/.
6 This figure comes from industry; Brazil, like many other countries, does not have official statistics on transgenic crops.
7 Similar conflicts between producers and companies are also emerging in relation to transgenic cotton (Valor Econômico, 2011a).
8 That is, twenty years after the initial filing of the patent application abroad—in this case, August 31, 1990.
9 For a detailed discussion of the National Campaign for a GM-Free Brazil, see Peschard (2010, chapter 6).

10 Emblematic of the intertwining of political and economic power is the figure of Blairo Maggi: governor (2003–10) and senator (2011–) of the state of Mato Grosso; he is widely considered the largest individual soybean producer worldwide.

11 According to Otero (2012, p. 282), alongside large agribusiness transnational corporations, the other components of the neoliberal regime are the state, which promotes international and national neoregulation that impose the neoliberal agenda, and biotechnology, the driver behind the modern agricultural paradigm.

References

Agência Brasil. (2010) "Agricultores reclamam que Monsanto restringe acesso a sementes de soja convencional," *Agência Brasil*, May 18, available at http://agenciabrasil.ebc.com.br/noticia/2010-05-18/agricultores-reclamam-que-monsanto-restringe-acesso-sementes-de-soja-convencional, accessed August 23, 2013.

Agência Estadual de Notícias. (2007) "Glifosato não elimina ervas daninhas que atacam transgênicos," *Agência Estadual de Notícias*, May 3, available at http://www.paisagismobrasil.com.br/index.php?system=news&news_id=903&action=read, accessed August 23, 2013.

Agência Estadual de Notícias. (2009) "Monsanto vai reajustar em 26% royalties para a soja transgênica," *Agência Estadual de Notícias*, August 21, available at http://www.historico.aen.pr.gov.br/modules/noticias/article.php?storyid=49797, accessed August 23, 2013.

Altafin, I. G. (2012) "Produtores querem limites nos royalties para transgênicos," *Agência Senado*, May 10, available at http://www12.senado.gov.br/noticias/materias/2012/05/10/produtores-querem-limites-nos-royalties-em-transgenicos, accessed August 23, 2013.

Aprosoja. (2013) "Comunicado da APROSOJA Brasil sobre royalties," press release, March 1, available at http://aprosojabrasil.com.br/?p=728, accessed August 23, 2013.

Aprosoja/BrasilWorks. (2013) "Mato Grosso soy growers reject Monsanto's royalty agreement," January 24, available at http://www.brazil-works.com/mato-grosso-soy-growers-reject-monsantos-royalty-agreement/, accessed August 23, 2013.

Ascom Aprosoja and Ascom Famato. (2012) "Nota Oficial: Justiça de MT determina que pagamento de royalties seja feito em juízo," December 5, available at http://www.aprosoja.com.br/noticia/nota-oficial-justica-de-mt-determina-que-pagamento-de-royalties-seja-feito-em-juizo/, accessed April 3, 2013.

Assessoria e Serviços a Projetos em Agricultura Alternativa (ASPTA). (2009a) "Soja transgênica produz 9% menos," *Boletim Por um Brasil Livre de Transgênicos*, no 448, July 3, http://antigo.aspta.org.br/por-um-brasil-livre-de-transgenicos/boletim/boletim-448-3-de-julho-de-2009/, accessed June 22, 2012.

Assessoria e Serviços a Projetos em Agricultura Alternativa (ASPTA). (2009b) "Brasil é o maior consumidor mundial de agrotóxicos; há quem comemore," *Boletim Por um Brasil Livre de Transgênicos*, no 443, May 29, http://antigo.aspta.org.br/por-um-brasil-livre-de-transgenicos/boletim/boletim-443-29-de-maio-de-2009/, accessed June 22, 2012.

Assessoria e Serviços a Projetos em Agricultura Alternativa (ASPTA). (2010) *Boletim por um Brasil livre de transgênicos*, no 497, July 9, http://antigo.aspta.org.br/

por-um-brasil-livre-de-transgenicos/boletim/boletim-497-9-de-julho-de-2010/, accessed June 22, 2012.

Associação de produtores de sementes de Mato Grosso (APROSMAT). (2010) *Press Release*, August 5, available at http://pratoslimpos.org.br/?p=1566, accessed June 22, 2012.

Athayde, P. (2009) "Polêmica nos campos de soja," *Carta Capital*, April 24.

Bauer, M. W. (2006) "Paradoxes of resistance in Brazil," Pp. 228–49 in G. Gaskell and M.W. Bauer (eds) *Genomics and Society. Legal, Ethical and Social Dimensions*, Earthscan, London.

Canal Rural. (2009) "Sojicultores gaúchos vão à Justiça contra a Monsanto," *Canal Rural*, March 19, available at http://agricultura.ruralbr.com.br/noticia/2009/03/sojicultores-gauchos-vao-a-justica-contra-a-monsanto-2445549.html, accessed August 23, 2013.

Congresso em Foco and DIAP. (2007) "*O que esperar do novo congresso. Perfil e agenda da legislatura 2007–2011*," Congresso em Foco e DIAP, Brasília.

Correa, C. M. (2007) "The Monsanto vs. Argentina dispute on GMO soybeans," *Third World Resurgence*, no 203–4, pp. 13–16.

Correio do Povo. (2010) "Royalties: começa perícia sobre soja contaminada," *Correio do Povo*, October 4, available at http://pratoslimpos.org.br/?p=1757, accessed August 23, 2013.

Diário de Cuiabá. (2011) "Monsanto cobra royalties duas vezes dos produtores," *Diário de Cuiabá*, February 12, available at http://pratoslimpos.org.br/?p=2223, accessed August 23, 2013.

European Court of Justice. (2010) *Monsanto Technology LLC v. Cefetra BV, Cefetra Feed Service BV, Cefetra Futures BV and State of Argentina*, C-428/08, judgment of July 6.

Faria, C. (2010) "Produzir soja convencional é mais barato, diz estudo da Embrapa," *Correio do Estado (MS)*, October 6, available at http://pratoslimpos.org.br/?p=1765, accessed August 23, 2013.

Farsul. (2013) "Declaração de princípios Soja RR1," *Assessoria de Imprensa Sistema Farsul*, February 8, available at Declaração de princípios Soja RR1, accessed August 23, 2013.

Folha de São Paulo. (2010) "Produzir soja transgênica custa mais em MT—mas sementes convencionais sumiram do mercado," *Folha de São Paulo*, June 16.

Folha de São Paulo. (2011) "Soja transgênica preocupa produtor de MT," *Folha de São Paulo*, April 5.

Gazeta do Povo. (2010) "Soja Livre: Mato Grosso quer conter expansão de transgênicos," *Gazeta do Povo*, December 14, available at http://www.sistemafaep.org.br/estatico/boletimdiario/bol_diario_141210.htm, accessed August 23, 2013.

Gomes, L. (2010) "Procura-se soja convencional," *Gazeta do Povo*, November 12, available at http://agro.gazetadopovo.com.br/arquivo/procura-se-soja-convencional/, accessed August 23, 2013.

Government of Brazil. (1997) *Institui a Lei de Proteção de Cultivares e dá outras providências* (Plant Variety Protection Act), Federal Law no 9.456, April 25.

Harvey, D. (2003) *The New Imperialism*, Oxford University Press, New York.

Heimes, R. S. (2010) "Post-sale restrictions on patented seeds: Which law governs?," University of Southern Maine, available at http://works.bepress.com/rita_heimes/2, accessed June 22, 2012.

Hisano, S. and Altoé S. (2008) "Brazilian farmers at a crossroads: Biotech industrialization of agriculture or new alternatives for family farmers?," Pp. 243–65 in G. Otero (ed.) *Food for the Few: Neoliberal Globalism and Biotechnology in Latin America*, University of Texas Press, Austin.

Inácio, A. (2011) "Céleres prevê 25,8 milhões de hectares cultivados com OGMs na safra 2010/11," *Valor Econômico*, January 18, available at http://pratoslimpos.org.br/?p=2132, accessed August 23, 2013.

Inácio, A., and Barros, B. (2010) "Produtores acusam a Monsanto de 'segurar' semente convencional," *Valor Econômico*, July 15, available at http://www.ihu.unisinos.br/noticias/noticias-arquivadas/34360-produtores-acusam-a-monsanto-de-segurar-semente-convencional, accessed August 23, 2013.

James, C. (2012) "Global status of commercialized Biotech/GM Crops: 2012," *ISAAA Brief*, no 44, ISAAA, Ithaca, NY.

Jepson, W. E., Brannstrom, C., and de Souza, R. S. (2008) "Brazilian biotechnology governance: Consensus and conflicts over genetically modified crops," Pp. 217–42 in Otero, *Food for the Few*.

Jornal Hora do Povo. (2010) "Embrapa lança no MT 'Soja Livre' para combater transgênicos da Monsanto," *Jornal Hora do Povo*, November 17, available at http://www.horadopovo.com.br/2010/novembro/2915-17-11-2010/P2/pag2e.htm, accessed August 23, 2013.

Kloppenburg, Jack R., Jr. ([1988] 2004) *First the Seed: The Political Economy of Plant Biotechnology, 1492–2000*, University of Wisconsin Press, Madison.

Macedo, D. (2010) "Agricultores reclamam que Monsanto restringe acesso a sementes de soja convencional," *Agência Brasil*, May 18, available at http://agenciabrasil.ebc.com.br/noticia/2010-05-18/agricultores-reclamam-que-monsanto-restringe-acesso-sementes-de-soja-convencional, accessed August 23, 2013.

Marques, L. C. (2010) "Custo menor favorece semente convencional," *Gazeta Digital*, May 30, available at http://www.gazetadigital.com.br/conteudo/show/secao/2/materia/241600, accessed August 23, 2013.

Maschio, J. (2009) "Soja transgênica deve chegar a 67% da produção brasileira," *Folha de São Paulo*, December 22.

McAfee, K. (2003) "Neoliberalism on the molecular scale. Economic and genetic reductionism in biotechnology battles," *Geoforum*, vol 34, no 2, pp. 203–19.

Monsanto. (2008) "Monsanto technology/stewardship agreement (limited use license)," Monsanto, http://www.monsanto.com/SiteCollectionDocuments/tug_sample.pdf, accessed June 22, 2012.

Monsanto. (2013) "Monsanto presents farmer solution while RR1 soybean term correction matter is ongoing in Court," press release, February 26, available at http://www.monsanto.com/newsviews/Pages/monsanto-presents-farmer-solution-while-rr1-soybean-term-correction-matter-is-ongoing-in-court.aspx, accessed August 23, 2013.

O Estado de São Paulo. (2013) "A Guerra pela semente da Monsanto," *O Estado de São Paulo*, February 25, available at http://www.estadao.com.br/noticias/impresso,a-guerra-pela-semente-da-monsanto-,1001130,0.htm, accessed August 23, 2013.

Oh, C. (2000) "TRIPS and pharmaceuticals: A case of corporate profits over public health," *Third World Network*, August/September, available at http://www.twnside.org.sg/title/twr120a.htm, accessed August 23, 2013.

Otero, G. (2008) *Food for the Few: Neoliberal Globalism and Biotechnology in Latin America*, University of Texas Press, Austin.
Otero, G. (2012) "The neoliberal food regime in Latin America: state, agribusiness transnational corporations and biotechnology," *Canadian Journal of Development Studies/Revue canadienne d'études du développement*, vol 33, no 3, pp. 282–94.
Parayil, G. (2003) "Mapping technological trajectories of the Green Revolution and the Gene Revolution from modernization to globalization," *Research Policy*, vol 32, no 6, pp. 971–90.
Pelaez, V. and Da Silva, L. R. (2008) "Social resistance to biotechnology: Attempts to create a Genetically Modified-Free territory in Brazil," *International Journal of Technology and Globalisation*, vol 4, no 3, pp. 207–22.
Peschard, K. (2010) "Biological dispossession: An ethnography of resistance to transgenic seeds among small farmers in southern Brazil," PhD thesis, McGill University, Canada, available at http://digitool.library.mcgill.ca/R/-?func=dbin-jump-full&object_id=86636¤t_base=GEN01, accessed June 22, 2012.
Rajan, K. S. (2006) *Biocapital: The Constitution of Postgenomic Life*, Duke University Press, Durham, NC.
Randeria, S. (2007) "The state of globalization. Legal plurality, overlapping sovereignties and ambiguous alliances between civil society and the cunning state in India," *Theory, Culture & Society*, vol 24, no 1, pp. 1–33.
Reis, M. R. (2005) "Propriedade intelectual, sementes e o sistema de cobrança de royalties implementado pela Monsanto no Brasil," www.territoriosdacidadania.gov.br/o/1206399, accessed June 22, 2012.
Reis, M. R. (2009) "Na câmara dos deputados, sojicultores transgênicos começam a expor o drama de se tornarem reféns da Monsanto. Estão acordando, mas um pouco tarde," *Boletim da Campanha por um Brasil livre de transgênicos*, no 459, September 18, available at http://aspta.org.br/campanha/boletim-459-18-de-setembro-de-2009/, accessed August 23, 2013.
Reis, M. R. (n.d.) "Considerações acerca do impacto da propriedade intelectual sobre sementes na agricultura camponesa," http://www.ebah.com.br/content/ABAAAAu3IAG/consideracoes-acerca-impacto-propriedade-intelectual-sobre-sementes-na-agricultura-camponesa, accessed June 22, 2012.
Reuters/Brasil Online. (2010) "Erva daninha resistente a herbicidas já preocupa agricultores no Brasil," Reuters/Brasil Online, June 10, available at http://ambienteja.com.br/ver_cliente.asp?id=164203, accessed May 22, 2011.
Revista Ótima S/A. (2010) "Segundo Conab, custo de produção da soja transgênica é maior," *Revista Ótima S/A*, October, available at http://aspta.org.br/campanha/boletim-518-03-de-dezembro-de-2010/, accessed August 23, 2013.
Riveras, I. (2009) "Biggest Brazil soy state loses taste for GMO seed," Reuters, March 13, available at http://www.reuters.com/article/2009/03/13/us-soy-brazil-gmo-idUSTRE52C5AB20090313, accessed August 23, 2013.
Romeu, A. (2011) "Plantio de soja convencional é questão de segurança nacional, defende presidente da Aprosoja," *Plantão News*, February 3, available at http://www.plantaonews.com.br/conteudo/show/secao/45/materia/28243, accessed August 23, 2013.
Ruralbr e Canal Rural. (2012) "Monsanto embarga decisão da Justiça do RS sobre cobrança de royalties da soja RR," *Ruralbr e Canal Rural*, April 12, available at http://agricultura.ruralbr.com.br/noticia/2012/04/monsanto-embarga

-decisao-da-justica-do-rs-sobre-cobranca-de-royalties-da-soja-rr-3724825.html, accessed August 23, 2013.

Salgado, R. (2009) "E os lucros secaram . . . ," *Revista Veja*, August 8.

Santilli, J. (2009) *Agrobiodiversidade e direitos dos agricultores*, Peirópolis & Instituto Internacional de Educação do Brasil, São Paulo.

Scoones, I. (2008) "Mobilizing against GM crops in India, South Africa and Brazil," *Journal of Agrarian Change*, vol 8, no 2–3, pp. 315–44.

Só Notícias. (2010) "Sorriso: custo da soja convencional—menor que o da transgênica," *Só Notícias*, February 12, available at http://www.sonoticias.com.br/noticias/2/100715/sorriso-custo-para-produzir-soja-convencional-e-menor-que-a-transgenica, accessed August 23, 2013.

Sousa, C. B. de (2012) "Mitos transgênicos," *Valor Econômico*, May 18, available at http://www.ihu.unisinos.br/noticias/509642-mitos-transgenicos, accessed August 23, 2013.

Stone, G. D. (2010) "The anthropology of genetically modified crops," *Annual Review of Anthropology*, vol 39, pp. 381–400.

Última Instância. (2012) "Ação bilionária sobre royalties cobrados pela Monsanto terá abrangência nacional," *Última Instância*, June 13, available at http://ultimainstancia.uol.com.br/conteudo/noticias/56545/acao+bilionaria+sobre+royalties+cobrados+pela+monsanto+tera+abrangencia+nacional.shtml, accessed August 23, 2013.

Valor Econômico. (2011a) "Algodão: royalty de semente transgênica também começa a gerar divergências," *Valor Econômico*, January 3, available at http://pratoslimpos.org.br/?p=2090, accessed August 23, 2013.

Valor Econômico. (2011b) "Venda de defensivos bate recorde no Brasil," *Valor Econômico*, February 17, available at http://www.valor.com.br/arquivo/873077/venda-de-defensivos-bate-recorde-no-brasil, accessed August 23, 2013.

Valor Econômico. (2012) "Decisão judicial suspende cobrança de royalties pela Monsanto," *Valor Econômico*, April 9, available at http://www.valor.com.br/legislacao/2604806/decisao-judicial-suspende-cobranca-de-royalties-pela-monsanto, accessed August 23, 2013.

Valor Online. (2012) "STJ dá ganho de causa a agricultores contra royalties da soja da Monsanto," *Valor Online*, June 13, available at http://www1.valor.com.br/agro/2713536/stj-da-ganho-de-causa-agricultores-em-acao-contra-monsanto, accessed August 23, 2013.

Zanatta, M. (2009) "Monsanto eleva em 26% royalties da soja," *Valor Econômico*, August 21, available at http://www.valor.com.br/arquivo/630339/monsanto-eleva-em-26-royalties-da-soja, accessed August 23, 2013.

Zanatta, M. (2010) "Produtor de soja ameaça recorrer ao Cade contra a Monsanto," *Valor Econômico*, May 19, available at http://www.valor.com.br/arquivo/825143/produtor-de-soja-ameaca-recorrer-ao-cade-contra-monsanto, accessed August 23, 2013.

10

US AGRI-ENVIRONMENTAL POLICY

Neoliberalization of Nature Meets Old Public Management

Steven A. Wolf

Introduction

If we are to make sense of state, market, and society relations, we need to have a sufficient conceptual understanding of our status and trajectories and the associated implications. To the extent that empirical study problematizes, or perhaps contradicts, theoretical framings of the situation, analysts need to take note. Few will object to empirical research aiming to specify what is meant by invoking the much debated term neoliberalism. It is possible that we do not have a satisfactorily coherent, encompassing, shared way to talk about "the state we are in" (Hutton, 1995) or what it is we object to and strive for as citizens and publicly engaged agri-food sociologists.

Through an empirical analysis of agri-environmental policy (AEP) in the United States, I contest totalizing narratives of neoliberalism as a program of institutional change in which market relations come to substitute for bureaucratic calculus and politically derived administrative controls. Rather than emphasize the now familiar insight that the "roll out" of market governance is predicated on an expanded role for state actors rather than a reduced roll (e.g., Heynen et al., 2007), the research highlights the resilience of state-centered governance in the face of reform efforts premised on market logic. I document continued commitment to administrative controls in an era when entrepreneurship and contract theory as represented in the New Public Management (NPM) are ascendant. While focusing on constraints to implementation of neoliberal policy designs, my objective is not to defend the status quo. I seek to problematize the legitimacy of bureaucratic controls derived from what could be considered outdated political alignments and a form of corporatism I will reference here as the "old public management."

In making the empirical claim that a policy domain that has been centered around a relatively closed policy network and state-bureaucratic controls for twenty-five years is, in fact, state-centered, there is a risk of being accused of circular reasoning or a lack of analytical ambition. Therefore, I want to clarify the empirical object under consideration. The focus here is on the fate of market-based policy designs currently being put into circulation by policy entrepreneurs that seek to expand commitments to environmental conservation. These reform proposals are consonant with the ethos of NPM, broad currents in conservation policy toward transactional approaches to restructuring access to land and associated ecosystem services, and long-standing critiques of traditional modes of soliciting public goods from farmers. What I find remarkable is that despite a macro historical context in which we observe significant shifts toward market-rule and sharp and persistent critiques of state-centered controls within the particular context of AEP in the United States, we observe a limited appetite for market discipline. Importantly, the analysis demonstrates that neoliberalism and bureaucratic governance are not bivariate categories that are either obtained or not obtained. AEP has, in the past, incorporated market calculus into policy designs and operating procedures. This process continues, albeit unevenly and indeterminately. Similarly, AEP as it exists and changes over time reflects the varied knowledge, organizational cultures, and political economic standing of a range of actors that define the relevant policy network.

The analysis suggests that the future, like the past, lies in institutional hybridity. Rigid separation of markets from other institutional modes of coordination is a reflection of a stale ideological debate and a product of a false model of social life. The way forward for analysis and social problem solving is a layering and blending of distinct institutional orders. Reclaiming the virtues of politics and bureaucracies would dilute the discursive and political authority of neoliberalism. Making use of the allocative efficiencies that emerge from incorporation of markets into responses to social problems would expand our ability to address information and collective action problems. The crisis of neoliberalism is an institutional crisis, and we need to figure out how to realize synergies by combining markets, state bureaucracies, and community-based forms to address social problems at and across a range of scales. Institutional hybrids, coordination strategies that draw on diverse justifications, and coordination mechanisms, offer promise in terms of constructing governance, accountability, and legitimacy needed to yield coherence. Coherence can be defined in reference to the traditional, static, criteria for assessing performance of institutions: effectiveness, efficiency, and equity. A fourth criterion, evolution, should also be considered, as learning over time is the basis of dynamic coherence.

In the next section, I review leading contemporary efforts to rationalize AEP in the United States through introduction of market-based policy designs. Noting that these neoliberal ideas are *not* being integrated into policy and practice on any significant scale at this time, we are left with important questions about the scope of neoliberalism and constraints to its realization. Why is AEP resistant to strong forms of marketization, and what does this resistance tell us about neoliberalism at this moment? To make headway on these questions, in the next section,

I describe efforts of policy entrepreneurs to introduce heightened market discipline into AEP. I then move on to situate these contemporary projects of policy reform in a historical context through assessment of the origins and evolution of AEP. Discussion then shifts to the theory of NPM to understand the intellectual roots of criticisms of AEP and to specify what we mean when we identify a new policy design as neoliberal. Based on empirical research conducted in Washington, DC, during the current Farm Bill cycle, I identify a set of explanations as to why appeals to imposition of more market discipline have not gained traction within AEP in the United States. In drawing conclusions, I reflect on the concept of institutional hybridity and the potential to realize integration of market-based and bureaucratic approaches to sociotechnical coordination.

Neoliberal Policy Entrepreneurship in US Agri-environmental Policy

Agriculture is the number one cause of water pollution, a significant source of greenhouse gas (GHG) emissions, and as the largest category of land use in the United States, a major determinant of biodiversity and environmental quality more generally. Among a set of environmental nongovernmental organizations (NGOs), think tanks, and academic fields, there is a long-standing and ongoing effort to rationalize design and implementation of US AEP (e.g., Hansen and Hellerstein, 2006; see Withey et al., 2012 for an example of this environmental economics discourse in conservation more generally).[1] Given dim prospects for expanding the amount of money Congress allocates to AEP, the focus is on changing how existing allocations are spent. Actors interested in effectiveness and efficiency are engaged in promotion of policy ideas that will introduce heightened accountability into the way society solicits environmental public goods from farmers. These policy ideas would change how the US Department of Agriculture (USDA) engages farmers and the ways in which farmers engage in conservation, and in this sense, added discipline would be applied at the level of both policy and practice.

Some of the most articulate and energized reform efforts are focused around incentivizing farmers through strategies closely linked to the logic of payment for ecosystem services (PES) (Potter and Wolf, 2013). Ecosystem services are "benefits people obtain from ecosystems" (Millennium Ecosystem Assessment, 2005), and the implied policy argument—that is, self-interest of people, firms, and nations should be at the core of conservation planning and policy, and pricing nature is a means to correct for market failure and achieve desirable allocation of natural resources—is an important focus of the "neoliberalism of nature" critique (Heynen et al., 2007). PES is a transactional conception of conservation in which buyers and sellers engage in exchange (Vatn, 2010). This exchange rests on specification of the objects put into circulation, valuation techniques, and accounting principles. The ways in which PES contributes to marketization/commodification of nature, and the implied disembedding and disenchantment, have given rise to a substantial critical literature (e.g., Sullivan, 2009; Robertson,

2012). While PES is increasingly understood as an ideal type—a species not found in nature—the effort to expand conservation through adoption of transactional discourse is quite real and potentially transformative (Potter and Wolf, 2013).

Relevant AEP policy designs are a departure from paying farmers to implement "best management practices" and a move toward paying for ecosystem service outcomes—for example, improvements in water quality deriving from reductions in nitrogen or phosphorous introduced into the environment relative to some historical baseline (Weinberg and Claassen, 2006). This PES or "pay-for-performance" (Winsten and Hunter, 2011) approach would enhance the productivity of investments in conservation by directing funding to those places in the landscape and to those farmers willing and able to produce the most environmental conservation at the least aggregate cost. Reliance on markets (i.e., social coordination based on interactions among self-interested farmers, each of whom is best positioned to know their resources and constraints) rather than administrative rules regarding program eligibility (i.e., social coordination based on planners' analyses and political economic interactions) is seen as a mechanism to enhance geographic targeting of investments. Farmers would in effect compete with one another to be low-cost producers of public goods. Conservation becomes a strategic profit center rather than a cost category, thereby unleashing the entrepreneurial capacities of farm enterprises. By moving away from a "polluter pays" argument and embracing a "provider gets" approach (Noble, 2010), PES is squarely aligned with the voluntaristic, incentive-oriented, self-organization ethos of neoliberalism.

Beyond tapping into the interests and heterogeneous capabilities of actors and capitalizing on emergent properties derived from markets (allocative efficiency), PES is very much about discipline. The principles of *conditionality* and *additionality* are almost always invoked in specification of what constitutes PES (e.g., Wunder, 2005). Conditionality holds that payments are only awarded when outcomes are achieved. Additionality holds that payments are contingent on evidence of creation of value above and beyond what would accrue under a counterfactual scenario. In other words, the change in status of the environmental variable of interest (e.g., GHG emissions, nitrogen pollution levels, habitat quality) is assessed relative to statistical controls. The relevant policy reform proposals emphasize accountability, as there is a dominant perception that conservation payments up to this point have not insisted on value-for-money. Moving to a transactional model of conservation founded upon market discipline is seen as capable of resolving this problem, thereby bolstering Congressional and taxpayer support for maintenance of public payments to farmers. My suggestion here is that invoking or performing the market is a response to the erosion of legitimacy of paying farmers for conservation. The tests, competition, and selection mechanisms implied by markets are powerful means of discipline.

The disciplinary impulse represented by PES goes beyond extracting more value from those farmers who choose to participate in voluntary conservation programs. There is a specific intention to discipline the public sector actors who fund and administer AEP. The introduction of PES as a focus of reform of AEP derives significantly from criticisms waged on the basis of weak accountability to taxpayers and weak defense of the environment. As perceived, the introduction of policy designs that privilege the productivity of investments will serve to constrain Congress and agency employees in ways that advance cost-effective conservation. As suggested above, there is an established sense that up to this point that payments by USDA to farmers in the name of environment have not always been predicated on commitments to public payments in exchange for public goods. By focusing attention on results and providing public agencies with tools to target the subset of farmers and places on the landscape that justify investment, these market-based approaches are seen as increasing accountability applied to actors soliciting ecosystem services from farmers.

Other contemporary programs of policy entrepreneurship in AEP are nested within the more radical technical and political project of creating markets for provision of ecosystem services (MES) (Daily and Ellison, 2002; Robertson, 2012). As discussed above, within the AEP context, PES represents a framework for contracting between a public agency and a farmer. PES ambitions focused around private sector purchases of conservation and MES are quite different. Parallel with the basic structure of GHG offsets under the Kyoto protocol, the vision is to have farmers and other vendors (e.g., managers of public forests, municipalities) respond to demand from multiple public and private sector buyers of ecosystem services—for example, water quality, climate change mitigation, biodiversity conservation. In such a world, farmers would not have a monopoly on supplying conservation to taxpayers, and USDA would not be the sole commissioner. Such changes would constitute a radical shift in environmental policy and management (Potter and Wolf, 2013). Importantly, unless we imagine firms, entire economic sectors, or commodity chains committing to purchasing environmental offsets on a large scale based on self-interest, creation of private sector willingness to pay could only arise from a process of political mobilization leading to legislating such demand into existence (Buttel, 2003). The establishment of sulphur dioxide trading under the Clean Air Act amendments and the European Trading System established under the Kyoto Protocol are examples of markets created through environmental regulation.[2]

While such a tectonic shift is difficult to imagine in the current US context, Congress did flirt seriously with establishing a cap-and-trade system for GHG in 2009 (i.e., the American Clean Energy and Security Act was passed by the house and defeated in the Senate). This political moment was the impetus for the USDA's creation of the Office of Ecosystem Services and Markets (OEM).[3] This new office is described as having "a unique role in the Federal government's efforts to develop uniform standards and market infrastructure that will facilitate market-based approaches to agriculture, forest, and rangeland conservation. OEM is bringing experts and stakeholders together with government agencies to build a robust, accessible, and scientifically credible market system that will protect and enhance America's natural capital into the future" (USDA-OEM, 2008). While this office is not particularly active at present, the creation of this organization highlights the salience of market discourse in AEP. Agriculture and state agencies very much want to remain part of the evolving discussion around if and how markets will be used to manage the environment.

While grounded explicitly in arguments about creation of public goods—a notion that is contested or at least very narrowly circumscribed by some strands of neoliberalism—these reformist (i.e., rationalize public payments to farmers through expanded reliance on market discipline) and radical (i.e., emergence of ecosystem service markets) policy ambitions reflect the power and status of neoliberal thought (see Busch's contribution to this volume for a specification). Neoliberalism, as a reference to a successful (i.e., widely diffused) political economic philosophy, has transformed many aspects of state-society relations and popular culture. My aim in this paper is to analyze how it is that neoliberal policy designs bolstered by well-established academic disciplines and circulated by well-connected policy entrepreneurs *have not been enacted* within AEP.

AEP under the Old Public Management

The PES-inspired policy ideas outlined above are importantly different from the historically structured and politically embedded mode of AEP that has been in place since the mid-1980s. To appreciate these differences, we must reflect on this formative period. Conservation programs in agriculture are multifunctional due to the political circumstances in which they were enacted. Introduction of payments to farmers for conservation in the watershed Farm Bill of 1985 made it possible to sustain subsidies to farmers' incomes while responding to demand to liberalize agricultural production (i.e., cut commodity payments or at least decouple production and conservation payments) and demand to address environmental degradation (Potter, 1998). AEP arose out of this political conjuncture, and the character of the current administrative procedures and the distribution of payments continue to reflect this history.

As argued elsewhere, Potter and Wolf (2013, p. 5) state:

> The resulting agri-environmental policy and rural development programs reflect profoundly the historical-institutional context in which they were made (Sheingate, 2000; Potter, 1998). The major source of funding for environmental management on private lands, they stand as classic examples of the second-best, politically compromised arrangements that policymakers and those who lobby them invent in order to address a range of policy problems—in this case rural environmental degradation, the support of farmers' incomes and the overproduction of agricultural commodities. Indeed, from the beginning, agri-environmental payments to farmers have served purposes other than conservation, and their wide scope and the trade-offs they embody are important explanations of their durability. For example, the Conservation Reserve Program (CRP), the most expensive federal agri-environmental program enrolling roughly

10% of all cropland, serves both environmental conservation and supply control goals while contributing $2 billion annually to farm incomes (Reichelderfer and Boggess 1988).

Because the programs that make up AEP were not designed to be cost effective relative to environmental protection goals, they are not. For example, the Environmental Quality Incentives Program (EQIP), the core conservation initiative focused on land in production, is required by Congress to spend 60 percent of program funds on livestock operations, which in practice results in public payment for manure containment facilities for large-scale concentrated animal feeding operations (CAFOs) (Noble, 2010). While the program is based on a bidding procedure whereby farmers nominally compete for available funds, the allocation to animal agriculture was an outcome of a political dynamic (Martin, 2008). Such rigidities explicitly limit capacity of frontline bureaucrats from optimizing allocation of money in their budgets. Additionally, under the 2008 Farm Bill, EQIP prohibits USDA from awarding conservation contracts to farmers who are willing to accept lower payment than that paid to other farmers for similar contracts. Again, we see explicit instances of market-defying logic embedded in policy and program implementation.

A final example derives from the original manifestation of the CRP. In 1985, USDA began to take farmland out of production through voluntary ten-year leases on what amounted to a first-come, first-serve basis. This approach stands in contrast to strategies that would prioritize farmland that could be rented at lowest cost or farmland that would produce the greatest environmental benefits. While the original program featured eligibility requirements, these coarse criteria came to be recognized as deficient. In response to critics, the USDA went on to develop their Environmental Benefits Index (EBI), a scoring algorithm that allowed them to rank acreage and enroll land more strategically. The EBI was an ad hoc bureaucratic planning tool that was updated more than a dozen times—adding new ranking criteria and reweighting them—depending on common wisdom and political winds. Some years, "prairie potholes" and duck hunting in South Dakota emerged as a priority. Other years, the formula favored bottomland hardwood wetlands in southern states. While the idea of leasing land to address environmental problems and the bidding procedures USDA developed over time to increase the productivity of these investments are neoliberal in character, I want to emphasize the agency of the state agency. USDA, in partnership with farm organizations and conservation NGOs and under the oversight of Congress, administered the CRP and the other AEP programs according to a politically negotiated bureaucratic logic.

Critique of AEP derives from many quarters, and it is not new. Batie (2009) offers a review and update of the traditional critique founded upon the work of an entire generation of agricultural/environmental economists, and she voices little optimism regarding institutional innovations that would significantly expand economic rationalization and effectiveness. Since the inception of contemporary AEP in the landmark 1985 Farm Bill, the continuation, specific aims, and size of the programs that make up the AEP portfolio have been actively debated. Yet, more than twenty-five years later, the basic structure of AEP remains in place. The justifications, general format, and contradictions have proven to be highly durable. This continuity is noteworthy for what it says about agriculture, the political economic status of the project of conservation on private lands, and the capacity of neoliberal thought to transform historically structured domains of knowledge, practice, relationships, and privilege. The embedded character of AEP has preserved central aspects of a tradition of old public management.

New Public Management, in Theory

We can understand the emergence of policy designs aimed at reforming AEP through reflection on broader tendencies in public policy as represented by new public management (NPM). This body of ideas emerged in the fields of political science and public administration in the early 1990s, and Hood (1991) is generally regarded as the key intellectual.

Mio Valbo (2009) offers a useful institutional analysis of NPM. In this account, NPM thinking combines new institutional economics, specifically emphasizing information economic and incentive structures, with contemporary conceptions of management. New institutional economics is reflected in NPM's insistence on "the purchaser-provider split, service specifications and contracts" (Valbo, 2009, p. 3, quoting Hardy and Wistow, 1998). Each of these three principles is of direct relevance here. With respect to the first criteria, the introduction of PES-like schemes into AEP would disrupt the solidarity USDA has enjoyed with farmers and agri-food more generally. Rather than serve in a role of modernizer, booster, and political representative, USDA would be responsible for disciplining farmers—making payments contingent on outcomes—and this would disrupt relations USDA have cultivated over decades. Conservation programs are relentlessly presented as voluntary, and there is a steadfast commitment to a relationship with farmers premised on incentive payments and technical assistance. USDA does perform the role of regulator in some specific instances—for example, enforcing "cross-compliance" (i.e., by law, eligibility of farmers for production subsidies is contingent on their meeting minimum standards for soil and wetland conservation). Yet, it is clear that USDA prefers to avoid enforcement functions. Competition within the state, specifically with the Environmental Protection Agency, an agency committed to regulation, can help us understand how it is that USDA has accepted limited enforcement duties over time.

Valbo's second and third principles similarly speak to interest in extracting efficiencies and ensuring outcomes through discipline. Commitment to specifying expected outcomes in greater detail within contracts is a key basis of accountability. In order to insist on value-for-money and in order to make a credible threat of withholding payment in the event effort and/or performance are lacking, it is necessary to define and measure what is being transacted. This facet of NPM logic can be identified in the rise of mandatory testing of students in public schools and linkages between school funding and students' performance on standardized tests. Similarly, we see antirecidivism programs being structured on a pay-for-performance basis. Contract theory is a central reference within contemporary public administration.

The modernist conception of management within NPM thinking stresses the capacity of managers to create value alongside and in tension with traditional Taylorist controls. Empowering managers (i.e., freeing them from controls of bureaucracies) and subjecting them to incentives and risks is viewed as a critical source of productivity gains. According to Valbo, "the model focus(es) on outputs instead of in-put and thereby entails an implicit critique of the Weberian rule-bound bureaucracy. The underlying idea is that decreasing emphasis on ex ante and processual controls over public sector managers would be balanced by increased emphasis on ex post evaluation of results, creating more discretionary space for managers to add value to public services" (2009, p. 5). Managers in the context of my analysis refers to USDA Natural Resource Conservation Service staff (NRCS) who interpret the policies legislated by Congress, produce the administrative rules, and implement conservation programs. Of course, these actions are done in consultation with actors in other public agencies and in commercial and nonprofit organizations occupying positions in the AEP policy network.

The same emphasis on empowering managers to be low-cost, high-quality producers of ecosystem services also applies to farmers and land managers in PES-like schemes. Rather than specify the technical practices farmers must follow in order to qualify for incentive payments, PES-like schemes emphasize flexibility, context specificity, creativity, and above all, entrepreneurship:

> Picture this. It's late winter, and Mr. Brown, an Iowa farmer with 283 ha (700 ac) of corn and soybeans and a midsized hog operation, is planning for the upcoming crop season. He wants to participate in a conservation program, but he's not looking through a long list of predefined practice standards, trying to find a few that would fit on his farm. Instead, he's analyzing his operation, identifying where soil and excess nutrients might be leaving his fields, and thinking of changes he could make to stop those losses. Why? Mr. Brown is not preparing for a traditional conservation program. Instead, he's preparing for a pay-for-performance conservation program, which pays farmers by the pound for quantifiable reductions in sediment and nutrient losses. (Winsten and Hunter, 2011, p. 111A)

AEP is not Neoliberal (or at Least as Neoliberal as It Could Be)

A core belief of neoliberalism and NPM is that given the right information and incentives, people and organizations "can deliver socially desirable outcomes without reliance either on direct political control or on any assumption that they are committed to some concept of public service" (Quiggin, 2003). In linking the two concepts, it is important to recognize that NPM is about achieving efficiencies in public administration by freeing up public sector managers, while neoliberalism is classically about shackling the public sector to preserve personal freedoms (Busch, 2011). While quite different in inspiration, they have each become important resources in projects to introduce market and quasi-market strategies into public policy. But when we examine AEP in the United States, we find that these overarching tendencies have been thwarted.

The project of introducing competition among farmers, shifting USDA's relation with farmers from booster to principle (vis-a-vis a principle-agent transactional framing) and breaking the monopoly USDA holds on contracting with farmers for provision of environmental public goods has, up to this point, largely been stifled. It is clear that the AEP of the past thirty years has not been a program of regulation, and it has, over time, incorporated significant elements of market calculus (e.g., bidding procedures employed by NRCS in allocating conservation contracts to farmers). We could, on this basis, conclude that AEP is neoliberal in character. I contend, however, that a more useful reading of the case lies for my purposes in focusing on the continuity and resilience of planning procedures and administrative controls of a sprawling state agency, USDA NRCS. In other words, the case of AEP in the United States is one in which neoliberalism has in significant ways gone unrealized. Perhaps we can even say that market discourse has been resisted and co-opted by public agencies and other actors in the relevant state-centered policy network. According to Selznick, "Cooptation is the process of absorbing new elements into the leadership or policy determining structure of an organization as a means of averting threats to its stability or existence" (Selznick, [1949] 1965, p. 13).The AEP policy network has selectively performed market governance largely on its own terms.

Rather than conclude that US AEP is an isolated backwater or that it highlights that neoliberalism is not deserving of popular and academic pronouncements regarding its status as a *fait accompli*, I argue that the case helps us understand institutional hybridity. As conceptualized in neoclassical terms, market rule is a myth (see Busch this volume; also Allaire and Wolf, 2004). Like all myths, it is powerful. It provides a useful way to think about and talk about the world. But it is not a comprehensive or coherent account of the world. AEP retains strong traditions of bureaucratic control and negotiations within and across policy networks. This mode of governance can be understood as corporatism, or what Dryzek (1997) has analyzed as "democratic pragmatism," a process of semi-inclusive deliberation among stable networks of elites.

Gaps in Knowledge

Market rule is not possible in this setting because property rights and satisfactory equivalencies—the foundation of market transactions—cannot be defined due to weak alignment of actors' interests and significant gaps in scientific knowledge.[4]

For example, in the scenario above, it is possible to imagine reliance on models, rather than direct measurement, to calculate nitrogen pollution outcomes. The model would need to determine the ecological responses to a range of specific changes in cropping practices, such as planting vetch as a winter cover crop or side dressing (i.e., not applying all fertilizer early in the season and waiting until the corn is well established to provide final feeding), accounting for soil type, landscape position, cropping history, yield, weather, and a range of additional variables. These relationships are not fully understood by scientists, particularly across the range of heterogeneous conditions found within and across landscapes. Because there is no broadly shared agreement about what counts and how to count what is thought to count, there is no clear basis for defining rights and responsibilities. Lack of agreement regarding how to account for variation in quality—quality of land, quality of practices, quality of verification strategies—stands in the way of efforts to place AEP on a transactional basis.

Property Rights

Market transactions are predicated on property rights and contracts. Technical and conventional understanding does not exist at this time to support moving AEP to a transactional basis. The interrelated problems of knowledge gaps and specification of property rights are highlighted when we consider that land generates multiple ecosystem services. Planting a cover crop or changing the timing of fertilizer treatments can generate climate change mitigation, water quality, and biodiversity benefits. Devising accounting procedures and contract design to deal with this "stacking" of ecosystem services is proving to be a significant constraint to realization of PES and MES. If an electric utility pays a farmer to offset their GHG emissions, and the farmer plants a nitrogen-fixing cover crop to reduce their reliance on inorganic nitrogen fertilizer, this change in cropping system could generate water quality and biodiversity benefits. How will these spillovers be addressed? In principle, the beneficiaries of each of these ecosystem services should share responsibility with the electric utility for paying for the production of ecosystem services. The knowledge base to support such transactions—and the platforms to allocation of costs and benefits and how to brook technical uncertainties—does not exist.

Local Political Economy and Organizational Friction

A program of subjecting farmers to competition and the accountability of outcome-based policies, and the more ambitious vision of creating markets for ecosystem services from agricultural landscapes, is not implementable in the current context because of political constraints. Most fundamentally, members of Congress are not willing to contemplate risks to reelection prospects that would accompany a geographic redistribution of conservation payments stemming from a move to outcome-based policies. Creation of ecosystem service markets and paying-for-performance (Winsten and Hunter, 2011) would almost certainly reshape the distribution of which Congressional districts, which commodities, and which farms receive payments. The pattern of allocation of federal conservation payments is generally well established, and this income stream is incorporated into land values (Wu and Lin, 2010). These are structural impediments to reform that counterbalance impulses toward market rule.

At the departmental level, USDA—and NRCS in particular—are not eager to contemplate the fallout of subjecting farmers to heightened accountability pressures. The voluntary nature of conservation programs leads to a self-selection dynamic whereby a certain class of farmers chooses to work with NRCS. NRCS employees have personal and professional relationships with these farmers in the individual counties where they live and work. Changing the rules and moving to a more transactional mode of conservation will strain local relationships. More broadly, farmers have indicated their opposition to the surveillance (i.e., loss of privacy) associated with on-farm monitoring and the data reporting needed to make payment conditional on outcomes and to run the imagined models that would generate proxies for on-farm measurement. The arm's-length approach to verification in place at present suits the interests of NRCS and farmers quite well.

USDA also confronts the internal frictions that would accompany a shift in organizational culture required to move from a "boots-on-the-ground" mentality to that of data-intensive, science-based models needed to contract for ecosystem services. The knowledge base on which conservation policy rests is, in important ways, derived from practical experience. The much touted Conservation Effectiveness Assessment Project (CEAP) is an effort to verify and improve the science base on which USDA conservation programs are predicated. This research effort can be seen simultaneously as feeding into efforts to move toward more outcome-based AEP and a rear-guard defensive action to derail criticisms wages on the basis of effectiveness.

Lastly, it appears that the US public, and more specifically Congress, are not prepared to regulate polluters so as to require them to purchase environmental offsets from farmers for water quality degradation, GHG emissions, biodiversity losses, and so forth. Yet nothing short of such legislation will lead to meaningful private sector demand for ecosystem services. There are voluntary programs in place, but corporate social responsibility must be viewed critically as a significant source of public goods creation. Some NGOs and private foundations are funding experimental projects to induce supply of ecosystem services (Majanen et al., 2011). Again, in the current political reality, this investment stream cannot constitute an important substitute, or even complement, to public funding of conservation.

Of these three complementary explanations as to why market rule has not been implemented in AEP, the first two are widely acknowledged in the literature (Batie, 2009). Transaction costs and information problems are anticipated by economic theory. A range of pilot projects and empirical applications confirm that costs of contracting and monitoring and lack of clarity regarding rights and responsibilities are difficult hurdles to overcome. The third argument is less well understood. Institutional inertia, switching costs, the power of incumbents in policy networks, and marked indifference of actors inside the state to opportunities for rationalization have not been emphasized in discussions of development and implementation of market-based environmental policy instruments. In the literature on PES, room to maneuver and resistance strategies of farmers and local people have been addressed (e.g., Higgins et al., 2012), but the role of policy actors has not attracted much attention (see Wynne-Jones, 2012 for an exception). We view these as exciting lines of inquiry.

Examining AEP through a historical lens allows us to appreciate the political economic context of these commitments to conservation. As noted above, these policies have not been designed to maximize productivity of investments in conservation because political support for them does not derive strictly from actors motivated by ecological considerations. Examining AEP through an organizational lens allows us to appreciate how local resource considerations and culture mediate a hypothetical embrace of market discipline. NRCS and the broader AEP policy community, including the relevant farmers, constitute an epistemic community, or a community of practice. In this community, bureaucratic administrative tradition runs deep. Despite vigorous promotion of PES-like schemes by able policy entrepreneurs, active experimentation by USDA on outcome-based approaches to commissioning the provision of conservation and the overarching discourses of neoliberalism and new public management, AEP continues to be structured and practices largely as it has been for twenty-five years. The justifications, assumptions, and knowledge base remain intact. The federal government is the sole buyer of conservation. Farmers have a monopoly on provision. Payments are not based on individuals' willingness to accept (i.e., price discovery does not occur according to market logic). Solidarity among farmers and between farmers and the USDA is emphasized. This continuity in the face of marketization pressures is remarkable.

Discussion and Conclusion

On one hand, neoliberalism has been rolled out in some domains of agri-food system regulation, and it has failed to deliver the goods (e.g., food security, public health). And on the other, neoliberal reforms have not been implemented rigorously in domains of production (i.e., production subsidies persist, an issue not addressed in this paper) and environmental management, and in this sense, neoliberalism has not lived up to its promises, again.

Despite the success of neoliberal ideology and the sleek elegance of concepts from the NPM, transaction costs, science gaps, and political constraints limit possibilities of bringing market discipline to bear on farmers, agency personnel, and policy makers. From the perspective of accountability to taxpayers (efficiency) as well as ecological conservation (effectiveness), realization of the ideas of paying for outcomes, putting farmers into competition with one another as rivals in production of low-cost ecosystem services, freeing up/disciplining policy actors to ensure productivity of investments are maximized, and bringing private sector money into conservation are very attractive propositions. Such institutional changes could potentially produce more environmental public goods at lower aggregate cost than the current system of political and bureaucratic governance. This statement is not a celebration of markets, and it is certainly not a statement of faith about markets as democratic, natural, or self-correcting. Rather, it is a recognition that the status quo—policy instruments reflecting political negotiations of twenty-five years ago and reliance on an outdated knowledge base—is hard to defend. Opposition to market reforms does not constitute a positive program.

So, rather than champion market discipline or celebrate state controls, I suggest development of a critical, pragmatic, adaptive approach that recognizes and makes use of state, market, and collective coordination mechanisms. In rare cases, these modes of action can be applied unalloyed or undiluted, I assume, although I am hard pressed to think of a single example. Generally, these institutional orders are combined in various permutations and through application of varied weights (Hollingsworth and Boyer, 1997). This process of hybridization is, in fact, how governance works in practice (Allaire and Wolf, 2004; Wolf and Hufnagl-Eichiner, 2007). By demonstrating the existence and the desirability of hybrid institutional arrangements, it is possible to move beyond neoliberalism as a claim about the superiority of markets as allocative instruments and as a normative prescription for how we should pursue sustainable development.

Acknowledgment

This research benefited from contributions by Seth Shames of EcoAgriculture Partners.

Notes

1 AEP is the set of national programs legislated in the Conservation Title of the Farm Bill, the agricultural policy that is revisited by Congress roughly every five years. These programs are administered by the USDA Natural Resources Conservation Service, which has a presence in almost all of the three thousand counties in the country. While one could include the relevant policies of state departments of agriculture, I do not do so in this paper.

2 Science never provides complete and fully integrated information, yet disciplinary closure and stabilization of beliefs are observed in some of the cases and not in others. In considering environment, broadly, it is interesting to reflect on how it is that the ambiguities of carbon accounting have been negotiated away to an extent that markets exist. Does political necessity override scientific gaps? Is knowledge closure around carbon flows more complete than around questions of, say, nitrogen or native pollinators? If and why equivalencies are proving difficult to negotiate in US AEP is a question that merits further research.

3 Since renamed the USDA Office of Environmental Markets (OEM).

4 There is, of course, significant potential for fraud and nonsensical scientific claims, as has been reported in the GHG offset market created under the neoliberal Kyoto Protocol (Lohmann, 2008). While it would be naive to believe that that these three developments would perform as advertised, one can imagine them improving productivity of investments and outcomes relative to the status quo.

References

Allaire, G. and Wolf, S. (2004) "Cognitive representations and institutional hybridity in agrofood innovation," *Science, Technology* & *Human Values*, vol 29, no 4, pp. 431–58.

Batie, S.S. (2009) "Green payments and the US Farm Bill: Information and policy challenges," *Frontiers in Ecology and the Environment*, vol 7, no 7, pp. 380–8.

Busch, L. (2011) *Standards: Recipes for Reality*, MIT Press, Cambridge, MA.

Buttel, F. (2003) "Environmental sociology and the explanation of environmental reform," *Organization & Environment*, vol 16, no 3, pp. 306–44.

Daily, G. and Ellison, K. (2002) *The New Economy of Nature*, Island Press, Washington, DC.

Dryzek, J. (1997) *The Politics of the Earth*, Oxford University Press, Oxford.

Hansen, L. R. and Hellerstein, D. (2006). *Better Targeting, Better Outcomes*, US Dept. of Agriculture, Economic Research Service, Washington, DC.

Heynen, N., McCarthy, J., Prudham, S., and Robbins, P. (2007) *Neoliberal Environments: False Promises and Unnatural Consequences*, Routledge, New York.

Higgins, V., Dibden, J., and Cocklin, C. (2012) "Market instruments and the neoliberalisation of land management in rural Australia," *Geoforum*, vol 43, pp. 377–86.

Hollingsworth, J. and Boyer, R. (1997) "Coordination of economic actors and social systems of production," Pp. 1–49 in *Contemporary Capitalism: The Embeddedness of Institutions*, Cambridge University Press, Cambridge.

Hood, C. (1991) "A public management for all seasons?," *Public Administration*, vol 69, pp. 3–19.

Hutton, W. (1995) *The State We're in: Why Britain Is in Crisis and How to Overcome It*, Jonathan Cape, London.

Lohmann, L. (2008) "Carbon trading, climate justice and the production of ignorance: Ten examples," *Development*, vol 51, pp. 359–65.

Majanen, T., Friedman, R., and Milder, J. (2011) *Innovations in Market-Based Watershed Conservation in the United States: Payments for Watershed Services for Agricultural and Forest Landowners*, EcoAgriculture Partners, Washington, DC.

Martin, A. (2008) "In the Farm Bill, a creature from the Black Lagoon?" *New York Times*, January 13, http://www.nytimes.com/2008/01/13/business/13feed.html, accessed May 27, 2013.

Millennium Ecosystem Assessment. (2005) "Ecosystems and human well-being: Biodiversity synthesis," available at http://www.millenniumassessment.org/documents/document.354.aspx.pdf, accessed May 27, 2013.

Noble, M. (2010) "Paying the polluters: Animal factories feast on taxpayer subsidies," Pp. 221–31 in D. Imhoff (ed.) *CAFO Reader: The Tragedy of Industrial Animal Factories*. Earth Aware, San Rafael, California.

Potter, C. (1998) *Against the Grain: Agri-environmental policy in the US and the EU*, CABI, Wallingford.

Potter, C. and Wolf, S. (2013) "Payments for ecosystem services in relation to US and UK agri-environmental policy: Disruptive neoliberal innovation or hybrid policy adaptation?," unpublished manuscript, Imperial College, London.

Quiggin, J. (2003) "Word for Wednesday: Managerialism (again)," John Quiggin, http://johnquiggin.com/2003/07/16/word-for-wednesday-managerialism-again/, accessed July 16, 2003.

Robertson, M. (2012) "Measurement and alienation: Making a world of ecosystem services," *Transactions of the Institute of British Geographers*, vol 37, pp. 386–401.

Selznick, P. ([1949] 1965) *TVA and the Grassroots: A Study in the Sociology of Formal Organization*, University of California Press, Berkeley.

Sullivan, S. (2009) "Green capitalism, and the cultural poverty of constructing nature as service provider," *Radical Anthropology*, vol 3, pp. 18–27.

USDA-OEM. (2008) "USDA Office of Environmental Markets," USDA Forest Service, http://www.fs.fed.us/ecosystemservices/OEM/index.shtml, accessed May 28, 2013.

Valbo, M. (2009) "New Public Management: The neoliberal way of governance," http://thjodmalastofnun.hi.is/sites/default/files/skrar/working_paper_4–2009.pdf, accessed May 28, 2013.

Vatn, A. (2010) "An institutional analysis of payments for environmental services," *Ecological Economics*, vol 69, pp. 1245–52.

Weinberg, M. and Claassen, R. L. (2006) *Rewarding farm practices versus environmental performance*, US Dept. of Agriculture, Economic Research Service, Washington, DC.

Winsten, J. and Hunter, M. (2011) "Using pay-for-performance conservation to address the challenges of the next farm bill," *Journal of Soil and Water Conservation*, vol 66, no 4, pp. 111–17.

Withey, J., Lawler, J., Polasky, S., Plantinga, A., Nelson, E., Kareiva, P., Wilsey, C., Schloss, C., Nogeire, T., Ruesch, A., Ramos, J., Jr., and Reid, W. (2012) "Maximizing return on conservation investment in the conterminous US," *Ecology Letters*, vol 15, pp. 1249–56.

Wolf, S. and Hufangl-Eichiner, S. (2007) "External resources and development of forest landowner collaboratives," *Society and Natural Resources*, vol 20, no 8, pp. 675–88.

Wu, J. and Lin, H. (2010) "The effect of the Conservation Reserve Program on land values," *Land Economics*, vol 86, no 1, pp. 1–21.

Wunder, S. (2005) *Payments for Environmental Services: Some Nuts and Bolts*, CIFOR, Occasional Paper No. 42.

Wynne-Jones, S. (2012) "Negotiating neoliberalism: Conservationists' role in the development of payments for ecosystem services," *Geoforum*, vol 3, pp. 1035–44.

11

FOR COMPETITIVENESS' SAKE?

Material Competition vs. Competitiveness as a National Project

Anouk Patel-Campillo

Introduction

Based on neoliberal ideologies of market-led economic integration and the intensification of global commodity flows, national governments strive to increase the international competitiveness of their economic sectors as a means to generate revenue.[1] In the agricultural sector, the quest to create or strengthen agro-export industries has intensified apace with globalization. Within this global context, developing country agro-industries in world markets often compete based on low-cost, increased productivity and the flexibilization of employment practices. In developed countries, competitiveness often rests on a combination of increased productivity, innovation, and product differentiation. Since the 1990s, the rapid incorporation of national economic sectors into world markets increased competitive pressures whereby the low-cost and differentiation strategies practiced by economic actors have become increasingly insufficient for gaining or maintaining competitiveness internationally. While economic actors respond to rising competitive pressures by reorganizing their economic activities, competitiveness as a national project requires governments to deploy their functions and resources to advance the position of domestic industries in world markets. Here, the notion of competitiveness goes beyond the habitual interactions among rival firms or industries to become a national project in which competitiveness is equated with national economic well-being. Within a neoliberal regime characterized by the so-called retreat of the state, national governments often rely on the creation of mobilizing mechanisms to justify and legitimize active intervention in economic matters and the marketization of their economies. "Competitiveness" is one such mechanism.

Linking competitiveness with export-led growth governments and international institutions, such as the World Bank and the International Monetary Fund, instills a sense of urgency to lend legitimacy, to competitiveness as a national project and facilitate its diffusion. Nonetheless, because export-led development models have had mixed reviews at best, their worn-out discourse has become insufficient to encourage the uptake of neoliberal practices. Nevertheless, national governments continue to endeavor to tightly bind and deepen the insertion of their national economic sectors into world markets, despite evidence of the perils associated with export-oriented development strategies (Barham et al., 1992; Bonanno, 1996; Collins and Krippner, 1999; Schurman, 2001; Kritzinger et al., 2004; Selwyn, 2007; Patel-Campillo, 2010a; Shaw, 2011). For former colonial empires like Holland, Britain, and the United States, capitalist forms of governance largely evolving from their historical trajectories as world hegemons (Arrighi, 1994) may account for their economic prowess (regardless of geographic size) and their preference for free markets. With the intensification of global commodity flows, however, developed countries are also engaged in national projects of competitiveness to maintain or strengthen their position in world markets.

Competitiveness as a national project moves beyond material competition to establish more permanent norms and practices that crystallize and entrench neoliberal ideologies across space and scale. Therefore, the deepening of the neoliberal regime depends not only on the legitimacy commanded by competitiveness as a national project but its uptake across scales of governance for its materialization. In the study presented here, the agency of national governments partially lies in the ability to utilize resources under their purview to embed more pervasive forms of neoliberal practices across scales of governance. Here, the agency of subnational units of governance lies on the degree to which these units adopt or contest this national project. By adopting this project, subnational units of government materialize neoliberal incursions across scales. These incursions create, modify, or eradicate the institutional architectures that uphold (or potentially weaken) the norms and practices associated with the neoliberal regime. For instance, the redrawing of national architectures might involve the dissolution or modification of national departments of labor, the creation of export promotion entities, and budgetary distributions among governmental entities. Because state architectures sediment over time, the adoption of competitiveness as a national project provides relative durability to the neoliberal regime across time and scales of governance.

Taking into account contextual specificities, in this chapter, I examine the deployment of competitiveness as a national project by the Colombian and Dutch national governments. I use an incorporated comparison (McMichael, 1990)[2] of one commodity—cut flowers—in two differentiated contexts, Colombia and the Netherlands,[3] to explore how legitimation plays a role in the diffusion and uptake of neoliberal practices at the subnational scale. Because mobilizing mechanisms and responses to them are dynamic, the findings presented here provide only a snapshot of how national governments in Colombia and the Netherlands seek to materialize this project in relation to their respective flower agro-industries. In the next section, I differentiate material competition or the habitual interactions between and among industries and firms vis-a-vis their rivals and competitiveness as a national project involving the diffusion of neoliberal ideologies aimed at shaping state architectures across space. Making such a differentiation is of importance as it places the scope and range of material competition within the purview of industries and firms compared to that of governments. To situate this analysis, I engage in a brief discussion of classical economics and more recent perspectives on the competitiveness of national industries in world markets. Then, I briefly examine scholarly discussions on the state in the context of globalization to illustrate that far from retreating, state control over institutional architectures is essential to diffuse and entrench neoliberal norms and practices across scales. I end by analyzing how competitiveness as a national project is deployed by the Colombian and Dutch governments using the example of their respective cut flower agro-industries.

This chapter makes the following methodological and policy-related contributions. First, the use of an incorporated comparison method illuminates the ways in which the Colombian and Dutch cut flower agro-industries are interrelated through their world market relations. Because these market relations are dynamic, generalization of an incorporated comparison is "historically contingent because the units of comparison are historically specified . . . [as the] . . . comparison becomes the substance of the inquiry rather than its framework" (McMichael, 1990, p. 384). Second, this chapter makes a policy-related contribution by suggesting that comparative research opens the possibilities to locate points of fracture or entry (such as a crisis of legitimacy) where governments face a more difficult task in coordinating the neoliberal regime. Based on the narrow view that export-led development can be attained by producing particular high-value commodities (i.e., cut flowers), a third contribution of this chapter is to call into question the analytical assumption that particular commodities are necessarily associated with certain types of chain governance, and thus, development prospects. This analytical decoupling has policy implications, as national governments and international entities such as the World Bank often promote the production of high-value export commodities for northern markets as a means to move up the ladder of development and create a "trickle-down" effect. However, because historical and contextual specificities influence the ways supply chains are formed and evolve over time, "moving up the chain" or producing high-value export commodities does not automatically guarantee improved opportunities for development.

Globalization and the Competition State

The role of the state in an era of globalization and economic integration has spurred scholarly debate. On the one hand, there are those who proclaim the demise of the state in a borderless world (Ohmae, 1990; Horsman and Marshall, 1994). Others reaffirm the relevance of the state, conceding that state power has eroded but contending that we are far from a fully globalized world (Hirst and

Thompson, 1999). Still others highlight "the adaptability of states, their differential capacity, and the enhanced importance of state power in the new international environment" (Weiss, 1997, p. 4). Clearly, although economic integration creates different types of pressures on the state, rather than weakening, the state transforms (Shaw, 1997, p. 498). As such, the state is dynamic in that it assumes various—at times contradictory—roles. While often actively supporting economic interests, the role of the state is weakened as a mediator between economic and societal interests, leading to what Bonanno calls "the crisis of democratization" (Bonanno, 1998). As such, the state is in a "crisis of political representation," whereby it is "increasingly unable to represent the will of its people" (Bonanno, 1998, p. 232). Captured by the narrow interests of economic actors, the state transforms or modifies its institutional architectures to further the interests of economic actors.

Similarly, rather than witnessing the eclipse of the state, Evans (1997) argues that the perceived weakness of the state can be interpreted as a change in its nature as it plays a key role in the reconfiguration of economic activity through its material functions (i.e., domestic and international policies). Evans (1997) finds that in the context of increased global trade, the rise of transnational corporations, deregulation of financial capital, and the roles of the state are pivotal for minimizing risk and protecting the interests of economic actors. Aligning with scholars in the regulation tradition (Hollingsworth and Boyer, 1997), Evans concludes that "[the state] may actually be a source of competitive advantage in a globalizing economy" (1997, p. 69). Equally important is Evans's argument that changes in the role of the state also have an ideological dimension, which facilitates the global diffusion of hegemonic neoliberal Anglo-American ideology and the process of capital accumulation and concentration.

Cerny (2007) agrees that in the context of globalization the state transforms, producing what he calls the "competition state." He argues that in this process, significant contradictions emerge. First, in the refashioning of state-market relations, state intervention and regulation become even more pivotal "in the name of competitiveness and marketization" (Cerny, 2007, p. 251). And second, in the interaction between economic transformation and the role of the state, the state itself is restructured, generating tensions between its role in fostering the public good and the demands of the competition state. The function of the competition state is to "pursue increased marketization in order to make economic activities located within the national territory, or which otherwise contribute to national wealth, more competitive in international and transnational terms" (Cerny, 2007, p. 259). Therefore, the competition state relies on the identification of national industries with strong international market presence to justify government intervention to shape economic outcomes.

In the varieties of capitalism approach, Hall and Soskice (2001), further distil the relationship between the competition state and economic actors by arguing that particular state architectures facilitate specific types of firm coordination strategies, which influence competitiveness and national economic performance

(p. 5). Hall and Soskice go as far as to claim "that national political economies can be compared by reference to the way in which firms resolve the coordination problems they face" (2001, p. 8). They support this contention by comparatively analyzing the ways firms organize; they find that one set of "firms coordinate their activities primarily via hierarchies and competitive market arrangements" (Hall and Soskice, 2001, p. 8), a type of organizational behavior that they link to liberal market economies. Another set of firms depend "more heavily on non-market relationships to coordinate their endeavours" (Hall and Soskice, 2001, p. 8); they identify this firm coordination strategy as an example of a coordinated market economy.

Hall and Soskice (2001) offer the concept of "comparative institutional advantage" to indicate that particular state architectures tend to generate identifiable firm coordination strategies. They contend that particular institutional architectures lend firms the ability to engage in specific activities within that context so that "[F]irms can perform some types of activities, which allow them to produce some kinds of goods, more efficiently than others" (p. 37). Here, particular institutional architectures are more conducive for firms to pursue certain types of coordination strategies (Storper and Salais, 1997), whereas others do not. While these analyses vary in their appraisal of the degree of entrenchment of economic interests and what renders the state a means to advance them, what is clear is that governments are increasingly implicated in the diffusion, entrenchment, and materialization of firm/industry competitiveness. Having explored the role of the state within the context of globalization and economic integration, in the next section, I briefly trace the idea of competition and competitiveness to draw distinctions between what I recognize as material competition and competitiveness as a national project.

From Smith to Porter: Delineating Material Competition

Competition, comparative advantage, and competitiveness have long been utilized to explain and organize economic activity. Based on the examination of businesses and entrepreneurs, Adam Smith first developed the ideas of competition, resource allocation, and efficiency in the field of economics. In *The Wealth of Nations* (1776), Smith argues that competition among entrepreneurs and businesses, while rooted in self-interest, can ultimately lead to broader societal benefits by generating lower prices and incentivizing economic actors to engage in diversified economic activities with the potential to generate value added (Busch, 2011). Although economists following Smith endeavored to differentiate imperfect versus perfect competition, the assumption of perfect competition and its normative use has prevailed over imperfect competition, especially in policy-making circles. Building on Smith's ideas, David Ricardo ([1911] 2004) broadened the scope of analysis to encompass international trade in a free-market context by examining bilateral trade in wine and cloth between England and Portugal, two colonial powers. In the context of free trade, he argued, value would be best accrued in a trading system that promoted specialization of production to lower costs. Currently, in export sectors, cost reduction includes the use of labor-displacing technology, increased productivity per worker, and the flexibility of employment practices.

Taking a more prescriptive approach towards improving the competitiveness of national industries in self-correcting global markets, Michael Porter (1990) identifies four determinants of competitive advantage (in what he calls the diamond model).[4] Here, I focus on the first determinant: firm strategy, structure, and rivalry. Unlike many of the scholars cited previously, Porter does not consider government as a source of comparative advantage. Instead, Porter's model places industry and its related activities at the core of competitive advantage, while "chance" and "government" are both considered peripheral elements that influence the four determinants. Although there are several factors that distinguish it, Porter's model builds on Smith's ideas of competition among economic actors and the Ricardian emphasis on productivity and opportunity costs in international trade.

Although Porter's unit of analysis is the industry rather than the firm, he finds that competitiveness is generated by individual firms mostly due to their ability to coordinate their "value chain system" or an interdependent network of activities connected by linkages, whose performance affects the rest of the value chain (Porter, 1990, p. 41). It is the firm's ability to coordinate this system through outsourcing, relocating, eliminating, or reorganizing productive activities that generate competitive advantage.[5] This dynamic system or material competition differs from competitiveness as a national project in that the former is narrowly comprised by habitual firm/industry interactions associated with firm strategy, structure, and rivalry vis-a-vis their counterparts, whereas the latter conflates the interests of economic actors with those of states. In this paradigm, the role of government in material competition is somewhat tangential.[6]

Drawing from Porter's industry-level analysis, competitiveness in the Global Commodity Chain (GCC) approach (Gereffi and Korzeniewicz, 1994) stems from the intra-chain coordination of supplier firms involved in discrete production processes within a supply chain (Gereffi and Korzeniewicz, 1994, p. 6). To remain competitive in global markets, firms mobilize a combination of internal attributes including capabilities and their ability to perform and codify complex transactions (Gereffi et al., 2005). Here, a low-cost or differentiation strategy hinges on a lead firm's ability to orchestrate the internal attributes of their value chains to successfully perform a combination of activities. From this perspective, material competition among suppliers provides lead firms with the leverage to influence power relations within a supply chain. While this model accounts for the intra-chain organizational activities of lead firms and leverage between producers and buyers, it does not operationalize the role of the state nor does it account for rival GCCs' influence on each other in world markets (Patel-Campillo, 2010b).[7]

In the next section, I describe global market change in cut flowers to highlight material competition and chain governance in the Colombian and Dutch cut flower agro-industries—buyer- and producer-driven chains, respectively—and habitual industry interactions associated with strategy, structure, and rivalry.

Cut Flowers: The Netherlands and Colombia in World Markets

The global cut flower trade is a dynamic and complex system of import and export flows in which developing and industrialized countries take part. Ninety percent of demand is in industrialized nations in Europe, Asia, and North America and estimates of annual consumption value range from forty to sixty billion US dollars (International Trade Centre, n.d.). Germany, the United States, the United Kingdom, and the Netherlands accounted for just over half of world imports in 2011 (International Trade Centre n.d.). Within these markets, there are regional and country-specific differences regarding market patterns and consumer tastes. This is exemplified by lower per capita consumption of flowers in the United States as compared to that of Germany and other European countries. Per capita consumption in the United States was estimated at US$29 in 2007, notably lower than other countries like Switzerland, with per capita consumption of US$112 (International Markets Bureau, 2010, p. 2). Within the global trading system of cut flowers, the two main suppliers are Colombia and the Netherlands. Because of their geographic proximity, Colombian cut flower growers supply mostly the US market, while the Netherlands has traditionally supplied both European and North American markets.

Cut flower consumption is traditionally driven by life events, such as marriage, births, and deaths, but peak sales are affected by seasonal and special events like Valentine's or Mother's Day. Although roses are the most traded flower across the globe, more than 200 varieties are traded in world markets (SADC Trade, n.d., p. 5). In 2007, US total retail sales of floriculture items reached US$34.6 billion, but dropped to US$29.6 billion in 2009. Since then, however, sales have begun to return to previous levels and reach US$32.1 billion by 2011 (Society of American Florists, 2012). With a long tradition of flower production, in 2011, the Dutch accounted for over half (55 percent) of the global cut flower export market or five billion US dollars (in value exported). Its main export markets are Germany, the United Kingdom, and France (which together account for 59 percent of exports). Colombian cut flower production began in the late 1960s and has been rising since. Currently, Colombian cut flowers represent 14 percent of world exports with a value of US$1.2 billion. Cut flowers account for 26.5 percent of the value of Colombian agricultural exports.

While the dominance of the Dutch and Colombian cut flower agro-industries has been relatively stable over time, their positioning within world markets has changed. In the 1990s, with the support of international institutions including the World Bank, export-oriented cut flower production spread throughout countries in Latin America, Africa, and Asia. New cut flower suppliers contribute to rising competitive pressures among established suppliers as well as newcomers. In Latin America, Ecuador emerged as an important player in the US market; in Africa, the leading cut flower producers include Kenya, Zimbabwe, Zambia, South Africa, and Uganda; and in Asia, China, Thailand, and India are the biggest producers.

Material Competition: Structure, Strategy, and Rivalry in the Colombian Cut Flower Agro-Industry

Attracted by its geographic location, favorable climate conditions for year-round production, cheap labor, and access to natural resources, four American entrepreneurs established Floramérica, the first leading export-oriented flower company in the savanna of Bogotá. In a period of two years from 1970 to 1972, Floramérica's exports of carnations to the US market went from US$400,000 to US$2 million (Fairbanks and Lindsay, 1997, p. 3). Prior experience in the flower business, knowledge of the US market, and new production and distribution practices, such as the use of cheap labor and changes in work routines and workplace hierarchies, played an important role in the success of the firm (Fedesarrollo, 2007, p. 25). Incentivized by Floramérica's rapid take-off, wealthy Colombian landowning families started their own farms and learned Floramérica's production and management techniques by hiring members of its staff. Also, Floramérica established an importer-distributor company in Miami introducing the practice of forward integration into the Colombian flower business. In addition, Colombian grower/exporters also cater to supermarkets and retailers supplying them directly according to their requirements.

Colombian grower/exporters established their position in the US market in the early 1970s, with a US market share that grew from 11 percent in 1968 to 89 percent in 1978 (Patel-Campillo, 2010b, p. 90). Although currently there are efforts to create differentiation, the Colombian cut flower agro-industry has traditionally relied on a low-cost strategy to compete in world markets. Market penetration by Colombian growers changed the positioning of other cut flower suppliers, including that of their US and Dutch counterparts. For instance, the entry of low-cost Colombian carnations into the US market displaced Dutch supply of the same. Similarly, Colombian presence in US markets displaced important segments of US production. However, the emergence of Ecuador as a flower supplier shifted the position of the Colombian cut flower agro-industry, which lost nearly 25 percent of the US market (Patel-Campillo, 2010b). With the increasing number of rivals with a low-cost competitive strategy, the Colombian flower agro-industry is broadening their competitive strategies to include product diversification and value-added products (i.e., bouquets), and they are diversifying their markets to include Russia, Japan, the United Kingdom, Canada, and Spain.

A buyer-driven chain, the Colombian cut flower agro-industry is a major player in world markets. This agro-industry is characterized by the concentration of production and export among a handful of firms with a predominantly low-cost strategy. At the municipal scale, flower production has provided much-needed employment. However, the flexible employment practices of growers and the environmental impacts of production create pressure on limited municipal resources. That is, while the competitive strategies of growers has helped the agro-industry maintain their positioning in world markets, these strategies have had adverse effects on workers, and the environment in some flower producing municipalities. As such, the "trickle-down" effect has not been felt in many communities, creating a crisis of legitimacy among municipal leaders and casting doubt on the benefits of the national project to increase the competitiveness of the cut flower agro-industry.

Material Competition: Structure, Strategy, and Rivalry in the Dutch Cut Flower Agro-Industry

Dating to the late 1800s, Dutch grower cooperatives emerged as a means to protect small growers from the rising influence of buyers. Partnering to market their products exclusively through the cooperatives, small growers consolidated the supply of flowers and created an intermediary segment in their supply chain. Rather than dealing with buyers directly, the cooperatives served to formalize, control supply, and make buyers accountable to growers for their purchases. To set prices, grower cooperatives used a declining price mechanism or an auction clock to increase the likelihood of obtaining higher prices for their flowers. The establishment of grower-led auctions and the adoption of a declining price auction mechanism enabled Dutch growers to transform the governance of their commodity chain into a producer-driven chain, a feature that remains to date (Patel-Campillo, 2011). In contrast to their Colombian counterparts, where production is concentrated in a handful of large firms, in the Netherlands, flower production is carried out by approximately six thousand small flower growers. At the municipal scale, flower farms often hire migrant workers, putting pressure on municipalities to provide housing. In addition, because flower production is an intensive activity, municipalities often have to deal with traffic congestion and pollution, among other issues. Nonetheless, because flower production contributes significantly to municipal budgets and a cluster of related industries has developed around this economic activity, municipal and provincial planners consider flower production an important part of their economic well-being.

The emergence of developing country suppliers and the increased presence of retailers has influenced the position of Dutch flowers in world markets. In the 1990s, the emergence of flower suppliers based in Africa increased material competition as new entrants forged direct relationships with supermarkets bypassing the grower-led auctions. Faced with this threat, Dutch growers voted to market foreign flowers through their auctions, which had two interrelated effects for the positioning of their agro-industry. First, the inflow of foreign flowers through the auction led to a decline of Dutch production in particular segments of the market. In response, some growers changed their production strategy by shifting to other types of flowers or moving into other segments of the supply chain. Second, as the volume in foreign cut flowers funneled through the auctions increased, the Dutch grower auctions became not only cut flower suppliers but also importers. With a 9.5 percent share in the world imports, the top four supplying countries—Kenya, Belgium, Ecuador, and Ethiopia—represent 71.1 percent of Dutch cut flower imports. Currently, approximately 15 percent of their 55 percent world market share consists of reexports of foreign flowers, a trend that is expected to continue.

Rising retailer influence posed a challenge to producer-led auctions because of their power over consumer markets and their capacity to source their flowers directly from flower suppliers around the world. To maintain control over their supply chain, the growers' cooperative strategy relies on quality as well as bulk and product diversification as well as the expansion of physical and logistic services and operations. In addition, Dutch cooperatives have ramped up the establishment of regional offices overseas to promote auction services to foreign growers. Finally, Dutch cooperatives have created alternative purchasing methods in addition to the auction clock to meet the particular needs of buyers (Patel-Campillo, 2011). Dutch growers are also targeting emerging markets in Eastern Europe, including Russia, Poland, Lithuania, and Croatia, where cut flower consumption is growing (IPM ESSEN, 2012). Despite these changes, however, governance in the Dutch cut flower agro-industry continues to be producer-driven.

Common responses to rising competition in the 1990s by the Dutch and Colombian cut flower agro-industries were to diversify their markets and add value to their products and services. As the world's hub for the trade in flowers, Dutch growers responded to material competition by auctioning foreign flowers, in effect repositioning themselves as middlemen. Although this shift has led to some restructuring domestically, the repositioning of the grower-led auctions as reexporters built on their strengths as a producer-driven agro-industry and commercial hub. In addition, because Dutch growers have mostly relied on a competitive strategy based on differentiation, foreign flowers served to meet demand for bulk products, thus enabling the Dutch grower cooperatives to fulfill both strategies. The Colombian cut flower agro-industry is responding to material pressure by seeking to diversify their products, shifting into value-added segments of the supply chain yet maintaining a low-cost strategy to cater to retailers and importers (Patel-Campillo, 2010a).

This example of material competition makes a twofold contribution to this book. First, an incorporated comparison of the Dutch and Colombian cut flower agro-industries illustrates the dynamic and mutually conditioning relationship of two global supply chains in world markets. This highlights how the material competitive strategies of flower growers influence the positioning of their rivals as well as their own. However, these strategies are neither mutually exclusive nor infallible. Although the Netherlands and Colombia retain dominance in the world market, accounting for over two-thirds of world exports—55 percent and 14 percent of the world's cut flower market, respectively as of 2011—their position in world markets is not uncontested. These dynamic fractures point to the instability of material competition in world markets, which helps to make the points of fracture among the agents of capital more visible.

Unlike much of the literature on international competitiveness and global commodity chains, where chain governance is assumed to be linked to the commodity produced itself, a second contribution of this chapter is to show that in the realm of material competition, chain governance is neither static nor determined by the commodity produced. Rather, chain governance, as demonstrated by the Dutch (PDC) and Colombian (BDC) cases, transforms and evolves out of context-specific historical, political-economic, and social processes. From a policy perspective, this is of importance given the emphasis by international entities such as the World Bank as well as national governments to embark on an export-led development model that focus on particular commodities regardless of their social organization, strategy, and positioning in world markets. Encouraging developing countries to produce the same commodities may not only increase volatility but trigger a race to the bottom even in the case of high-value agricultural crops.

Diffusing the Competitiveness Agenda: The Other Side of the Coin

The 1990s marked a period not only of global economic change characterized by deepening economic integration but also a time of institutional restructuring. In the Netherlands, this shift emphasized interscalar coordination to increase the role of subnational units of government, including provinces and municipalities, while increasing central government control over economic activities. This sentiment is captured in the motto of the Fifth National Policy Document on Spatial Planning 2000–2020: Making Space, Sharing Space (Ministry of Housing, 2001) [hereafter the Fifth Plan]: "centralised where necessary, decentralised when possible" (Ministry of Housing, 2001, n.p.). In the Fifth Plan, the central government laid out the policy framework governing the economic and spatial future of the Netherlands, reiterating its commitment to economic growth by "strengthening the Netherlands' competitive position in the European arena" (Ministry of Housing, 2001, n.p.). This highlights Evans's point that the central state has reinforced its engagement in economic matters, albeit in a different fashion. In the Fifth Plan, not only did the central government focus on the creation of infrastructure to reinforce the Netherlands' position as an international commercial hub, but it explicitly set out to increase the competitiveness of specific Dutch industries, including cut flowers, illustrating Cerny's point (2007). Furthermore, this shows that as a competition state, the function of the Dutch central government was to increase the marketization of selected national sectors as it restructures:

> The national government will restrict itself to overseeing aspects deemed to be having national significance. The national government's role can be divided into three categories: execution, promotion and facilitation.

> Execution pertains to situations of great national or international interest, the end result of which is the ultimate responsibility of the national government. (Ministry of Housing, 2001, n.p.)

By identifying the cut flower agro-industry as a matter of national interest, the central government legitimized competitiveness as a national project and allocated land and resources for the expansion and scaling up of flower production operations. Although seemingly trivial, in the context of the Netherlands, where land is scarce and there are competing claims for space for recreation, housing, and other purposes, the adjudication of land to this industry is highly significant. To advance competitiveness as a national project, the central government delegated the coordination of activities aimed at increasing the competitiveness of the flower agro-industry to flower producing provinces and municipalities. Provinces were in charge of coordinating activities with the national government as well as municipalities. The task of individual municipalities was to work with provinces to execute the plans for the physical restructuring of the cluster. Here, interscalar coordination between the central government, provinces, and municipalities was necessary to meet the objectives delineated in the Fifth Plan. Outweighing the adverse impacts of intensive flower production, uptake of the competitiveness project by provinces and municipalities was legitimized by the employment and economic benefits of flower production, which municipal governments consider an important component of their local economies (Patel-Campillo, 2010a). In this instance, the competitiveness project moves forward because it commands the legitimacy necessary to gain the support from provinces and municipalities to carry out the necessary tasks.

In Colombia, the 1990s were marked by national government efforts to increase market liberalization through the process of "*apertura*," or market opening, a project aligned with neoliberal principles of political and economic governance. Through this process, the national government sought to deepen its insertion in world markets by liberalizing domestic markets, relaxing regulatory frameworks, and reinvigorating exports. In the revised 1991 constitution, the national government sought to legitimize the deepening of neoliberal practices by reasserting its power over national economic matters while mandating the creation of municipal economic and social plans.

Like its Dutch counterpart, the Colombian government sought to diffuse the competitiveness project subnationally by deploying a combination of top-down intervention and interscalar coordination. The government aimed to create a network of governmental entities at the national scale, such as the Consejo Superior de Política Económica y Social (CONPES), to formulate national plans for competitiveness and create regulatory frameworks to materialize them across scales. Led by the president and the ministers of Agriculture, Commerce and Trade, and Planning, the CONPES produces legally binding documents, charting the vision and direction of competitiveness as a national project. In 2004, CONPES (number 3297) reinforced the institutional foundation for the active pursuit of the competitiveness agenda to be guided by: "the country's aggressive strategy of economic integration" [3439, p. 1]. Here, competitiveness as a national project entails increasing (labor) productivity, upgrading infrastructure, and creating the institutional architectures to make the competitive project durable.

In CONPES (3297), the cut flower agro-industry was singled out as a matter of national interest, justifying the use of national funds and resources to increase its international competitiveness. A covenant between the government and the cut flower agro-industry committed government support for international marketing and export promotion of Colombian cut flowers and large infrastructure projects. Nonetheless, top-down government efforts to engage subnational units of government in increasing the competitiveness of the cut flower agro-industry have not been entirely uncontested. Unlike Dutch provinces that coordinate between the central state and municipalities, their Colombian counterparts have traditionally played a weak, almost nominal role. Their relative lack of power and political presence prevents the national government from using this scale of governance to coordinate activities to improve the competitiveness of Colombian flowers. More importantly, municipalities have historically been characterized by having comparatively more influence, making them pivotal for the materialization of competitiveness as a national project. Because the benefits of cut flower production have failed to trickle down to some flower producing municipalities, their responses to the national project to increase the competitiveness of this agro-industry was met with reticence, if not rejection.

Push back from some of the most important flower growing municipalities relied on the 1991 constitutional national mandate requiring them to create their own economic and social plans governing their activities and territories. Using this newly acquired power, some flower producing municipalities, rather than encouraging the expansion of flower production, prohibited the construction of new green houses and curtailed the growth of existent ones (Patel-Campillo 2010a). Municipal planners point to the low-cost competitive strategy of flower firms as creating adverse social, economic, and environmental impacts locally as a reason for push back. Also, although flower farms can be the main source of employment in some of these municipalities, the flexibility of employment practices to increase labor productivity in flower farms has caused employment instability, food insecurity, and circular migration (Patel-Campillo, 2010a). This has stretched municipal resources and services, including housing, educational facilities, and hospitals beyond their capacity. In this instance, the worn-out promise of export-led growth has contributed to a crisis of legitimacy, disincentivizing the uptake of the competitiveness project at the municipal scale.

In this study, the national competitiveness project is contested by some Colombian flower producing municipalities due to a crisis of legitimacy rooted in worn-out development promises of export-led growth and the absence of a trickle-down effect. The highly concentrated structure of this supply chain among a few large firms at the production end partly accounts for the crisis in legitimacy. In the Netherlands, competitiveness as a national project holds more legitimacy because the benefits associated with flower production are acceptable enough for municipalities and provinces to work to improve its position in world markets. Here, the structure of the chain at the production end is evenly distributed among small growers, disallowing the concentration of capital and power.

Conclusion

While the neoliberal regime continues to expand apace with globalization, it does so persistently yet somewhat haphazardly. While the actors implicated in the entrenchment of neoliberal norms and practices are many, the adoption of competitiveness as a national project has far reaching implications. Authority over national resources as well as the power to shape regulatory frameworks makes competitiveness as a national project a means to deepen the entrenchment of the neoliberal regime. Avenues for further research entail comparative studies of commodity-based global supply systems, their structure, strategy, and positioning in world markets. Such research may be instrumental in gaining a better understanding of whether and how these factors influence the socioeconomic well-being of workers and communities. Also, comparative studies examining supply chain governance and structure of like commodities may provide important insights whether and how the production of particular export commodities (i.e., high value) are necessarily more conducive to development than others, as international institutions like the World Bank claim. In relation to broader perspectives on the neoliberal regime, its incursions and fractures, this analysis points to the need for further research on how mobilizing mechanisms such as "competitiveness" are used to legitimize the entrenchment of undemocratic forms of governance that conflate narrow economic interests with those of governments and people.

Notes

1 Although economic actors, regional entities, and policy makers, among others, are also largely implicated in the adoption of policies and practices to increase competitiveness, here I choose to focus on national governments because of the greater authority they hold over national resources and regulation, among others.
2 An incorporated comparison method is based on the premise that the examination of international organization is dynamic and conditioned by historical processes, which are illuminated through the analysis or comparison of its component parts (McMichael, 1990, p. 384).
3 This research is based on participant observation and interviews with cut flower growers, workers, trade unions, auction representatives, auction traders, government officials, and planners, among others, spread over a period of three years in the Netherlands and Colombia. Critical analysis of primary and secondary data was also used.
4 These are firm structure, strategy and rivalry, demand conditions, related and supporting industries, and factor conditions.

5 It is not uncommon for trust-based relationships among firms to not be present. As Bonanno et al. point out, one of the limits to the power of transnational firms is that they are often fraught with friction and distrust (2000), and that far from amiable and trust-based, relationships among economic actors are fragile and often times require "assistance from the state even as they attempt to bypass it" (Bonanno et al., 2000, p. 455).

6 In later work, however, Porter and van der Linde acknowledge that environmental regulation may spur firm innovation (1995, p. 6) by providing incentives for firms to innovate production processes and practices. With friendly regulation, companies are more likely to invest and use improved environmental practices in place because this lowers their production costs as well as adds value to their products (i.e., premiums for "green" products) (Porter and van der Linde, 1995, pp. 1–4). To illustrate the point, they argue that the international competitiveness of the Dutch cut flower agro-industry largely depends on technological (close-loop production system) changes they implemented as a result of friendly regulation. While a change to a close-loop system of flower production did occur, this is just but one aspect of the environmental impact associated with greenhouse gas emissions and energy intensive and year-round cut flower production in the Netherlands. From this perspective, relatively narrow changes in environmental practices do not fully account for the Dutch cut flower agro-industry's competitiveness in world markets. Rather, it is the historical trajectory, regulation, and actor strategies that have relatively more explanatory power as to why the Dutch continue to be competitive in world markets despite their natural disadvantages (Patel-Campillo, 2011). Similarly, a study of ecocertification in Ecuadorian banana production indicates that resource productivity or the adoption of sustainable practices to lower production costs or to target "green" markets is neither even within an industry nor cost effective (Melo and Wolf, 2007). Instead, Melo and Wolf (2007) find that while large farms (as opposed to small farms) have a higher rate of adoption and thus make greater strides toward sustainability, even large Ecuadorian banana producers would have difficulties implementing the ecocertification standards required without northern resources and support. Equally important is the argument that voluntary certification is not a substitute for state regulation and enforcement to achieve environmental improvements (Melo and Wolf, 2007, p. 258). Thus, from a sustainability perspective, without proper state regulation and enforcement as well as northern resources, the likelihood of adoption of certification schemes may be slim (Melo and Wolf, 2007). Pointing to similar constraints, other studies find that, in some instances, the adoption of transnational certification schemes by developing country growers and their associations is highly contested (Korovkin and Sanmiguel-Valderrama, 2007; Patel-Campillo, 2012).That is, when these schemes focus on the social (i.e., labor and gender rights) aspects of sustainable production, the outright rejection, denial, redefinition, or co-optation by producers detracts or undermines its social aims. Here, as Friedmann (2005) points out, green capitalism's responses to public pressure "is selective, choosing those demands that best fit with expanding market opportunities and profits" (pp. 230–1).

7 For a detailed analysis of the three generations of the Global Commodity Chain approach, see Patel-Campillo (2011).

References

Arrighi, G. (1994). *The Long Twentieth Century: Money, Power, and the Origins of Our Times*, Verso, London.

Barham, B., Clark, M., Katz, E., and Schurman, R. (1992) "Nontraditional agricultural exports in Latin America," *Latin America Research Review*, vol 27, no 2, pp. 43–82.

Bonanno, A. (1996) *Caught in the Net: The Global Tuna Industry, Environmentalism, and the State*, University Press of Kansas, Lawrence.

Bonanno, A. (1998) "Liberal democracy in the global era: Implications for the agro-food sector," *Agriculture and Human Values*, vol 15, no 3, pp. 223–42.

Bonanno, A., Constance, D. H., and Lorenz, H. (2000) "Powers and limits of transnational corporations: The case of ADM," *Rural Sociology*, vol 65, no 3, pp. 440–60.

Busch, L. (2011) "The private governance of food: Equitable exchange or bizarre bazaar?," *Agriculture and Human Values*, vol 28, no 3, pp. 345–52.

Cerny, P. G. (2007) "Paradoxes of the competition state: The dynamics of political globalization," *Government and Opposition*, vol 32, no 2, pp. 251–74.

Collins, J. L. and Krippner, G. R. (1999) "Permanent labor contracts in agriculture: Flexibility and subordination in a new export crop," *Society for Comparative Study of Society and History*, vol 41, no 3, pp. 510–34.

Consejo Nacional de Política Económica y Social, Republica de Colombia, Departamento Nacional de Planeación (2006) "Institucionalidad y principios rectores de política para la competitividad y productividad *Documento Conpes 3439*," https://www.dnp.gov.co/Portals/0/archivos/documentos/Subdireccion/Conpes/3439.pdf accessed August 24, 2013.

Evans, P. (1997) "The eclipse of the state? Reflections on stateness in an era of globalization," *World Politics*, vol 50, no 1, pp. 62–87.

Fairbanks, M. and Lindsay, S. (1997) *Plowing the Sea: Nurturing Hidden Sources of Growth in the Developing World*, Harvard Business School Press, Boston, MA.

Fedesarrollo. (2007) "The emergence of new successful export activities in Colombia," http://www.iadb.org/res/laresnetwork/projects/pr284finaldraft.pdf, accessed November 10, 2007.

Friedmann, H. (2005) "From colonialism to green capitalism: Social movements and emergence of food regimes," *Research in Rural Sociology and Development*, vol 11, pp. 227–64.

Gereffi, G., Humphrey, J., and Sturgeon, T. (2005) "The governance of global commodity chains," *Review of International Political Economy*, vol 12, no 1, pp. 78–104.

Gereffi, G. and Korzeniewicz, M. (eds) (1994) *Commodity Chains and Global Capitalis*, Greenwood Press, Westport, CT.

Hall, P. A. and Soskice, D. W. (2001) *Varieties of Capitalism: The Institutional Foundations of Comparative Advantage*, Oxford University Press, Oxford, UK.

Hirst, P. and Thompson, G. (1999) *Globalization in Question*, Polity Press, Malden, MA.

Hollingsworth, J. R. and Boyer, R. (eds) (1997) *Contemporary Capitalism: The Embeddedness of Institutions*, Cambridge University Press, Cambridge, UK.

Horsman, M. and Marshall, A. (1994) *After the Nation State: Citizens, Tribalism, and the New World Disorder*, HarperCollins, London.

International Markets Bureau. (2010) *Consumer Trends: Cut Flowers in the United States*, ITC, Geneva, Switzerland.

International Trade Centre. (n.d.) *Comtrade Tradema p*, ITC, Geneva, Switzerland.

IPM ESSEN. (2012) "Developments and trends in the flower and plant markets," http://www.schweissen-schneiden.com/media/presse_1/ipm_5/2012_17/englisch/Market_Figures.pdf, accessed May 3, 2013.

Korovkin, T. and Sanmiguel-Valderrama, O. (2007) "Labour standards, global markets and non-state initiatives: Colombia's and Ecuador's flower industries in comparative perspective," *Third World Quarterly*, vol 28, no 1, pp. 117–35.

Kritzinger, A., Barrientos, S., and Rossouw, H. (2004) "Global production and flexible employment in South African horticulture: Experiences of contract workers in fruit exports," *Sociologia Ruralis*, vol 44, no 1, pp. 17–39.

McMichael, P. (1990) "Incorporating comparison within a world-historical perspective: An alternative comparative method," *American Sociological Review*, vol 55, pp. 385–97.

Melo, C. J. and Wolf, S. A. (2007) "Ecocertification of Ecuadorian bananas: Prospects for progressive north-south linkages," *Studies in Comparative International Development*, vol 42, no 3–4, pp. 256–78.

Ministry of Housing, S. P. A. T. E. (2001) "Summary: Making space, sharing space" Fifth National Policy Document on Spatial Planning 2000/2020, http://international.vrom.nl/docs/internationaal/engelsesamenvattingnr.pdf, accessed March 5, 2007.

Ohmae, K. (1990) *The Borderless World: Power and Strategy in the Interlinked Economy*, McKinsey & Company, New York.

Patel-Campillo, A. (2010a) "Agro-export specialization and food security in a subnational context: The case of Colombian cut flowers," *Cambridge Journal of Regions, Economy and Society*, vol 3, no 2, pp. 279–94.

Patel-Campillo, A. (2010b) "Rival commodity chains: Regulation and agency in the US and Colombian cut flower agro-industries," *Review of International Political Economy*, vol 17, no 1, pp. 75–102.

Patel-Campillo, A. (2011) "Transforming global commodity chains: Actor strategies, regulation, and competitive relations in the Dutch cut flower sector," *Economic Geography*, vol 87, no 1, pp. 79–99.

Patel-Campillo, A. (2012) "The gendered production-consumption relation: Accounting for employment and socioeconomic hierarchies in the Colombian cut flower global commodity chain," *Sociologia Ruralis*, vol 52, no 3, pp. 272–93.

Porter, M. E. (1990) *The Competitive Advantage of Nations*, Free Press, New York.

Porter, M. E. and Van der Linde, C. (1995) "Green and competitive: Ending the stalemate," *Harvard Business Review*, September–October, pp. 120–34.

Ricardo, D. ([1911] 2004) *The Principles of Political Economy and Taxation*, Library of Congress, Washington, DC.

SADC Trade. (n.d.) "Trade information brief: Cut flowers and foliage," http://www.sadctrade.org/files/TIB1CutFlowersLong_0.pdf, accessed March 9, 2012.

Schurman, R. A. (2001) "Uncertain gains: Labor in Chile's new export sectors," *Latin America Research Review*, vol 36, no 2, pp. 3–29.

Selwyn, B. (2007) "Labour process and workers' bargaining power in export grape production, North East Brazil," *Journal of Agrarian Change*, vol 7, no 4, pp. 526–53.

Shaw, C. (2011) "Global agro food systems: Gendered and ethnic inequalities in Mexico's agricultural industry," *McGill Sociological Review*, vol 2, pp. 92–109.

Shaw, M. (1997) "The state of globalization: Towards a theory of state transformation," *Review of International Political Economy*, vol 4, no 3, pp. 497–513.

Smith, A. ([1776] 2006) *An Inquiry into the Nature and Causes of the Wealth of Nations*, Echo Library, Cirencester, UK.

Society of American Florists. (2012) "Society of American Florists," www.safnow.org, accessed December 9, 2012.

Storper, M. and Salais, R. (1997) *Worlds of Production: The Action Frameworks of the Economy*, Harvard University Press, Cambridge, MA.

Weiss, L. (1997) "Globalization and the myth of the powerless state," *New Left Review*, vol 225, pp. 3–27.

12

THE NEOLIBERAL FOOD REGIME AND ITS CRISIS

State, Agribusiness Transnational Corporations, and Biotechnology

Gerardo Otero

Introduction

Transgenic crops, the product of advanced genetic-engineering techniques based on recombinant DNA, started to be commercialized in the mid-1990s. Since well before their commercialization, biotechnology, in general, and transgenic crops, in particular, have been touted as miraculous technologies (Cage, 2008; Lee, 2008; Harvey and Parker, 2008); if only given a chance, they would make deserts bloom and do away with world hunger. The intensity of these assertions has not been tempered by the fact that most transgenic crops are not even geared for direct human consumption. Transgenic crops are sold in volatile global markets as raw materials to produce livestock feed, agro-fuels, cooking oil, and sweeteners, among other products. These are grown in huge industrial monocropping operations (which amount to two-thirds of global food production grown with biotechnology), like soybeans, corn (a fourth of global production), cotton, and canola.

The purpose of this paper is to present an overview of the neoliberal food regime and its crisis. If a food regime is the articulation of a set of regulations and institutions, making the accumulation of capital possible and stable in agriculture, the global food-price inflation crisis of 2008 is likely an indication that this food regime's contradictions are now out in the open. The food-price crisis came after almost one hundred years of continuously declining food prices. Although biotechnology has expanded agricultural productivity, the narrow range of crops that it has affected is now being fought over by a handful of economic agents with varying and antagonistic interests. This clash is at the root of the crisis, so this chapter pays particular attention to the role of biotechnology. In order to understand what is behind the crisis, we must also disentangle what are the central features of the neoliberal food regime: what are its chief dynamic factors, and how can they be modified to resolve the crisis in a progressive manner?

As we have argued elsewhere (Pechlaner and Otero, 2008, 2010), three key dynamic factors of the neoliberal food regime are the state and neoregulation, which provide the political, legislative, and policy context; agribusiness transnational corporations, which constitute the driving economic actors; and biotechnology as the main technological form chosen to expand profitability. The first two factors are discussed in the next section, followed by an analysis of biotechnology as the main technological form. I argue that biotechnology is the continuation of the modern agricultural paradigm, which started with the petrochemical, mechanical, and hybrid-seeds revolution of the 1930s. This paradigm has also been called "the industrial intensification of agriculture" (Wolf and Buttel, 1996, p. 1270). I offer an overview of how biotechnology has played out in reshaping Latin America's agrarian structures with the neoliberal turn. The third section offers an outline of the economic contradictions of the neoliberal food regime and how its global crisis has impacted a selection of Latin American countries. Food-price inflation has been much higher in Latin America than in advanced capitalist countries. The concluding section discusses the political consequences of naming the neoliberal food regime and its possible future.

The Neoliberal Food Regime

What was the food regime that emerged with neoliberalism in the 1980s and, more specifically, since the deployment of transgenic crops in the mid-1990s? A good starting point to address this question is the food-regime perspective, as conceptualized first by Harriet Friedmann and Philip McMichael (1989). A food regime is a temporally specific dynamic in the global political economy of food. It is characterized by particular institutional structures, norms, and unwritten rules around agriculture and food that are geographically and historically specific. These dynamics combine to create a qualitatively distinct "regime" of capital-accumulation trends in agriculture and food, which finds its durability in the international linking of agri-food production and consumption relations in accordance with global capital accumulation trends more broadly. Friedmann and McMichael (1989) identified two clearly demarcated food regimes: (1) the settler regime with British hegemony that lasted from the latter part of the nineteenth century until World War II. It was based on the expansion of the agricultural frontier for capital accumulation, as modern agriculture was still not present; and (2) the second food regime, dominated by the United States, emerged after a transitional period from World War I to World War II and lasted until the 1970s. The second food regime was based on a modern agricultural paradigm, which was centered on petrochemicals, machinery, and hybrid seeds that generated surpluses. Each food regime has been grounded in relatively stable (albeit typically unequal) international trade relations and has been associated with a specific technological paradigm, as discussed below.

Philip McMichael (2005) has elaborated the concept of the third food regime, which emerged after the crisis of Fordism in the United States. Central to the second food regime, Fordism was a regime of accumulation that was focused primarily on the national economies, mass production and mass consumption, and the welfare state. Its crisis led to trying to extend capital accumulation well beyond national borders on a global scale. The third food regime is thus part of a global political project, and McMichael argued that its central tension is between the globalization of corporate agriculture and countermovements informed by food-sovereignty principles and a national focus for agriculture. This characterization reflects, however, the broad brush of the regulation-school and world-systems theory in which the food-regime perspective was rooted. Such a macro view raised critiques by other scholars in regard to the original food-regime perspective's structuralism (Goodman and Watts, 1994), suggesting the need for more nuanced investigation and the call for a meso-level analysis (Bonanno and Constance, 2001, 2008). For instance, while the perspective does acknowledge the role of the state, especially in advanced capitalist countries in maintaining subsidies for their agricultures, McMichael overemphasizes the main beneficiaries of the food regime, corporations, and the South's dependency as a result. McMichael's formulation is clear that markets are politically constructed via states as members of the World Trade Organization (WTO). This means also that states are therefore subject to resistance from countermovements, which are part of the food-regime dialectic and transformation. The trouble is that McMichael's analysis remains at the level of the world economy, so my goal is to provide a friendly amendment that introduces some nuance at the national level of analysis, which is the point at which any food-sovereignty program may even be attempted.

While there is no doubt that corporations have become the dominant economic agents, especially after the neoliberal turn of the 1980s with its declining social-welfare aspect, I argue that we must continue to take full and explicit account of the specific role of the state. In contrast to McMichael's characterization of the "*corporate* food regime" (2005, 2009), therefore, I subscribe to the naming of this regime as the "*neoliberal* food regime" (Pechlaner and Otero, 2008, 2010). This characterization takes into account national-level states and local or domestic-level resistance struggles. This contrasts, for instance, with the prominence that McMichael gives to one particular social movement, Vía Campesina. Admittedly, this is the most important grassroots organization that wages much of its struggle at the transnational level. What must be recognized, however, is that the struggles of constituent organizations of Vía Campesina are firmly rooted at the national level (Desmarais, 2007, 2008); their objects of struggle are primarily their national states and the state's involvement both in local-level legislation as well as international regulations promoted and enacted by suprastate organizations.

While Vía Campesina and affiliated organizations have been quite successful in derailing the WTO's Doha Round of negotiations, the central goal of which has been to further liberalize agricultural trade, this sector was already substantially liberalized through the passage of the WTO's Uruguay Round in 1993. The extent to which such liberalization is materialized in each country's agriculture largely depends, however, on the interaction between states and domestic mobilization and resistance.

Thus, it is not simply that the "core principle" has been displaced from the state to the market with the move from the second to the third food regime, as McMichael (2009) posits. Rather, the state continues to play a central role, even if it has changed to favor the predominance of Agribusiness Transnational Corporations (ATNCs) in food production and distribution (i.e., supermarkets). Also, "the market" does not exist in the abstract; it is constructed in large part by states that also deploy some minimal rules of the market game and legislate, among other things, intellectual property rights, which are critical to biotechnology development. So, under neoliberalism, the state apparatus has indeed contracted and cut social programs. In this sense, there is a crisis of the progressive social state. But the state continues to be a central actor in facilitating corporate domination. The most dynamic elements of the neoliberal food regime discussed in this paper are thus: the state, which promotes neoregulation, a series of international agreements and national legislation that impose the neoliberal agenda; large agribusiness multinationals, which have become the central economic actors; and biotechnology as the chief technological form that continues and enhances the modern agricultural paradigm contained in the earlier Green Revolution. Supermarkets are another key driver (Reardon et al., 2003; Brunn, 2006; Burch and Lawrence, 2005, 2007) of the neoliberal food regime but are not discussed here. Let us turn to the first two dynamic elements.

The State and Neoregulation

Given the national state's key role in promoting the new set of policies and regulations associated with neoliberal globalism, I use the term *neoregulation* rather than *de*regulation, as is common in the food regime and other literatures (Ó Riain, 2000; Weiss, 1997). In spite of the free-trade rhetoric, the US state has made concerted policy and regulatory efforts to facilitate the development of its biotechnology-based industry (Kloppenburg, 1988a; Kenney, 1986). Although ATNCs have become the principal economic actors in the production and dissemination of agricultural inputs (see next subsection), this rise to dominance took place in close association with the US government through its Department of Agriculture and the Land Grant Universities, heavily supported by the state at both federal and state levels (see the chapter by Lawrence Busch in this book). The latter produced the science with public funds, and private firms developed the inputs for modern agriculture, including biotechnology (Pavitt, 2001). The US state was also very active in pressuring other states to homogenize patent laws so that its biotechnology companies would have enhanced intellectual-property-rights protection in the global economy.

Government or public support has not been limited to research and development funding. It has included the issuing of new policy and legislation to protect intellectual-property rights, as will be discussed below. Although US farmers also participated in this alliance, they never played a determining role with regard to what technologies were to be developed or produced; they were simple recipients of technological innovations that responded to the profit-maximizing logic of ATNCs (Pechlaner, 2012).

While other international agreements relevant to agricultural biotechnologies exist, to date, the most significant supranational regulatory body remains the WTO. Agriculture has featured prominently in WTO negotiations since it replaced and absorbed the General Agreement on Tariffs and Trade (GATT) in 1995 (Pechlaner and Otero, 2010). The issue of reducing trade distortion in agriculture has become increasingly important in succeeding rounds of negotiation. Negotiations have not advanced, however, with developing countries arguing that agreements to date have supported the protectionist practices of developed countries, such as the United States, while "development" goals fell by the wayside. Nonetheless, a number of agreements reached during the GATT's Uruguay Round of negotiations (1987–93) have had a significant impact in the (neo)regulation of agricultural biotechnology. Most notable are the Sanitary and Phytosanitary Measures Agreement (SPS) and the Agreement on Trade-Related Aspects of Intellectual Property Rights (TRIPS), whose implementation began in 1995 along with the WTO (Pechlaner and Otero, 2010).

The most relevant point for our discussion here, then, is that states continue to be central agencies in the deployment of neoregulation and policies that enhance neoliberalism. Whether in agreeing (or not) to participate in suprastate agreements or developing national legislation, states have been the key actors implementing neoregulation. James Klepek (2012) documents a counterexample for Guatemala, which has resisted aspects of neoregulation: It has been able to resist the adoption of transgenic corn, given the great biodiversity of maize residing within its borders. Such resistance comes from the bottom up in the form of peasant and indigenous social movements. Conversely, Elizabeth Fitting (2008, 2011) has shown that the Mexican antitransgenic corn network has mobilized similar symbolism around maize. It has garnered so much international media coverage and transnational nongovernmental organization (NGO) involvement because it is the first case of "genetic pollution" in a crop's center of origin (see also McAfee, 2008). Yet, neoregulation has moved right along in Mexico, and transgenic corn was soon to be commercially deployed as of early 2013.

Until 2006, the Brazilian case was emblematic of a state that had resisted the marketing of transgenic soybeans for years (Jepson et al., 2008; Hisano and Altoé, 2008). Eventually, through the pressure of large landowners, who had in fact been smuggling transgenic seeds from Argentina and Monsanto's lobbying efforts, the Lula government caved in to this pressure from top economic actors in 2006. Surprisingly, however, the large landowners, once the very promoters of the legalization and adoption of transgenic crops (Herring, 2007), are now in uproar against their resulting dependency on Monsanto, one of the world's leading ATNCs; such dependency has become economically disadvantageous, as Monsanto creams off the bulk of the profits. Karine Peschard's chapter in this book documents this resistance from Brazil's traditionally very strong agrarian bourgeoisie. It remains to be seen whether the Brazilian state reverses neoregulation in view of these new pressures.

Agribusiness Transnational Corporations (ATNCs)

Five agrochemical companies dominate biotechnology product development and production, while their customers are primarily mid- to large-size farmers, well endowed with capital, whose main production logic is geared by the profit motive. As highlighted singly in McMichael's (2009) characterization of the food regime, corporations are indeed the central economic actor. But states have regulated the markets even if such regulation has been mostly to the corporations' own advantage. Thus, corporations as an explicit feature of the neoliberal food regime have come to dominate such markets; there is an increasingly *limited* number of horizontally and vertically integrated corporations dominating agricultural production. This concentrated—oligopolistic—market structure squeezes producers between few input sellers, processers, and food retailers, and also limits consumer options (Hendrickson and Heffernan, 2005). As the US Agribusiness Accountability Initiative put it sharply and succinctly:

> This [ATNCs-controlled] system isn't working for farmers. The power of large agribusinesses on the buying and selling sides means that farmers have less and less control over what they produce, how they produce it, where they can sell it, and what price they can get for it. The system isn't good for consumers and rural communities either: we are all affected when agribusinesses squeeze the rural economy or put profit above environmental and health concerns, community values, or fair wages. (n.d., p. 1)

Furthermore, the pervasiveness of ATNCs in the agri-food system is important both with respect to their influence over neoregulation and with respect to their ability to deflect resistance to any socially undesirable features of the new regime—such as some groups consider agricultural biotechnologies. The lack of labeling for genetically engineered (GE) content in North America is a case in point. It could be argued that farmers of all classes, including the agrarian bourgeoisie, have become subsumed under agribusiness capital.

Biotechnology, Modern Agriculture, and Neoliberalism

The biotechnology revolution coincided with the neoliberal reformation of capitalism and thus exacerbated and deepened the socioeconomic effects of the prior agricultural revolution in Latin America, the so-called Green Revolution (1940s–1970s). As defined by the Canadian Food Inspection Agency, "'**Modern biotechnology**' is used to distinguish newer applications of biotechnology, such as genetic engineering and cell fusion from more conventional methods such as breeding, or fermentation" (2012, bold in original). According to the same source, for instance, "Mutagenesis" involves "the use of methods to physically change or 'mutate' the genetic sequence, without adding DNA from another organism." I focus on transgenic seeds, which involve the introduction of foreign genetic material into plant varieties but acknowledge that there are other forms of modern biotechnology that do not involve such genetic alteration.

The Green Revolution was the incarnation of what had earlier emerged as the modern agricultural paradigm in US agriculture. The technological paradigm of modern agriculture involves a specific package of inputs made up of hybrid and other high-yielding plant varieties, mechanization, agrochemical fertilizers and pesticides, and irrigation. The "Green Revolution" is the name adopted by this technological package when it was exported to developing countries. While the Green Revolution technically began in Mexico in 1943, with a program promoting high-yielding wheat varieties (Hewitt de Alcántara, 1978), its origin and initial development were located in the agriculture of the United States, dating from the 1930s (Kloppenburg, 1988a). This exported package then became the "technological paradigm" for modern agriculture throughout the twentieth century (Otero, 2008).

The Green Revolution has been called a "technological paradigm" (Otero, 2008) in the sense that the range of solutions to problems emerging in agricultural production tends to be solved within a narrow variety of options shaped by the paradigm. In analogy with Thomas Kuhn's "scientific paradigm," Giovanni Dosi (1984) suggested that technological paradigms move along technological trajectories shaped by the "normal" solution to problems. Such technological paradigms not only select solutions but also have exclusionary effects of alternative solutions that do not pertain to the paradigm. The technological paradigm, then, defines both the agendas for research and development and the technologies that are excluded from the frame of vision and technological imagination of engineers and, in our case, plant breeders and other agricultural researchers. It should be emphasized that there is not technological determinism here. It is not about merely responding to a social need; rather, it is mostly the scientists and technologists, institutions and policy makers who promote particular technologies.

Problems emerging in agriculture will likely be solved along the lines determined by this technological trajectory. Other new technologies emerging in the 1990s, like precision farming based on global positioning satellite systems (GPS), for instance, are also geared to optimize the use of "industrial intensification agriculture," made up mostly by "chemical fertilizer; synthetic pesticides; large-scale, tractor-based mechanization; and fertilizer-responsive, higher-yielding, genetically uniform crop varieties" (Wolf and Buttel, 1996, p. 1270). Large scale and monocropping have also become key features of modern agriculture. It is not surprising, then, that the application of agrochemicals has increased dramatically with the extension of this paradigm from its place of origin, the United States, to most regions of the world that practice capitalist agriculture. It should not be surprising either that the ATNCs involved in their production have become dominant economic actors in world agriculture, as seen above.

In what ways is biotechnology part of the modern agricultural paradigm? From its start at the laboratory stage in the 1980s, proponents described agricultural biotechnology, in general, and genetic engineering, in particular, as potent tools for sustainable development and efforts to end world hunger, food insecurity, and malnutrition. It is well known that these problems are disproportionately concentrated in developing countries, which also have larger proportions of their population dedicated to agriculture. But the technological profile of modern agriculture centers, above all, on improving the productivity of large-scale operations—those that are highly specialized on a single crop and are very intensive in the use of capital inputs rather than labor. Compared with this productive and technological model and bias, therefore, the majority of smallholder peasant cultivators in developing countries have been rendered "inefficient."

In the Food and Agriculture Organization's (FAO) calculation, twenty to thirty million peasants were displaced by new policies and technologies in the 1990s (Araghi, 2003). Some of these peasants were transformed into wageworkers for large, capitalized farms, while countless joined the unemployed. Many of these people have contributed to the growing trend toward internal and international migration, separating them from their communities and families for prolonged periods of time—or even permanently. Hence, Castells and Miller (2003) have called neoliberal capitalism "the age of migration." In Mexico, for instance, hundreds of thousands became redundant in agriculture (Corona and Tuirán, 2006), while the rest of its macro economy was incapable of absorbing them (Otero, 2004, 2011). As a result, Mexico became the number one sending nation in international migration; between 2000 and 2005, Mexico economically expelled two million people, mostly to the United States but also increasingly to Canada (González and Brooks, 2007). By comparison, as reported by the World Bank, China and India sent fewer migrants abroad during this period even though they are more than ten times bigger than Mexico in terms of population (González and Brooks, 2007). My contention as to why Mexico has expelled so many migrants is that its government has followed one of the most aggressive neoliberal policies in the world (Otero, 2011; Cypher and Delgado-Wise, 2010; Moreno-Brid and Ros, 2009).

The neoliberal reforms that started in the 1980s had deep consequences, many of them negative, for the agricultural sectors of Latin America and for a large proportion of agricultural producers. The ideological preamble of these reforms is constituted by what has been called neoliberal globalism (Otero, 2004, 2008). This ideology, which vilified state intervention and glorified the private sector and free trade, emerged during the almost simultaneous administrations of Margaret Thatcher in the United Kingdom and Ronald Reagan in the United States. For Latin America, economic liberalization generally involved the unilateral end of protectionist policies, the opening of agricultural markets with the reduction or elimination of tariffs and import permits, privatization or dismantling of government agencies for rural credit, infrastructure, marketing, or technical assistance, the end or reversal of agrarian reforms, or the reorientation of food policies centered on domestic markets toward an agricultural economy geared toward exports. Yet, the neoliberal reform was implemented in advanced capitalist countries' agriculture only partially, as they continue to subsidize and protect their agricultural sectors with billions of dollars per year, placing Latin American producers at great competitive disadvantage.

"Neoliberal globalism" is considered an ideology in the sense that the thought and policies associated with it are not inevitable. They can be modified with a different perspective, which must be backed up by alternative social and political forces, such as bottom-up social movements demanding that states implement a food-sovereignty program.

What is the problem, however, with the emerging domination of ATNCs, if they can produce food more efficiently for a growing population? Or can they, really? For millennia, peasants have been directly responsible for the preservation of huge plant biological diversity. In fact, given the vagaries of nature, developing countries possess the greatest plant biological diversity on the planet (Fowler and Mooney, 1990), as well as the greatest problems with soil erosion and ecological degradation (Montgomery, 2007), some of which is related to global warming created mostly by the advanced capitalist countries since the onset of the industrial revolution (Foster, 2000, 2009; Jarosz, 2009). As Sreenivasan and Christie (2002, p. 1) put it, "All biodiversity is richer in the South than in the North. . . . This is as true for agricultural biodiversity as for wild or biological diversity."

Capitalized and intensive farmers cannot preserve biological diversity, given modern agriculture's monocropping bias toward high-yield, hybrid, or transgenic plant varieties. That is to say, in order for large producers to stay viably in the market, they must specialize, devoting large areas of land to a single crop. Ironically, plant breeders, who require plant biological diversity as their raw material to keep the process of crop improvement going, depend on the availability of plant genetic diversity afforded by small peasant cultivators. It is in these materials that plant breeders find the desirable traits to improve crops. If peasants disappear, therefore, the same fate awaits the raw materials for future plant breeding (Fowler and Mooney, 1990; Kloppenburg, 1988b). Combined with neoliberal globalism, agricultural biotechnology can only exacerbate the tendencies to social polarization and ecological degradation, given its immersion in the modern agricultural paradigm.

Beyond ecological concerns, the productive logic of modern agriculture contrasts with that of smallholder petty-commodity producers. Rather than producing to generate profits, petty-commodity producers are geared toward self-consumption and producing for local, regional, and national markets. By definition, petty-commodity producers are content to produce quality use values for human consumption that generate enough revenues for the simple reproduction of their household units. Such production may occasionally generate income above and beyond simple reproduction needs. In this case, such income may contribute to improve their living standards or even set the conditions for bourgeoisification. Most of the time, however, petty-commodity producers are in economic-survival mode given the structural constraints they operate under (Chayanov, 1974; Wolf, 1966; Otero, 1999). The European Union considers the support of petty-commodity producers as a form of safeguard of the environment, thus recognizing them at the institutional level.

The point here is to transcend a dichotomous way of classifying agricultural producers. As I have argued elsewhere (Otero, 1998), there is the possibility for petty-commodity producers to become peasant entrepreneurs, successfully incorporated into modern markets. These are family farms and farmers whose activities can include export-oriented monocropping as well as mixed farming oriented towards the local, regional, and even national markets. These producers are clearly embedded in the market and not geared toward self-consumption, though without being capitalist corporations. Van der Ploeg (2008) has also offered a three-way categorization of agricultural producers, which includes the category of "entrepreneurial farming" between peasant and capitalist farming. Entrepreneurial farmers may be best suited for engaging in a food-sovereignty program, as they can also be ecologically sustainable. Their production is oriented to the market, but their logic of production is still imbued with a moral economy (Van der Ploeg, 2008, p. 140). In this moral economy, the market will no doubt represent an ongoing and harsh context and only a few will "win." Because entrepreneurial farmers are content with recovering costs and gaining the equivalent of a self-attributed wage, however, their numbers could be much greater than if only capitalist farms prevail; they seek simple rather than expanded reproduction, as in capital accumulation.

Capitalist farmers, by contrast, must produce primarily exchange values based on salaried workers, for human use or otherwise. Their chief goal is to produce a profit above and beyond their simple-reproduction needs in order to stay competitive (Van der Ploeg, 2008, p. 2). Thus, while capitalist farmers must also operate on a continuous economic-survival mode, their productive logic allows them—indeed compels them—to look beyond producing use values for human consumption. To the extent that ATNCs increase their domination of agricultural research and production, the exchange-value and profit logic has come to prevail, whether it is to produce food or agro-fuels (Bello, 2009, p. 15).

The capitalist mode of producing food crops is thus not the most adequate to fulfill human needs. It is neither ecologically nor socially sustainable. Nevertheless, the global food price inflation that started in 2007 unleashed a return to the rhetoric that places all faith in biotechnology and transgenics. This technology has been posed once again as the necessary solution to the food crisis (Cage, 2008; Lee, 2008; Harvey and Parker, 2008). According to this perspective, world hunger can be eradicated in poor countries only with greater yields and cheaper and more efficient crops, and transgenics are supposed to bear out this promise. A particularly relentless observer said that it would be criminal to sidestep the hope offered by biotechnology to the worse-fed people in the world (Lomborg,

2009). Critics of the position that biotechnology offers a solution to the world's poor, however, have multiplied their voices with empirically based research in the social sciences (Glover, 2010a, 2010b, 2010c, 2010d; Hisano, 2005; Jansen and Gupta, 2009; Scoones, 2002, 2008).

The idea that we can solve the food crisis simply by increasing yields is problematic in a world in which hunger is present in the midst of plenty; the world produces enough food for everyone on the planet, but the hungry simply cannot afford it. The core issue is one of inequality and lack of access to food rather than that of not producing enough of it. Since the 1960s, the world has seen a reduction in the numbers of people affected by famine even as the numbers of food insecure has risen relentlessly. We thus have the phenomenon in which obesity is combined with hunger on a planetary scale (Patel, 2007). Importing transgenic crops produced at low costs failed to protect Mexicans from high corn prices once the crisis struck, given Mexico's inequality; when price inflation for corn was 15 percent in December of 2007, consumption plummeted by 30 percent (Notimex, 2009).

Whatever level of transgenic crops is adopted in Mexico or other Latin American countries, it is doubtful that this will help feed their peoples. On the contrary, US-based ATNCs sell their seeds to farmers each agricultural cycle, always as part of a technological package that includes herbicides and other agrochemicals and all under contract (see Peschard's chapter in this book). More direct adoption of transgenic crops leads to greater dependency on the import of capital-intensive inputs, lowering demand for labor, and thus further threatening peasant agriculture. This trend will exacerbate the socially polarizing effects brought about by the Green Revolution (Pearse, 1980; Hewitt de Alcántara, 1978) and further expel workers from the countryside (Otero, 2011; Cypher and Delgado-Wise, 2010).

Adopters of transgenic crops to produce soybeans or corn for the export market, which may be more lucrative than the domestic market in Latin America, are large-scale farmers with substantial capital endowments. Export agriculture will, of course, take agricultural land away from food production for the domestic market. Hence, we had the Argentine paradox at the turn of the twenty-first century (Teubal, 2008)—having been the second soybean exporter after the United States (until it was displaced by Brazil), and one of the leading agricultural exporters in the world, there was nonetheless a substantial growth in the number of people going hungry.

Although peasant agriculture is not highly productive in economic terms—that is, in terms of generating profits for producers, at the very least, it can produce food and subsistence for those who depend on it for their livelihoods, for local, regional, and national markets. On the other hand, peasants have scarce alternative job opportunities in an economy that does not offer sufficient or adequately paid employment to urban migrants, nor the rights or dignity to international migrants (Cypher and Delgado-Wise, 2010; Otero, 2011). Therefore, substituting peasant farming for export-oriented industrial agriculture forces many in the rural population to depend on dollar remittances from migrant relatives and thus increases food insecurity at the family level, even if the nation's total agricultural production is increased. Besides producing subsistence, peasants offer gratis to society the service of plant biological conservation, as they do not focus on monocropping as large-scale cultivators (Bartra, 2004).

The social polarization trends brought about by the Green Revolution, and then by biotechnology and the neoliberal reform, were exacerbated with the food-price inflation crisis of 2007–8, which resurfaced in 2010–11. Excluding a process of bottom-up technological innovation, one that builds on the actual needs of smallholder farmers, the reaction of some suprastate institutions, like the World Bank, has been to promote industrial agriculture (Akram-Lodhi, 2012). Based as it is on the profit motive, this approach can hardly help mitigate the food crisis (see next section). Several studies have shown the limits of ATNC-promoted biotechnology, including strong doubts about its effective economic performance, the strong tendency to favor ATNCs, and the limited benefits for smallholder producers or the hungry (Friends of the Earth International, 2009; McAfee, 2008; Otero et al., 2008; Otero and Pechlaner, 2005, 2009; Pechlaner and Otero, 2008, 2010).

The Crisis of the Neoliberal Food Regime and Global Food-Price Inflation

The neoliberal food regime entered a period of protracted crisis in 2007, at which point there was a reversal of nearly a century of diminishing food prices. The crisis was generated primarily by the presence of new economic actors focused on the exchange value rather than the use value of food—for example, finance capital in agricultural-futures markets, or state policies concerned with the geopolitics of oil and energy dependency, such as the US and EU public policies geared to expand the production of agro-fuels, which divert land from food production (Bello, 2009; McMichael, 2009). But this is just the tip of the iceberg, as modern agricultural production has also been a central contributor to climate change by producing about 30 percent of greenhouse gas emissions.

A critical point of cleavage that arises with capitalist-dominated agriculture is that there are now several for-profit interests competing over the same crops. The most dramatic example is corn. Driven by hefty state subsidies on one side or another, corn is fought over by industrialists who process it to make ethanol for fuel or high fructose corn syrup, by livestock growers who use corn as feed, and finally, by financial speculators who invest in corn's futures. The latter group is interested in higher prices of corn, while all others are interested in lower prices, as it is used as a raw material. But whether corn is used to produce ethanol or food, it has vastly different consequences. In fact, producing ethanol rather than food also has a huge impact in overall food-price inflation (Turrent Fernández et al., 2012). Furthermore, although many think ethanol is green, they are wrong; ethanol produced with corn actually produces more greenhouse gas emissions than even gasoline (Otero and Jones, 2010), and it is unsustainable without subsidies. The 2012 drought in the United States will no doubt exacerbate the cleavages and tensions among these groups with high stakes on corn. Highly fragmented consumers of corn products are of course at the mercy of this highly oligopsonic competition over a vital food, which is also used to produce fuel or straight financial profits. Only the state can provide a forceful agency to regulate these forces in a different direction. In the United States, the question is whether it will continue to be controlled by strong lobby groups.

My goal here is not to engage in a detailed discussion of the global food-price inflation crisis set off in 2007–8, which has been discussed by many scholars (e.g., Bello, 2009; Holt-Giménez and Patel, 2009; McMichael, 2009). Rather, I want to highlight the role of food-import dependency on the extent to which the crisis impacted a sample of Latin American countries, plus a few other countries for further comparison and contrast. All data come from the UN Food and Agriculture Organization (see Figure 12.1 for source). In general, rich OECD economies experienced accumulated inflation rates no higher than about thirty-five points above 2000-level prices (i.e., 2000 = 100) through 2011. A comparison of the United Kingdom, France, and Germany clearly shows that the United Kingdom was the most sharply affected with a price hike of about 35 percent by 2010, likely due to its early wholehearted adoption of the neoliberal reforms, in contrast to Germany (18 percent) and France (23 percent). In spite of this inflation level, the UK as well as the rest of Western European countries—except Spain and Portugal—were considered to experience a low food security risk (Carrington, 2011).

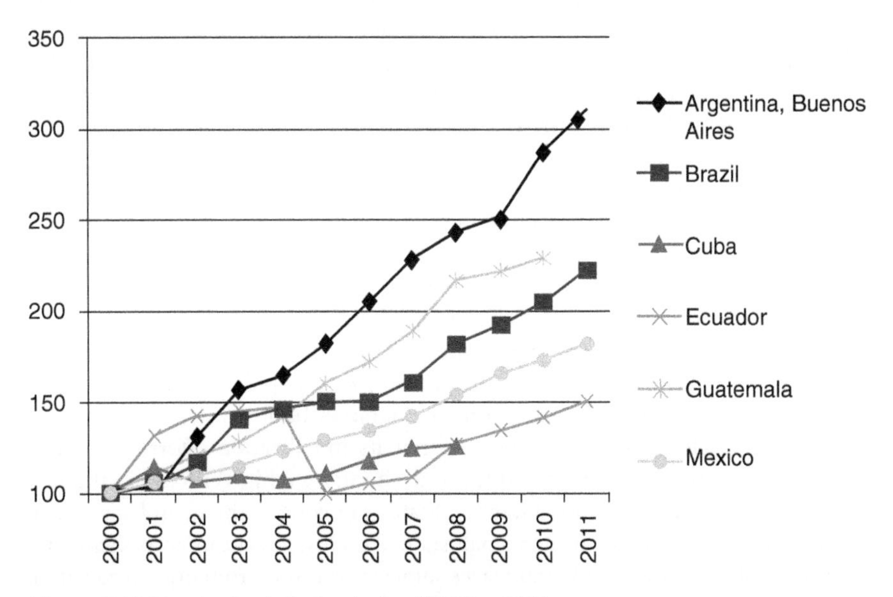

Figure 12.1 Food price inflation index (2000 = 100).

In contrast, there is a sharp disparity in food-inflation rates for Latin American countries with those for advanced capitalist countries, which is accompanied by a higher food-security risk in all Latin American countries (Carrington, 2011). Argentina's food inflation skyrocketed to over 200 percent in 2006 and reached over 300 percent accumulated food-price inflation by 2011. Brazil also experienced considerable food-price inflation, but its rate in 2011 was about one hundred points below that of Argentina. In the American continent, the United States and Canada were at the bottom with the lowest accumulated food-inflation rates since 2000 of about 36 to 37 percent, respectively, while all the developing countries had rates three to four times higher. The notable exception is Cuba's inflation, which has been below that of even advanced capitalist countries, likely reflecting the fact that its government has pursued a concerted food-sovereignty policy.

Conclusions

Building on previous work (Pechlaner and Otero 2008, 2010), I have argued that the chief dynamic factors of the neoliberal food regime are the state, ATNCs, and biotechnology. States continue to be the chief agency to implement neoregulation—that is, the type of legislation best suited to protect the intellectual-property rights of ATNCs and their coming to prominence in the economy; ATNCs clearly dominate the markets shaped by the state, and biotechnology is the central technological form, enabling the continuation of the modern agricultural paradigm (Otero, 2008).

The state is posited to be a central agent in the neoliberal food regime both for its deployment and transcendence. Even if neoliberalism has involved a reduction of the state's direct intervention in the economy and its social-policy aspects, the state continues to play a critical role in providing hefty subsidies and setting up the conditions under which the private sector enters the market through neoregulation. Having a fuller picture of the key dynamic factors puts us in a better position to develop strategies for resistance and perhaps transcending the neoliberal food regime with a focused targeting of the state. Thus, emphasizing only its "corporate" aspect detracts from a more dialectical engagement with the analysis of the state and how it operates not just to impose the logic of capital but also to respond to mobilization and pressure from below. Need we be reminded that the state itself is a relationship with contradictions (Jessop, 2007)? Subordinate classes can thus use these contradictions to advance a popular-democratic cause, such as food sovereignty. In fact, even the agrarian bourgeoisie would benefit from having a clear picture of where to put its mobilization efforts—not just against ATNCs but also against and through the state. While ATNCs are the key economic agents, they are not almighty. States can control them, and states can be influenced from below. Even a small state like Guatemala can keep them and their technologies at bay if the state faces sufficient pressure from organized social movements (Klepek, 2012).

Biotechnology emerged in the 1980s as an industry in its own right, driven as it was by the association of venture capitalists and academics with promising products based on molecular biology and genetic engineering (Kenney, 1986). Yet, biotechnology was eventually absorbed by the preexisting giant firms in the chemical and pharmaceutical industries and turned it into an "enabling" technology (Otero, 2008). Biotechnology thus enabled these industries to extend the modern agricultural paradigm in which they were invested.

While the technology itself may indeed theoretically contain unsuspected promise to alleviate human and ecological issues, the question is who actually drives technological development, as a function of what and who's technological "problems," and in who's interests. The research agendas driven by the modern agricultural paradigm and their main economic actors—large ATNCs—have served primarily the goal of maximizing their own profits. It is not clear that even large capitalized farmers can benefit from the technology, let alone peasant or entrepreneurial farmers. Given the role of these ATNCs in agri-food production, it would seem that the vast majority of cultivators that come into the orbit of biotechnology throughout the Americas have become their contract managers to deploy biotechnology products. Even the agrarian bourgeoisie is becoming subsumed under agribusiness capital.

Food-price inflation, which reemerged again in 2010, is likely to motivate the strengthening of resistance movements like Vía Campesina. I have suggested that the state can be a critical target to steer the food regime in a progressive direction. At a time when the cheap-food era seems to be over, enhanced food dependency resulting from the neoliberal food regime has made most developing countries in Latin America much more vulnerable to food-price volatility. The food-sovereignty program, strongly advocated by Vía Campesina (Desmarais, 2007), is the safest policy route to take for developing countries, raising small and entrepreneurial peasants to a central productive and environmental role. The neoliberal food regime's crisis was brought on as a consequence of neoregulation favoring the private sector. Progressive forces may continue to press the states to transcend it into a post-neoliberal era.

Acknowledgments

The author was a Highfield Fellow at the Centre for Advanced Studies at the University of Nottingham (January–April 2012) in the United Kingdom when he wrote the first draft of this paper. Thanks very much to Adam Morton (Politics and IR) and Wyn Morgan (Economics) for nominating me and for providing me the exciting intellectual experience. I gratefully acknowledge the Social Sciences and the Humanities Research Council of Canada, which has awarded me several research grants since 1994. Christina Holmes, James Klepeck, and Karine Peschard provided useful comments on a previous version of this chapter. John Harriss and four anonymous reviewers for the *Canadian Journal of Development Studies*, which published an earlier version, also offered very valuable input. Any remaining limitations are the author's sole responsibility.

References

AAI (Agribusiness Accountability Initiative). (n.d.) "Corporate power in livestock production: How it's hurting farmers, consumers, and communities—and what we can do about it," available at http://www.ase.tufts.edu/gdae/Pubs/rp/AAI_Issue_Brief_1_3.pdf, accessed April 26, 2012.

Akram-Lodhi, H. (2012) "Contextualising land grabbing: Contemporary land deals, the global subsistence crisis and the world food system," *Canadian Journal of Development Studies*, vol 33, no 2, pp. 114–42.

Araghi, F. (2003) "Food regimes and the production of value: Some methodological issues," *Journal of Peasant Studies*, vol 30, no 2, pp. 41–70.

Bartra, A. (2004) "Rebellious cornfields: Toward food and labor self-sufficiency," Pp. 18–36 in G. Otero (ed.) *Mexico in Transition: Neoliberal Globalism, the State and Civil Society*, Zed Books, London.

Bello, W. (2009) *Food Wars*, Verso, London.

Bonanno, A. and Constance, D.H. (2001) "Globalization, Fordism, and Post-Fordism in agriculture and food: A critical review of the literature," *Culture and Agriculture*, vol 23, no 2, pp. 1–15.

Bonanno, A. and Constance, D.H. (2008) *Stories of Globalization: Transnational Corporations, Resistance, and the State*, Pennsylvania State University, University Park, PA.

Brunn, S. (ed.) (2006) *Wal-Mart World: The World's Biggest Corporation in the Global Economy*, Routledge, New York.

Burch, D. and Lawrence, G. (2005) "Supermarket own brands, supply chains and the transformation of the agri-food system," *International Journal of Sociology of Agriculture and Food*, vol 13, no 1, pp. 1–18.

Burch, D. and Lawrence, G. (2007) *Supermarkets and Agri-Food Supply Chains: Transformations in the Production and Consumption of Foods*, Edward Elgar, Cheltenham, UK.

Cage, S. (2008) "Food prices may ease hostility to gene-altered crops," Reuters, July 9.

Canadian Food Inspection Agency (CFIA). (2012) "Modern biotechnology: A brief overview," CFIA, http://www.inspection.gc.ca/plants/plants-with-novel-traits/general-public/fact-sheets/overview/eng/1337827503752/1337827590597, accessed June 13, 2012.

Carrington, D. (2011) "Food is the ultimate security need, new map shows," *Guardian*, August 31, available at http://www.guardian.co.uk/environment/damian-carrington-blog/2011/aug/31/food-security-prices-conflict#, accessed September 9, 2012.

Castells, S. and Miller, M.J. (2003) *The Age of Migration: International Population Movements in the Modern World*, Guilford Press, New York.

Chayanov, A.V. (1974) *La organización de la unidad económica campesina*, Nueva Visión, Buenos Aires.

Corona, R. and Tuirán, R. (2006) "Magnitud aproximada de la migración mexicana a Estados Unidos después del año 2000," Papeles de población, vol. 14, núm. 57, pp. 9–38, Universidad Autónoma del Estado de México.

Cypher, J.M. and Delgado-Wise, R. (2010) *Mexico's Economic Dilemma: The Developmental Failure of Neoliberalism*, Rowman & Littlefield Publishers, Lanham, MD.

Desmarais, A.A. (2007) *La Via Campesina: Globalization and the Power of Peasants*, Fernwood Press, Halifax.

Desmarais, A.A. (2008) "The power of peasants: Reflections on the meanings of La Via Campesina," *Journal of Rural Studies*, vol 24, no 2, pp. 138–49.

Dosi, G. (1984) *Technical Change and Industrial Transformation*, Macmillan, London.

Fitting, E. (2008) "Importing corn, exporting labor: The neoliberal corn regime, GMOs, and the erosion of Mexican biodiversity," Pp. 135–58 in G. Otero (ed.) *Food for the Few: Neoliberal Globalism and Biotechnology in Latin America*, University of Texas Press, Austin.

Fitting, E. (2011) *The Struggle for Maize: Campesinos, Workers, and Transgenic Corn in the Mexican Countryside*, Duke University Press, Durham, NC.

Foster, J.B. (2000) *Marx's Ecology: Materialism and Nature*, Monthly Review Press, New York.

Foster, J.B. (2009) *The Ecological Revolution: Making Peace with the Planet*, Monthly Review Press, New York.

Fowler, C. and Mooney, P. (1990) *Shattering: Food, Politics, and the Loss of Genetic Diversity*, University of Arizona Press, Tucson.

Friedmann, H. and McMichael, P. (1989) "Agriculture and the state system: The rise and decline of national agricultures, 1870 to the present," *Sociologia Ruralis*, vol 29, no 2, pp. 93–117.

Friends of the Earth International. (2009) "Who benefits from GM crops? Feeding the biotech giants, not the world's poor," report, February, issue 116, http://www.foeeurope.org/GMOs/Who_Benefits/full_report_2009.pdf, accessed August 22, 2013.

Glover, D. (2010a) "Exploring the resilience of Bt cotton's 'Pro-Poor Success Story,'" *Development and Change*, vol 41, no 6, pp. 955–81.

Glover, D. (2010b) "Is Bt cotton a pro-poor technology? A review and critique of the empirical record," *Journal of Agrarian Change*, vol 10, no 4, pp. 482–509.

Glover, D. (2010c) "GM crops: Still not a panacea for poor farmers," *Appropriate Technology*, vol 37, no 3, pp. 19–29.

Glover, D. (2010d) "The corporate shaping of GM crops as a technology for the poor," *Journal of Peasant Studies*, vol 37, no 1, pp. 67–90.

González, A.R. and Brooks, D. (2007) "México: el mayor expulsor de migrantes, dice el BM," *La Jornada*, http://www.jornada.unam.mx/2007/04/16/index.php?section=politica&article=003n1pol, accessed October 4, 2009.

Goodman, D. and Watts, M. (1994) "Reconfiguring the rural or fording the divide? Capitalist restructuring and the global agro-food system," *Journal of Peasant Studies*, vol 22, no 1, pp. 1–49.

Harvey, F. and Parker, G. (2008) "Top UK scientist pushes for GM crops," *Financial Times*, July 8.

Hendrickson, M. and Heffernan, W. (2005) "Concentration of agricultural markets," Department of Rural Sociology, University of Missouri, February, available at www.agribusinesscenter.org/docs/Kraft_1.pdf, accessed February 22, 2012.

Herring, R. (2007) "Stealth seeds: Bioproperty, biosafety, biopolitics," *Journal of Development Studies*, vol 43, no 1, pp. 130–57.

Hewitt de Alcántara, C. (1978) *Modernización de la agricultura mexicana*, Siglo XXI Editores, Mexico City.

Hisano, S. (2005) "Critical observation on the mainstream discourse of biotechnology for the poor," *Tailoring Biotechnologies*, vol 1, no 2, pp. 81–105.

Hisano, S. and Altoé, S. (2008) "Brazilian farmers at a crossroads: Biotech industrialization of agriculture or new alternatives for family farmers?," Pp. 243–65 in G. Otero, *Food for the Few*.

Holt-Giménez, E. and Patel, R., with Shattuck, A. (2009) *Food Rebellions: Crisis and Hunger for Justice*, Food-First Books, San Francisco.

Jansen, K. and Gupta, A. (2009) "Anticipating the future: 'Biotechnology for the poor' as unrealized promise?," *Futures*, vol 41, no 7, pp. 436–45.

Jarosz, L. (2009) "Energy, climate change, meat, and markets: Mapping the coordinates of the current world food crisis," *Geography Compass*, vol 3, no 6, pp. 2065–83.

Jepson, W.E., Brannstrom, C., and de Sousa, R.S. (2008) "Brazilian biotechnology governance: Consensus and conflict over genetically modified crops," Pp. 217–42 in G. Otero, *Food for the Few*.

Jessop, B. (2007) *State Power: A Strategic-Relational Approach*, Polity, Cambridge, MA.

Kenney, M. (1986) *Biotechnology: The University-Industrial Complex*, Yale University Press, New Haven, CT.

Klepek, J. (2012) "Against the grain: Knowledge alliances and resistance to agricultural biotechnology in Guatemala," *Canadian Journal of Development Studies*, vol 33, no 3, pp. 310–25.

Kloppenburg, J.R., Jr. (1988a) *First the Seed: The Political Economy of Plant Biotechnology, 1492–2000*, Cambridge University Press, New York.

Kloppenburg, J.R., Jr. (1988b) *Seeds and Sovereignty: The Use and Control of Plant Genetic Resources*, Duke University Press, Durham, NC.

Lee, J. (2008) "GM crops may be answer to food crisis: Ecologist," Reuters, June 30.

Lomborg, B. (2009) "Another 'Green Revolution,'" *National Post* (Canada), March 25.

McAfee, K. (2008) "Exporting crop biotechnology: The myth of molecular miracles," Pp. 61–90 in G. Otero, *Food for the Few*.

McMichael, P. (2005) "Global development and the corporate food regime," *Research in Rural Sociology and Development*, vol 11, pp. 269–303.

McMichael, P. (2009) "A food regime analysis of the 'world food crisis,'" *Agriculture and Human Values*, vol 26, no 4, pp. 281–95.

Montgomery, D.R. (2007) *Dirt: The Erosion of Civilizations*, University of California Press, Berkeley.

Moreno-Brid, J.C. and Ros, J. (2009) *Development and Growth in the Mexican Economy: A Historical Perspective*, Oxford University Press, Oxford.

Notimex. (2009) "Subieron 15% los alimentos básicos y cayó 30% el consumo," Notimex, January 7.

Ó Riain, S. (2000) "States and markets in an era of globalization," *Annual Review of Sociology*, vol 26, pp. 187–213.

Otero, G. (1998) "Atencingo revisited: Political class formation and economic restructuring in Mexico's sugar industry," *Rural Sociology*, vol 63, no 2, pp. 272–99.

Otero, G. (1999) *Farewell to the Peasantry? Political Class Formation in Rural Mexico*, Westview Press, Boulder, CO.

Otero, G. (2011) "Neoliberal globalization, NAFTA and migration: Mexico's loss of food and labor sovereignty," *Journal of Poverty*, vol 15, no 4, pp. 384–402.

Otero, G. (ed.) (2004) *Mexico in Transition: Neoliberal Globalism, the State and Civil Society*, Zed Books, London.

Otero, G. (ed.) (2008) *Food for the Few: Neoliberal Globalism and Biotechnology in Latin America*, University of Texas Press, Austin.

Otero, G. and Jones, H. (2010) "Biofuels or biofools: A socio-ecological critique of agrofuels," paper presented at the International Congress of the Latin American Studies Association, October 6–9, Toronto, Ontario, Canada.

Otero, G. and Pechlaner, G. (2005) "Food for the few: Neoliberal globalism and the biotechnology revolution in Latin America," *Canadian Journal of Development Studies*, vol 26, no 4, pp. 867–87.

Otero, G. and Pechlaner, G. (2009) "Is biotechnology the answer? The evidence from North America," *NACLA Report on the Americas*, vol 42, no 3, pp. 27–31.

Otero, G., Poitras, M., and Pechlaner, G. (2008) "Latin American agriculture, food, and biotechnology: Temperate dietary pattern adoption and unsustainability," Pp. 31–60 in G. Otero, *Food for the Few*.

Patel, R. C. (2007) *Stuffed and Starved: Markets, Power and the Hidden Battle for the World Food System*, Portobello, London.

Pavitt, K. (2001) "Public policies to support basic research: What can the rest of the world learn from US theory and practice? (And what they should not learn)," *Industrial and Corporate Change*, vol 10, no 3, pp. 761–79.

Pearse, A. C. (1980) *Seeds of Plenty, Seeds of Want: Social and Economic Implications of the Green Revolution*, Clarendon Press, Oxford.

Pechlaner, G. (2012) *Corporate Crops: Biotechnology, Agriculture, and the Struggle for Control*, University of Texas Press, Austin.

Pechlaner, G. and Otero, G. (2008) "The third food regime: Neoliberal globalism and agricultural biotechnology in North America," *Sociologia Ruralis*, vol 48, no 4, pp. 351–71.

Pechlaner, G. and Otero, G. (2010) "The neoliberal food regime: Neoregulation and the new division of labor in North America," *Rural Sociology*, vol 75, no 2, pp. 179–208.

Reardon, T., Timmer, C. P., Barrett, C. B., and Berdegué, J. (2003) "The rise of supermarkets in Africa, Asia, and Latin America," *American Journal of Agricultural Economics*, vol 85, no 5, pp. 1140–6.

Scoones, I. (2002) "Can agricultural biotechnology be pro-poor? A sceptical look at the emerging 'consensus,'" *IDS Bulletin*, vol 33, no 4, pp. 114–19.

Scoones, I. (2008) "Mobilizing against GM crops in India, South Africa and Brazil," *Journal of Agrarian Change*, vol 8, no 2–3, pp. 315–44.

Sreenivasan, G. and Christie, J. (2002) "Intellectual property, biodiversity, and the rights of the poor," Ottawa, Canadian Council for International Co-operation, http://dspace.cigilibrary.org/jspui/bitstream/123456789/18726/1/Intellectual%20Property,%20Biodiversity,%20and%20the%20Rights%20of%20the%20Poor.pdf?1, accessed June 13, 2012.

Teubal, M. (2008) "Genetically modified soybeans and the crisis of Argentina's agriculture model," Pp. 189–216 in G. Otero, *Food for the Few*.

Turrent Fernández, A., Wise, T., and Garvey, E. (2012) "Achieving Mexico's maize potential," Global Development and Environment Institute Working Paper No. 12–03, Tufts University, Boston.

Van der Ploeg, J.D. (2008) *The New Peasantries, Struggles for Autonomy and Sustainability in an Era of Empire and Globalization*, Earthscan, London.
Weiss, L. (1997) "Globalization and the myth of the powerless state," *New Left Review*, no 225, pp. 3–27.
Wolf, E. (1966) *Peasants*, Prentice-Hall, Englewood Cliffs, NJ.
Wolf, S. and Buttel, F. (1996) "The political economy of precision farming," *American Journal of Agricultural Economics*, vol 78, December, pp. 1269–74.

13

"JUST ANOTHER ASSET CLASS"?

Neoliberalism, Finance, and the Construction of Farmland Investment

Madeleine Fairbairn

Introduction

In 2008, the US housing bubble burst, setting off a severe national recession whose effects reverberated throughout the global economy. However, not all real estate markets are created equal. While the residential real estate market was stagnating, US farmland markets began to boom. Meanwhile, developing countries throughout Latin America, Southeast Asia, Eastern Europe, and Africa became the target of a rash of large-scale land acquisitions by foreign investors. This "global land grab" (GRAIN, 2008) is taking place on several fronts. The price of basic grains rose throughout 2007 and spiked sharply in 2008, leading several grain producing countries to institute export restrictions and generating fears about food security among food importing countries. Governments in some of these countries, including Saudi Arabia, China, and South Korea, began acquiring land abroad, bypassing markets to ensure a direct link to future food supplies. Meanwhile, the concurrent financial crisis revealed the fragility of financial assets, like the now notorious mortgage-backed security, leading to a search for more concrete investment opportunities. This demand for real assets, combined with continued high grain prices, stimulated an unprecedented interest in farmland among private investors.

The land rush has been interpreted both as a sign that the internal contradictions of neoliberalism are intensifying and as evidence of continued adherence to neoliberal prescriptions, even in the face of these contradictions. Soaring grain prices and export restrictions call into question the ability of free markets to ensure food security, and the increasing prevalence of financial crises and banking scandals casts retrospective doubt on the wisdom of financial deregulation. McMichael (2012, p. 681) interprets the land rush as evidence of a crisis in the "neoliberal globalization project," arguing that it is "a short-term attempt to resolve the contradictions of rising agro-industrial costs on the one hand, and rising (food) costs of reproduction of labor on the other, but under conditions of agribusiness as usual that will only accelerate ecological and social contradictions" (p. 684). However, the proposed solutions to this crisis may actually extend neoliberalization into new arenas. For instance, "green grabs" for the purposes of carbon trading or other environmental services hinge on the commodification of nature and the premise that market forces can remedy complex environmental problems (Fairhead et al., 2012). This chapter argues that neoliberal transformations also helped lay the groundwork for the rapid emergence of a financialized and increasingly global market in land.

The chapter focuses on one significant dimension of the current land rush: the increasing appetite for farmland evinced by the financial sector. Wealthy individuals, families, and some institutions have long viewed land as a safe and stable means to store wealth. Beginning in 2007–8, however, land started to attract an unprecedented amount of attention from institutional investors. These are organizations with huge amounts of capital, such as pension funds, endowments, insurance companies, and hedge funds. The US pension giant TIAA-CREF, for example, began buying farmland in 2007 and now owns US$2.8 billion worth of land (Minaya and Ourso, 2012). Farmland also gained unaccustomed cachet thanks to well-publicized investments by celebrity financiers like George Soros and Warren Buffet (O'Keefe, 2009). Even with this heightened status, the amount of capital committed to farmland is still miniscule compared to that in other asset classes; the new institutional farmland investors are generally investing less than 1 percent of their portfolios (Carter, 2010). But given the enormous pools of capital under institutional control, this has the potential to add up quickly. Estimates of total institutional investment in farmland currently range between ten and forty billion US dollars globally (Highquest Partners, 2010; Wheaton and Kiernan, 2012), which translates into a lot of land.

Understanding the financial sector's sudden affinity for farmland requires an examination of the particular role of finance during the economic reorganizations of the past four decades. Harvey (2010) argues that advanced capitalism is characterized by a constantly escalating search for new investment opportunities to absorb surplus capital. Since the 1970s, the "financialization" of the global economy has provided a major new outlet for this excess liquidity. Financial deregulations, buttressed by neoliberal economic doctrine, have opened up a shimmering universe of derivatives and other financial instruments. Investment in real estate is another good way to absorb surplus capital (Harvey, 2010), and when combined with geographic expansion, it contributes a temporary "spatial fix" to the accumulation problem (Harvey, 2010; McMichael, 2012). In addition to these new spatialities, neoliberalism has contributed to changing temporalities, as financial actors' search for new asset classes takes on an ever more frenetic pace. However, this process is not without costs. The US housing bubble, which burst in 2008, was just the latest in a succession of asset price bubbles that have contributed to the increasing frequency of economic crises since the 1970s. Though farmland is a comparatively tiny asset class, its increasingly warm reception by financial institutions suggests it is at least being weighed as a promising candidate in this ongoing search for new investment outlets.

This chapter problematizes the sudden popularity of farmland investment in financial circles by asking *how* farmland is being shaped into a desirable and investible asset class. There is a tendency, particularly within political-economic scholarship, to reify neoliberalism, treating as monolithic a phenomenon that is, in fact, highly contradictory and contingent (Brenner et al., 2010). Brenner et al. (2010, p. 208, original italics) suggest that this mistake can be partially remedied through a focus on "*neoliberalization processes*" rather than on "some fully formed, coherently functioning, self-reproducing and 'regime-like' state of neoliberal*ism*." In this chapter, I analyze the current land rush through the lens of both political-economic and discursive processes (Fairhead et al., 2012). For the latter, I draw on the considerable body of work within economic sociology and related disciplines that views finance as socially constructed. This work disrupts common-sense understandings of finance in which it is naturalized as a self-evident field or a politically and culturally neutral science. It instead examines how finance is continuously constituted through the institutional arrangements and daily cultural practices of those employed in the sector (Abolafia, 1996; Ho, 2009), as well as by the financial science and economic models that purport to describe it (De Goede, 2005; MacKenzie, 2006). My data is drawn from interviews with actors along the farmland investment chain, mostly based in North and South America, as well as ethnographic research at farmland investment conferences.

The three body sections of this chapter explore the current farmland rush from vantage points that can be roughly classified as structural, institutional, and discursive. First, I provide a structural backdrop to the current farmland boom by exploring the financialization of food and agriculture during the past three decades of neoliberal reorganization. Second, I examine how a financial industry in farmland investment is being hastily assembled through the establishment of new institutions, particularly investment conferences and farmland funds. Finally, I examine the narratives that help to construct farmland as an investible asset class. Farmland fund managers and other actors promoting investment in land minimize the differences between land and other types of asset classes and foster a boom mentality in which timing is crucial.

Structural Shifts: The Financialization of Agriculture

Neoliberalism and Financialization

Beginning in the 1970s, the United States and global economies underwent a suite of changes that are commonly captured under the term "financialization." At its broadest, financialization can be conceptualized as "the increasing role of financial motives, financial markets, financial actors and financial institutions in the operation of domestic and international economies" (Epstein, 2005, p. 3).

Over the last four decades, finance has taken on an increasingly central role in the economy, politics, and everyday life. Financial sector profits have swelled, and even companies in other sectors have increasingly turned to financial investments as a way to supplement the profits they make from their normal, productive activities (Krippner, 2011). In short, finance, whose purpose used to be the timely supply of capital to those involved in production, has increasingly become an end in itself.

Financialization can be traced, at least in part, to the economic policy shifts associated with neoliberal restructuring (Kotz, 2010), particularly the deregulation of the financial sector. Throughout the 1980s and 1990s, US banks churned out new financial instruments, such as over-the-counter derivatives and credit default swaps, but rather than creating new regulations to keep pace with these innovations, the government repealed long-standing laws that had formerly limited the scope of banking activities. Restrictions on credit card interest rates were removed, and the Glass-Steagall Act, which was set up to regulate finance in the wake of the Great Depression, was gradually dismantled. Restrictions on bank mergers were removed, as were prohibitions on cross-industry activity that had formerly prevented banking, insurance, and investment activities from taking place within the same firm (Tomaskovic-Devey and Lin, 2011). The same period saw an enormous increase in global capital mobility, beginning in 1971 with the end of the Bretton Woods system of fixed exchange rates and continuing throughout the 1980s and 1990s as the International Monetary Fund (IMF) pressured developing countries to liberalize their capital markets (Eichengreen, 2008). Global stock and financial trading markets were integrated, allowing capital to more easily flow around the globe in search of the highest rates of return (Harvey, 2010). The rise of neoliberal economic and political thought provided an ideological justification for these policy changes, leading Tomaskovic-Devey and Lin (2011, p. 556) to remark that, "If neoliberalism is a policy and intellectual movement away from state regulation, financialization is perhaps its most fundamental product."

One major aspect of the financialization process has been the growth in both size and power of institutional investors. By the 1980s, institutions had amassed enormous amounts of capital and were rapidly buying up corporate stock (Useem, 1996). The amount of company stock owned by institutions grew from under 10 percent in the 1950s to over 70 percent by 2009 (Bogle, 2005). This concentration of investment power contributed to the rise of the "shareholder value" model of corporate governance, under which companies are thought to exist primarily for the purpose of enriching shareholders (Fligstein, 2001). Critics argue that this model leads to a pervasive short-termism. "Impatient finance" engages in "asset churning" in pursuit of quick capital gains rather than investing in the long-term potential of a corporation (Crotty, 2005).

The financialization era has also seen a rise in the frequency and severity of asset price bubbles. Bubbles have a long and storied history that includes such famous episodes as the Dutch "tulipmania" of the 1630s and the South Sea Bubble of 1720 (Kindleberger and Aliber, 2005). During these episodes, asset prices lose their grounding in actual asset value. Investor expectations become distorted, leading them to believe that prices will continue to rise and to make speculative investments based on this belief. In the financialization era, such inflated investor expectations have become a central pillar of the economy. Sweezy and Magdoff (1987) argue that financialization was able to act as a (temporary) solution to the accumulation crisis of the 1970s precisely because soaring investor expectations allowed for far higher returns than could be made had asset prices stayed grounded in the stagnating real economy. Since then, the boom and bust cycles associated with bubbles have intensified, often contributing to financial crises (Kindleberger and Aliber, 2005). Parenteau (2005) links the increasing prevalence of bubbles to financialization via both institutional and political shifts. Asset managers have come under increasing pressure from institutional investors, leading them to focus on extremely short-term profits. Meanwhile, he argues, the financial sector has gained political clout, leading the Federal Reserve and politicians to prioritize its health over bubble prevention (Parenteau, 2005).

Food, Agriculture, and Finance

Financialization has rapidly reshaped many sectors. Asset managers are constantly scouring the economy for the next big asset class into which to pour capital. As the capital market darling of the late 1990s, the information technology sector was buoyed by enormous amounts of financing. Its fortunes quickly reversed, however, when the dot-com bubble burst and investors fled the scene. The residential real estate market is another well-known beneficiary-cum-casualty of capital market favor.

In recent years, the agricultural sector has increasingly been integrated into financial circuits via the creation of new avenues for investment in food, agriculture, and related industries. One such avenue is investment in agricultural commodity derivatives like soybean futures. Traditionally, these derivative markets were dominated by commercial traders (i.e., farmers, grain elevator operators, food processors), who used them to hedge their productive activities by ensuring a certain price or level of supply at a later date. In the 1990s, however, deregulation allowed professional investors to become increasingly active in these markets. With the bursting of the dot-com bubble in 2000, and particularly beginning around 2005, their involvement started to increase dramatically (Mayer, 2010). Unlike the hedging activities of commercial traders, these financial investors use commodity derivatives either as a way to balance out their investment portfolios or to speculate on commodity price movements (Clapp and Helleiner, 2012). These noncommercial investors, particularly the enormous commodity index funds, have come under fire for contributing to high and volatile global grain prices in recent years (Masters, 2008), although many economists argue that this link is unsubstantiated (Irwin and Sanders, 2011). Once again, financial deregulation—most notably the Commodity Futures Trading Commission's (CFTC) creation of position-limit exemptions for several large traders—has enabled this financialization to take place (Clapp and Helleiner, 2012).

The increasing penetration of finance into food and agriculture extends well beyond commodity markets. Burch and Lawrence (2009) demonstrate that financialization is occurring throughout the agri-food system, and that financial investors are not the only ones driving this process. The "impact of finance capital has led other actors in the agri-food supply chain to act like the private equity companies, hedge funds and fund managers—in short, to behave like finance capital" (Burch and Lawrence, 2009, p. 276). This means, for instance, that Britain's Tesco supermarket chain is moving into banking, while international grain trader Cargill owns a major global hedge fund. The financialization of food and agriculture is now moving further upstream, to the land itself. McMichael (2012) and Harvey (2010) both suggest that land is increasingly being viewed as a source of potentially high financial returns and used as a way to sop up excess global capital—the newest backstop for the accumulation crisis. Farmland is unlikely to be the next dot-com boom, but it is developing considerable appeal in the eyes of some institutional investors.

However, creating a new asset class requires more than just the right political-economic conditions. In line with theoretical perspectives that view neoliberalization as a fundamentally creative and constitutive process (Larner, 2003), the rest of this chapter examines *how* this connection between farmland and finance is actually being crafted. First, turning farmland into an investible financial asset involves constructing the institutional architecture of the investment space. Markets cannot exist without the social networks, rules, and organizations that shape them (Fligstein, 2001). Second, it involves the development of new narratives (or sometimes, the revival of old ones) that construct farmland as a desirable investment. As De Goede (2005, p. 7) explains, finance is "a discursive domain made possible through performative practices which have to be articulated and rearticulated on a daily basis." An analysis of these institutional and discursive underpinnings reveals the prosaic processes through which the financialization of the agri-food sector unfolds.

Anatomy of the Farmland Investment Sector: Investment Strategies and Emerging Institutions

Investors are increasingly turning their gaze to farmland, but transforming this interest into investment requires the establishment of new institutions. The capacity to receive and deploy institutional capital in global farmland markets is still being developed. A fund manager based in Uruguay explained that, "my impression of the whole industry, if I compare it to the professionalism in the [real estate] industry, is that nobody knows what they are doing. I mean, yeah, there's people making a lot of money, but it's not defined. There is no structure in the industry." This sentiment was echoed by an executive at a US-based asset management firm with international farmland holdings: "What we're trying to do is impose

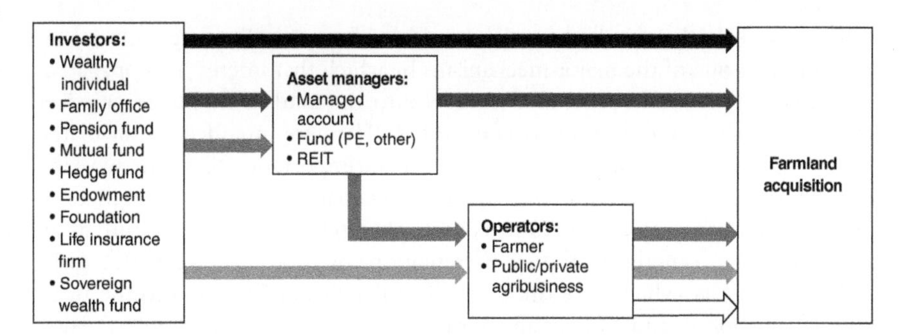

Figure 13.1 Farmland investment chain. Boxes contain examples of entities involved and arrows represent flows of capital from investors to farmland markets.

an institutional framework on a market that is not even retail." Insofar as it does exist, however, the farmland investment sector primarily involves three types of entities: the investors, who provide the capital; asset management firms, who create investment vehicles offering exposure to farmland; and agricultural operating companies, who farm the land and may also own it (see Figure 13.1). Somewhat confusingly, in the case of institutional investors like pension funds, the investor is also a type of asset manager.

Investment Conferences

One significant new development has been the emergence of high-profile conferences focused on agricultural investment. These began appearing in 2008 and quickly flourished, illustrating the extent of the emerging interest in agricultural investment. According to the organizer of one conference series, "when we started putting on [our conference] five years ago I was not aware of any other conferences, anywhere around the country that were dedicated to investment in land and now I get a new invite in the mail every week for a conference somewhere in the world about land investment." Global AgInvesting, a conference series put on by the agricultural research and consulting firm Soyatech, was first held in 2009 just as investor interest was beginning to take off. The conference, which took place at a Marriott hotel in mid-town Manhattan, attracted a healthy two hundred participants. The following year, the conference drew twice as many attendees. By 2011, it was up to six hundred attendees and had moved to the opulent Waldorf-Astoria hotel, with offshoot events in London and Singapore. Another recently initiated conference series, Agriculture Investment Summit, is hosted by the business event-coordinating firm Terrapinn. It is even more geographically dispersed, with conferences in Johannesburg, Singapore, San Francisco, Miami, and London. In addition, there are regional investment conferences, such as the Land Investment Expo, put on by rural real estate firm Peoples Company, which has been held annually since 2008 in Des Moines, Iowa. The rapid growth of these conferences is both an indicator of the sudden interest in direct agricultural investment and one of the major mechanisms by which that interest is propagated.

The conferences also provide a window into the kind of mundane practices through which financialization is constituted. They are attended by all of the major entities involved in this sector and are an important site for networking between them. Investors (or more accurately, the managers who control investor capital) attend these conferences to educate themselves about agriculture as an asset class. The conferences feature presentations by economists, climatologists, and other experts, which have titles like "The Benefits of Farmland to the Institutional Portfolio" and "Increasing Weather Risks and Climate Shifts: Driving Land Values to New Highs and Lows." Other presentations are given by firm executives and essentially constitute a sales pitch for their fund or company. These are sometimes clustered into panels by geographic area or investment vehicle, so that prospective investors can watch several talks, for instance, on farmland funds operating in Africa or on agricultural private equity.

For managers of farmland funds and operating companies, meanwhile, the appeal of these conferences lies primarily in the chance to recruit investors. Frequent coffee breaks, long lunches, and generous cocktail hours provide ample opportunity to network with potential investors. While raising capital for a new fund, asset managers attend many such conferences, shaking hands and giving out business cards over innumerable coffees and danishes. For a fee, managers can also spread the word about their fund by being one of the conference speakers or through official conference sponsorship, which buys their company conspicuous promotion throughout the conference. Companies that provide agricultural equipment or consulting might also attend such conferences, sometimes with information tables set up in the reception area, so that they can explain their services to mingling conference attendees.

Investment Vehicles

Another important step in the institutionalization and professionalization of farmland investment has been the creation of new farmland investment vehicles. Until recently, the primary options for those interested in farmland investment were either purchasing land directly or creating a separate, managed account through one of the very few asset management firms offering farmland investment as an option. In the United States, three asset managers offered this service to institutional investors: Hancock Agricultural Investment Group (HAIG), Prudential Agricultural Investments, and UBS Agrivest. These managers assemble a portfolio of farmland on behalf of their clients and manage it either directly or through a partner land management company. These tend to be relatively long-term, low-risk investments, and generally have a minimum investment of around fifty million US dollars, making them inaccessible to all but the wealthiest investors. Alternatively, investors content with indirect land ownership could also invest in a land-owning farmland operating company, some of which are publicly traded.

Since 2008, however, growing interest in farmland as an asset class has spawned a host of new funds aimed at capturing investor capital and funneling it toward agricultural land. Many of these take the form of private equity (PE) funds. Information on these funds is limited, but one estimate suggests that there are now roughly 190 PE funds operating in agriculture and farmland (IIED, 2012). The typical modus operandi of PE funds is to buy a company or portfolio of companies, make them more profitable, and resell them for more than the purchase price. PE funds dealing in farmland acquire either an operating company that owns farmland or a portfolio of farmland itself (Daniel, 2012). These funds are administered by asset managers who are paid a management fee (typically 2 percent of the fund's net asset value), as well as a performance fee (typically 20 percent of the fund's total profits). Compared to managed accounts, these new agricultural PE funds generally have lower minimum investments and are more diverse in their geographic targets. These funds generally have a fixed term of either seven or ten years and an exit strategy that allows them to return capital to investors after this period. The exit generally involves either selling the farmland or operating company, taking the company public, or rolling the fund over into another fund. Another new option for investors is to acquire land via investment in a real estate investment trust (REIT), a "pass-through" entity that is exempt from paying corporate taxes by virtue of the fact that it distributes 90 percent of its income directly to investors. Although private and public timberland REITs have existed for many years (Gunnoe and Gellert, 2010), farmland REITs have only sprung up in the last few years and are not yet publicly traded.

While farmland used to be considered a relatively stodgy, low-risk, low-return investment, the establishment of these new funds increases its appeal to a broader spectrum of investors. Investors are diverse and so are their reasons for buying land. The type of investment made will depend on both the level of risk and return that the investor is looking for and the role that he wants farmland to play in his portfolio (see Table 13.1). An investor looking for a very low risk-return investment for

Table 13.1 Attributes of farmland investment projects and level of risk and return

Risk-return	*Ownership strategy*	*Country*	*Land type*	*Crop type*
Higher ↑	Lease and operate	Least developed (e.g., Africa)	Undeveloped land	
↕	Own and operate	Intermediate (e.g., Brazil, E. Europe)		Livestock and dairy
↓				Permanent crops
Lower	Own and lease out	Most developed (e.g., United States, Canada)	Mature farmland	Row crops

Source: Based on Stookey (2010).

the purposes of value storage and inflation hedging, for instance, can find an asset manager that specializes in long-term investments in US row crop land. Since there is more risk in the operation of the land than the ownership, the land will likely be leased out to a tenant farmer or operating company. On the opposite end of the spectrum, an investor may be willing to take on more risk in exchange for the potentially high capital gains to be made from appreciation. This investor can find a farmland fund or operating company dedicated to acquiring "undeveloped" land somewhere in Africa or a frontier region of a more developed country like Brazil. These investments often hinge on making "transformations," such as clearing the land or adding infrastructure, and may involve reselling the land after a few years. By catering to a range of risk appetites and investment strategies, the new funds make farmland investments attractive to a wide range of investors.

These institutional developments lay the necessary groundwork for farmland to become a serious asset class in the eyes of big investors. However, the following section argues that discourse is also crucial to making farmland comprehensible and attractive to finance.

The Discursive Construction of Farmland Investment

Economic sociologists have increasingly recognized the role of discourse in shaping the economic world. According to Leyshon and Thrift (1997, p. 289), "the world of money is a world of interpretive power struggles, where competing sets of scripts and discourses conjure up alternative plausible 'orderings' of the economic world for purposes of financial advantage." These discourses are inseparable from the institutions of the economy and can have profound material effects. Thrift (2001), for instance, found that entire asset price bubbles can be inflated with the hot air of rhetoric. Because agriculture was such a minor asset class until recently, its sudden rise has necessitated a hasty education campaign in which financial narrative takes center stage.

"Like a One-Year Duration, Income-Bearing Money Market Fund": Making Farmland Legible to Investors

In a 2010 interview, TIAA-CREF executive Jose Minaya explained how the pension giant views their enormous new investments in farmland. "This is just another asset class that has the potential of going the route that real estate, private equity, [and] hedge funds did in the past" (McFarlane, 2010). This statement captures one of the key narratives constructed around farmland investment. This narrative depicts farmland as essentially interchangeable with stocks, bonds, or commercial real estate. In doing so, it takes an inherently uneven, uncertain, and socially meaningful "fictitious commodity" (Polanyi, [1944] 2001) and constructs it as identical to any other financial asset. This makes farmland legible and comprehensible to potential investors and can help allay misgivings about agriculture being outside their purview.

One aspect of this narrative is the frequent parallel drawn between land and other asset classes. A particularly common comparison is between farmland and fixed-income securities like bonds. Investors are accustomed to dealing with bonds, which make up a considerable percentage of most standard investment portfolios. In addition to making farmland comprehensible to investors, putting it in the context of fixed income also casts it in a favorable light in terms of returns. Bonds are considered very safe, but with interest rates hitting rock bottom their returns (known as coupons) are currently very low. Or, as one US farmland fund manager cogently put it when asked about the current appeal of farmland, "you can't get income off anything these days". She went on to compare farmland favorably to certificates of deposit (CDs) and treasury bills. Another manager of a US farmland fund compared timberland and farmland to two different types of bonds:

> I tell people that I come from a bond background [and] to me timber is like a zero-coupon thirty-year bond. You get all your coupon at the end when you finally harvest the trees and nothing in between. Row crops are like a one-year duration, income-bearing money market fund because each year you are getting the rent or the share of the crop as it's harvested and each year you can change the crop, so to me it's a lower risk asset.

The fixed-income analogy may be particularly common among managers operating in the United States or other similar low-risk, low-return contexts. Another frequent analogy is between farmland and gold. Farmland is referred to as "black gold," or, more frequently, the gold and bond metaphors are combined and it is referred to as "gold with a coupon" (Land Commodities, 2009). The gold comparisons are telling because they draw attention to land's ability to store and increase in value (generating capital gains) rather than its productive capacity.

Telling investors how to think of farmland in terms of other asset classes is important because institutions need to be able to conceptualize farmland fitting into their broader portfolio strategy. In an interview with the manager in charge of natural resources for a major US-based charitable foundation, I asked why, despite attending agricultural investment conferences and being frequently solicited by managers of farmland funds, he had chosen *not* to invest in farmland. His explanation revealed doubts about how farmland fit into an institutional portfolio:

> So we're not really sure. Is farmland a real estate play or is it a commodity play? We want the commodity exposure. We don't want real estate, you know? . . . So we're not really sure, and so we're still trying to figure this out. Now if the real estate is going up because the price of commodities is going up, so that makes the real estate more valuable if it could be used for farmland, that makes sense to me. That's commodity exposure. But if

> it's just real estate because real estate happened to be doing well at that time, that's not the beta we're looking for.

His uncertainty stems from an inability to classify farmland. Like other real estate, it stores value and appreciates over time, but as a productive asset, it is also closely aligned with agricultural commodity prices. And as his comment suggests, the two roles are not entirely separable. Discursively equating farmland with other well-known asset classes helps to allay this uncertainty.

A second aspect of this narrative is that it minimizes the things that make farmland and agriculture, more broadly, unique. One way in which this occurs is through the treatment of uncertain investment outcomes associated with farmland as manageable risk. Although often conflated, the economic concepts of risk and uncertainty actually have distinct meanings (Froud, 2003). Both refer to events that might occur in the future. However, "risk" refers to situations in which the probability of those events occurring is calculable, while uncertainty refers to situations in which the future is fundamentally unknowable. Of the many things that could potentially go wrong with an agricultural investment, several can be adequately summed up with the word "risk." These dangers include crop loss due to weather or reduced agricultural commodity prices caused by a supply response. Their likelihood can be estimated to some degree through scientific and historical analysis, and they can therefore be insured against. There are many other potential pitfalls, however, that are more adequately captured with the concept of "uncertainty." Crop loss due to extreme climatic events caused by global climate change is arguably better described by uncertainty, since the unprecedented nature of climate change makes it impossible to accurately predict. Farmland investments are also particularly susceptible to political uncertainty due to anything from host country political unrest to Federal Reserve interest rate policy.

Those promoting agricultural investment do acknowledge its unique dangers but tend to frame them as manageable risk. According to one farmland investment report:

> From a portfolio planning perspective it is also important to recognise that natural disasters and manmade disruptions, such as wars and social or political unrest, often have negative effects on equity holdings in an investment portfolio. These same conditions have the opposite effects on agricultural commodity prices because production disruptions effect [sic] supply. In this respect, allocation in investment portfolios to farmland at the higher end of the quality spectrum and lower end of the risk spectrum can actually provide investors with a hedge against such extreme events. (Land Commodities, 2009)

What is notable about this passage is not the use of the word "risk" but the implication that the enormous uncertainties of farmland investment can be managed through a careful weighing of risks and returns and resultant portfolio allocation decisions. However, to a certain extent, the institutional changes described above may increasingly make this true. Large farmland funds are diversified by both geography and crop-type, which provides a hedge against political and climatic uncertainties. Managers of TIAA-CREF's enormous farmland fund, for instance, claim that this type of diversification insulated them from the US drought of 2012 (Minaya and Ourso, 2012).

A Boom (But not a Bubble): Timing in Farmland Investment Narratives

Another key aspect of this discursive construction of farmland investment is the investment boom narrative, with its heavy emphasis on timing. Some investors are buying farmland with the intent to hold it indefinitely as a store of value. They are therefore unconcerned about timing. For those drawn to the potentially high capital gains to be made from land appreciation, however, timing is paramount. The boom narrative appeals to these investors by suggesting that there is a limited window of opportunity in which to invest, but that it is still early enough to do so without getting caught by a bubble.

First, a sense of urgency is fostered through descriptions of macroeconomic and environmental trends. As Li (2012) observes—drawing on Tsing (2000)—narratives highlighting crisis and scarcity are essential to the spectacle and drama that constitute the global farmland boom. Farmland investment reports feature graphs of rising global population, rising food prices, increasing meat consumption in developing countries and other "market fundamentals." These images "seem to indicate that you can't go wrong with acquiring land. Whether you plan to use it, or hold it, or flip it soon for speculation, the graphs always go upwards—usually sharply, dramatically, spectacularly" (Li, 2012, p. 2). Although all of the demographic and environmental trends in question are extremely long term, the present is described as a crucial juncture in which these trends come together to make farmland an extremely valuable commodity. Some proponents of farmland investment, such as influential asset manager Jeremy Grantham, even suggest that we have reached a Malthusian crisis of global resource scarcity (Grantham, 2011).

This sense of urgency is also cultivated through the specter of looming inflation. Quantitative easing, public spending, and other potentially inflationary government policies are indicated as signals that the dollar could soon experience a major, or even catastrophic, loss in value. A speaker at one investment conference painted a stark image in which "central banks are not going to sit around and do nothing, they are going to continue to print money . . . they're throwing gasoline on a smoldering fire that's going to erupt into huge price increases and huge inflation and huge profits for the people who get it right." He was advocating investment in agriculture and natural resources generally, but farmland is often touted as particularly suited to wealth preservation under inflationary conditions.

Second, in this narrative, the agricultural sector is depicted as underinvested and inefficient but on the verge of being discovered. According to a speaker at one agricultural investment conference, "There's a wall of cash that's about to hit agriculture." At another investment conference, a panelist used a different, but equally evocative image, saying, "There is a tsunami of money that wants to enter the agricultural sector." A farmland investment report, meanwhile, explains that, "Current market conditions have created what might be described as a perfect storm in terms of buying opportunities" (Land Commodities, 2009). These metaphors depict an impending boom that is extremely close, extremely large, and almost violent in its intensity. The implication is that *now* is precisely the moment to invest in agriculture. There are huge arbitrage opportunities to be had as agriculture is brought into the twenty-first century—in the case of farmland through a rapid increase in land price—but because this flood of institutional money is about to hit the sector, those opportunities won't last long.

The boom narrative contains an inherent tension caused by the possibility of a farmland bubble. Acknowledging that land prices are rising fast because people believe that they will continue to rise is a powerful selling point for fund managers, but it also sounds disconcertingly like a bubble—a point not lost on savvy investors. By 2010, US farmland prices had already risen so steeply that the Federal Deposit Insurance Corporation (FDIC) considered it necessary to hold a symposium titled "Don't Bet the Farm" aimed at determining whether farmland prices were headed toward "another asset bubble similar to residential real estate or dot-com equities before it" (FDIC, 2010).

Two possibilities exist for investors negotiating the boom-bubble tension. The first is simply to deny the existence of a bubble. The second, and perhaps more effective, is to resort again to timing; bubble or no bubble, if you get in soon you can still make a profit before it bursts. Those promoting farmland investment often acknowledge that the meteoric rise in farmland prices is no longer in its earliest stages but emphasize that it still has a long way to go. Once again, the boom narrative depends on temporal position within the investment cycle. A publication by the financial advising firm Wellington West used a baseball analogy to describe this position, "We believe we are in the third or fourth inning of a significant farmland price appreciation cycle resulting from high farmer income combined with scarce arable land" (Wellington West, 2008). Three years later, a US-based farmland fund manager resorted to the same analogy in addressing the boom-bubble issue: "It's not the first inning of the game, [but] It's not the eighth inning either" (Gustke, 2011).

The boom narrative can also be maintained in the face of the sector's increasingly frothy appearance via recourse to the looming calamities mentioned above. The keynote speaker of the 2012 Land Investment Expo resolved the bubble concern through reference to the menace of inflation: "As we go into 2012, I don't care if the president of the Kansas City Fed says maybe cropland's in a bubble. Well, so what? Because if your government continues to print a trillion and a half a year, we ain't seen nothin' yet when we get a bubble going like that"

(Dotzour, 2012). He then went on to show a one hundred trillion Zimbabwean dollar bill and summed up "so that's why I can say at the same time cropland is perceived to be in a bubble and at the same time say it's possible it could go up dramatically in value."

Conclusion

Farmland is currently experiencing an investment renaissance of a kind not seen in the United States since at least the 1970s and globally perhaps ever. This chapter has examined the emerging farmland investment sector, not as a "natural" by-product of investor demand but as a historically situated development and an active site of institutional innovation and discursive work. The financial deregulation and the liberalization of global capital flows under neoliberalism have transformed many sectors of the economy. The denizens of Wall Street, the City of London, and other financial hubs are perpetually prospecting for new asset classes in which to deploy investor capital. This search has led to spectacular booms and equally dramatic busts in sector after sector, contributing to a period of unprecedented volatility in the global economy. Rather than passively identifying the next big investment trend, these financial actors are deeply involved in the process of making markets. Though finance has always been a social construction (De Goede, 2005), the financialization era has given financial discourse even more room to operate in shaping markets and determining asset "values." It is in this environment that capital markets have taken an increasing interest in the agri-food sector, generally, and are now targeting farmland itself. In this chapter, I have taken an initial look at the processes by which global farmland is being constructed as an accessible and desirable component of the institutional portfolio. Since 2008, a host of new agricultural investment conferences and farmland funds have charged out of the gate, providing the sector with the lineaments of an institutional framework. The nascent sector is bolstered by the discursive construction of farmland as an asset like any other and a boom narrative that exhorts people to invest before time runs out. This analysis suggests that, though the land rush may be symptomatic of a crisis of neoliberalism (McMichael, 2012), it also extends neoliberal logic by further incorporating land markets into financial circuits.

Financial sector interest in farmland may prove to be transient. The bubble (if indeed one exists) may burst; the "tsunami of money" may recede. In the event that the trend continues, however, several implications are worth considering. First, as Daniel (2012) observes in her examination of farmland PE funds operating in Africa, limited regulatory oversight and lack of transparency can lead to a dangerous power asymmetry between powerful investors and the developing countries with which they work. Commercial pressure is already leading to the displacement of many African communities at the hands of their governments (Wily, 2011). Another concern is that this trend could further the consolidation of land tenure in developing countries. In fact, consolidation is an explicit transformational strategy for some of the new farmland funds. Finally, the short investment periods and preplanned exit strategies of many PE funds may not be conducive to long-term maintenance of soil and water resources. A ten-year PE investment in farmland is long term from the perspective of many investors but seems fleeting compared to the multigenerational view of farmland ownership taken by many farm families. Farmland may soon return to being an investment wallflower, but as long as it remains in the spotlight, these social and environmental implications require scrutiny.

Acknowledgments

This chapter is based on work supported by a National Science Foundation (NSF) Graduate Research Fellowship under grant number DGE-1256259, as well as by a Social Science Research Council (SSRC) International Dissertation Research Fellowship.

References

Abolafia, M. (1996) *Making Markets: Opportunism and Restraint on Wall Street*, Harvard University Press, Cambridge, MA.

Bogle, J. (2005) "Individual stockholder, R. I. P.," *Wall Street Journal*, October 3, p. A16.

Brenner, N., Peck, J., and Theodore, N. (2010) "Variegated neoliberalization: Geographies, modalities, pathways," *Global Networks*, vol 10, no 2, pp. 182–222.

Burch, D. and Lawrence, G. (2009) "Towards a third food regime: Behind the transformation," *Agriculture and Human Values*, vol 26, no 4, pp. 267–79.

Carter, D. (2010) "Fertile ground for investment," *Pensions & Investments*, available at www.pionline.com/article/20100419/PRINTSUB/304199989, accessed May 23, 2012.

Clapp, J. and Helleiner, E. (2012) "Troubled futures? The global food crisis and the politics of agricultural derivatives regulation," *Review of International Political Economy*, vol 19, no 2, pp. 181–207.

Crotty, J. (2005) "The neoliberal paradox: The impact of destructive product market competition and 'modern' financial markets on nonfinancial corporation performance in the neoliberal era," Pp. 77–110 in G. Epstein (ed.) *Financialization and the World Economy*, Edward Elgar, Cheltenham, UK.

Daniel, S. (2012) "Situating private equity capital in the land grab debate," *Journal of Peasant Studies*, vol 39, no 3–4, pp. 703–29.

De Goede, M. (2005) *Virtue, Fortune, and Faith: A Genealogy of Finance*, University of Minnesota Press, Minneapolis.

Dotzour, M. (2012) "Economic outlook for investors and business decision makers," presentation at the Land Investment Expo, January 20, Des Moines, Iowa, available at http://youtube/41lzJqt5xiA, accessed August 28, 2013.

Eichengreen, B. (2008) *Globalizing Capital: A History of the International Monetary System*, Princeton University Press, Princeton, NJ.

Epstein, G. (2005) "Introduction: Financialization and the world economy," Pp. 3–16 in Epstein, *Financialization and the World Economy*, Edward Elgar, Northampton, UK.

Fairhead, J., Leach, M., and Scoones, I. (2012) "Green grabbing: A new appropriation of nature?," *Journal of Peasant Studies*, vol 29, no 2, pp. 237–61.

FDIC. (2010) "Don't bet the farm: Assessing the boom in US farmland prices," symposium held March 10, Arlington, Virginia, available at http://www.fdic.gov/news/conferences/2011-03-10.html, accessed August 28, 2013.

Fligstein, N. (2001) *The Architecture of Markets: An Economic Sociology of Twenty-First-Century Capitalist Societies*, Princeton University Press, Princeton, NJ.

Froud, J. (2003) "The private finance initiative: Risk, uncertainty, and the state," *Accounting, Organizations and Society*, vol 28, no 6, pp. 567–89.

GRAIN. (2008) "SEIZED! The 2008 land grab for food and financial security," GRAIN, Barcelona, www.grain.org/article/entries/93-seized-the-2008-landgrab-for-food-and-financial-security, accessed June 14, 2012.

Grantham, J. (2011) "GMO Quarterly Letter, April 2011," Scribd, www.scribd.com/doc/54681895/Jeremy-Grantham-Investor-Letter-1Q-2011, accessed January 14, 2012.

Gunnoe, A. and Gellert, P. (2010) "Financialization, shareholder value, and the transformation of timberland ownership in the U.S.," *Critical Sociology*, vol 37, no 3, pp. 265–84.

Gustke, C. (2011) "Digging into farmland," CNBC, www.cnbc.com/id/40784051/Digging_Into_Farmland, accessed January 3, 2012.

Harvey, D. (2010) *The Enigma of Capital: And the Crises of Capitalism*, Oxford University Press, Oxford.

Highquest Partners. (2010) *Private Financial Sector Investment in Farmland and Agricultural Infrastructure*, Organization for Economic Co-operation and Development, New York.

Ho, K. (2009) *Liquidated: An Ethnography of Wall Street*, Duke University Press, Durham, NC.

IIED. (2012) "Farms and funds: Investment funds in the global land rush," International Institute for Environment and Development, London, pubs.iied.org/17121IIED.html, accessed January 9, 2012.

Irwin, S. and Sanders, D. (2011) "Index funds, financialization, and commodity futures markets," *Applied Economic Perspectives and Policy*, vol 33, no 1, pp. 1–31.

Kindleberger, C. and Aliber, R. (2005) *Manias, Panics, and Crashes: A History of Financial Crises*, 5th edn, Wiley and Sons, Hoboken, NJ.

Kotz, D. (2010) "Financialization and neoliberalism," Pp. 1–18 in G. Teeple and S. McBride (eds) *Relations of Global Power: Neoliberal Order and Disorder*, University of Toronto Press, Toronto.

Krippner, G. (2011) *Capitalizing on Crisis: The Political Origins of the Rise of Finance*, Harvard University Press, Harvard, MA.

Land Commodities. (2009) "The land commodities global agriculture and farmland investment report 2009," Land Commodities Asset Management AG, Baar Switzerland, www.farmlandinvestmentreport.com/Farmland_Investment_Report.pdf, accessed January 9, 2012.

Larner, W. (2003) "Neoliberalism?," *Environment and Planning D*, vol 21, no 5, pp. 509–12.

Leyshon, A. and Thrift, N. (1997) *Money/Space: Geographies of Monetary Transformation*, Routledge, London.

Li, T. (2012) "What is land? Anthropological perspectives on the global land rush," paper presented at the International Academic Conference on Global Land Grabbing II, Cornell University, October 17–19, Ithaca, NY.

MacKenzie, D. (2006) *An Engine, not a Camera: How Financial Models Shape Markets*, MIT Press, Cambridge, MA.

Masters, M. (2008) "Testimony before US Senate Committee on Homeland Security and Governmental Affairs," Washington, DC, May 20.

Mayer, J. (2010) "The financialization of commodity markets and commodity price volatility," Pp. 73–98 in S. Dullien, D. Kotte, A. Márquez, and J. Priewe (eds) *The Financial and Economic Crisis of 2008–2009 and Developing Countries*, United Nations, New York.

McFarlane, S. (2010) "Pension funds to bulk up farmland investments," Reuters, uk.reuters.com/article/2010/06/29/uk-pensions-farmland-idUKLNE65S01K20100629, accessed March 29, 2012.

McMichael, P. (2012) "The land grab and corporate food regime restructuring," *Journal of Peasant Studies*, vol 39, no 3–4, pp. 681–701.

Minaya, J. and Ourso, J. (2012) "US drought shouldn't scorch long-term farmland investing," TIAA-CREF, www.tiaa-cref.org/public/advice-planning/market-commentary/market_commentary_articles/articles/mc_053.html, accessed January 15, 2012.

O'Keefe, B. (2009) "Betting the farm," *Fortune Magazine*, June 16, money.cnn.com/2009/06/08/retirement/betting_the_farm.fortune/, accessed March 22, 2012.

Parenteau, R. (2005) "The late 1990s' US bubble: Financialization in the extreme," Pp. 111–48 in Epstein, *Financialization and the World Economy*.

Polanyi, K. ([1944] 2001) *The Great Transformation: The Political and Economic Origins of Our Time*, 3rd edn, Rinehart, New York.

Stookey, H. (2010) "Ag portfolio decisions," slide presented at Global AgInvesting 2010, May 6–7, New York.

Sweezy, P. and Magdoff, H. (1987) *Stagnation and the Financial Explosion*, Monthly Review Press, New York.

Thrift, N. (2001) " 'It's the romance, not the finance, that makes the business worth pursuing': Disclosing a new market culture," *Economy and Society*, vol 30, no 4, pp. 412–32.

Tomaskovic-Devey, D. and Lin, K. H. (2011) "Income dynamics, economic rents, and the financialization of the US economy," *American Sociological Review*, vol 76, no 4, pp. 538–59.

Tsing, A. (2000) "Inside the economy of appearances," *Public Culture*, vol 12, no 1, pp. 115–44.

Useem, M. (1996) *Investor Capitalism: How Money Managers Are Changing the Face of Corporate America*, Basic Books, New York.

Wellington West. (2008) "Global farmland primer," Wellington West Capital Markets Inc., Toronto.

Wheaton, B. and Kiernan, W. (2012) "Farmland: An untapped asset class? Quantifying the opportunity to invest in agriculture," Macquarie Agricultural Funds Management, New York.

Wily, L. (2011) "The law is to blame," *Development and Change*, vol 42, no 3, pp. 733–57.

14

NEOLIBERALISM IN THE ANTIPODES

Understanding the Influence and Limits of the Neoliberal Political Project

Geoffrey Lawrence and Hugh Campbell

Introduction

It has become a well-worn trope of Australasian agri-food scholarship to cite the cases of Australia and New Zealand as early entrants into the neoliberalization of agriculture. Commencing with a process of radical political reform in the mid-1980s (and taking only mildly variant pathways), Australia and New Zealand broke away from the post-World War II norm of protectionist-inspired agricultural governance (Le Heron and Roche, 2009; Almås and Campbell, 2012; Campbell, 2012) and pursued a deregulatory pathway, articulated under the framework of the wider neoliberal political project.

In 2000, we commenced an analysis of the impacts of these neoliberal reforms on agriculture and rural society in Australia and New Zealand (see Campbell and Lawrence, 2003). It confirmed Australia and New Zealand as the paradigmatic exemplars of the impacts of neoliberal reforms in agriculture, reviewing the negative impacts on rural livelihoods, farming families, and the abandonment of any kind of rural policy. Using the French regulationist-influenced terminology of the time, the impact of neoliberal policies was argued to be influential in moving agriculture in both countries from a "conjunctural crisis" into a "structural crisis" (see also Drummond et al., 1999).

While the narrative of negative impacts on antipodean agriculture was supported by over a decade of prior rural sociological research in both countries, the wider framing of the earlier chapter posited that Australia and New Zealand were remarkable mainly in that "localized resistance and/or amelioration of the effects of the neoliberal experiment have been slow to materialize" (Campbell and Lawrence, 2003, p. 98). This was contrasted with a more equivocal framing of the wider impetus of the neoliberal political project in agriculture on a global scale.

As it is now twenty-five years since the beginning of the neoliberal revolution in Australia and New Zealand, this chapter provides the opportunity to revisit some of the original assumptions in Campbell and Lawrence (2003)—both in terms of how we theorize neoliberalism and how we understand its impacts and limits. We compare two very contrasting cases: the relative lack of resistance to neoliberalization accompanied by the financialization of the agri-food sector in Australia (and abroad) and the reassertion of the power of agricultural collectives as a new hybrid in the dairy and kiwifruit industries in New Zealand. In doing so, we are seeking to engage with the wider reframing of the "neoliberal regime" as articulated in this book collection—a reframing that is looking for evidence of both coherence in the neoliberal project as well as incursion, rupture, and contestation of these processes.

Part 1: Neoliberalism Embraced? Rural Restructuring, Supermarket Power, and the Financialization of Australian Agriculture

Contextualizing Australian Agriculture

After federation of the colonies in 1901, a series of federal and state governments of various political hues provided strong support and funding for agriculture. Experiment stations were built, research and extension were funded, new regional towns were planned and gazetted, irrigation works were commenced, railroads and ports for export of farm products were built, and farmers were provided with various incentives to increase production and efficiency (Lawrence, 1987, pp. 183–4). For three decades after World War II, a mantle of protectionism gave some stability to an industry that was, as a price taker in the international marketplace, quite vulnerable to the vagaries of supply and demand (Lawrence, 1987). The jolt of Britain joining the European Union in 1973, accompanied by the hike in oil prices and a decade of high interest rates in the 1980s, which impacted negatively on farming, led the Australian government to remove protection and "liberate" the economy by floating the dollar, removing tariffs, and embracing economic rationalism. Here was the calculus: what Australian farmers would lose in state support for agriculture, they would gain in new sales in a deregulated, free-trade-oriented, world market (Andrews et al., 2003). Today, Australia's farming sector receives the second lowest level of protection of any agricultural trading nation. The Producer Support Estimate (PSE), which measures the amount of farm income provided by government, is 2.6 percent for Australia but rises to 7.7 percent for the United States, 17.4 percent for China, and 60.8 percent for Norway (DFAT, 2012, p. 2). Australia prides itself on its free-market stance and wants all other countries to follow suit.

The Impacts of Neoliberalism on Australia's Food and Farming Industries

Neoliberalism in Australian agriculture has had two distinguishing features: marketization and the ideology of "self-help" (Cheshire, 2005; Cheshire and Lawrence, 2005; Lockie et al., 2005). Marketization (accompanied by privatization and outsourcing—see Peters, 2011) is the creation of, and support for, market mechanisms that are assumed to deliver the most efficient and cost-effective outcomes for society. Self-help is the endorsement of personal and community responsibility in economic and social matters (Beck and Beck-Gernsheim, 2002; Cheshire and Lawrence, 2005)—something the Foucauldians, such as Dean (1999) and Rose (1999), have referred to as "responsibilization." Together, marketization and "self-help" have justified and led to the removal of government-created support mechanisms, the strengthening of private property rights—allowing producers to pursue efficiency and productivity gains without the hindrance of earlier regulatory structures—along with the entrenchment of what Dibden and Cocklin (2005) have termed "competitive productivism."

Competitive productivism promotes resource and labor efficiency and the expansion of output as the fundamental imperatives of farming. Driven by these imperatives, Australian farms have become more specialized in production and more industrial (intensive). As the least-efficient farmers have left the industry, farm size has increased, and remaining farms have become more capitalized—relying upon the machinery, pesticide, fertilizer, seed, and other inputs from transnational agribusiness to achieve efficiency and productivity gains (Gray and Lawrence, 2001; Argent, 2002). However, the real costs of production are rarely calculated, and one of the so-called externalities of productivist agriculture has been massive environmental degradation, including irreversible salinization and acidification of soils, pollution of waterways, over-allocation and misuse of scarce water resources, and significant biodiversity losses (Lawrence et al., 2012, 2013). This is a prime example of Marx's "metabolic rift" between agriculture and its natural resource base and between people and nature (Schneider and McMichael, 2010). While the Australian Greens have criticized the environmental impacts of farming, neither of the major political parties—the Labor or the Liberal-National Coalition—have specific policies to alter the current trajectory of farming; they endorse the expansion of agriculture along industrial lines and according to a market logic.

As suggested, farmer attrition has also accompanied the expansion of productivist agriculture. Farm numbers are declining at the rate of one percent per year (there was a 25 percent reduction in the number of farmers between 1983 and 2003, and the trend continues—see Productivity Commission, 2005, p. 31). Production is also concentrated—the largest 10 percent of farmers contribute 50 percent of total production, while the smallest 50 percent contribute only 10 percent (Productivity Commission, 2005, pp. 31, 37).

Some of the larger producers have taken advantage of opportunities provided by free markets, global trade links, and new information technologies to directly market their products abroad. These "family farm entrepreneurs" are not typical farmers, nor are they corporations. They utilize flexible financial and legal structures and build upon personal contacts made during international travel to sell their products directly to overseas buyers, thereby effectively bypassing middle agents and increasing their profit margins (Pritchard et al., 2007; Cheshire and Woods, 2013). They are described as being "exceptionally entrepreneurial, market-sensitive, technologically-oriented, knowledge-seeking and highly capitalized" producers who deal readily and confidently with corporations and who are the polar opposite of Kautsky's small-scale farmers, who scrape through by exploiting both their own, and family, labor (Pritchard et al., 2007, p. 85).

The declining number of farmers has been linked to other global free-market pressures—in particular, a strong Australian dollar as a result of the recent mining boom. The strong dollar makes Australia's farm products more expensive than those of competitors on global markets and therefore limits sales and reduces the viability of some farms (Munro et al., 2013). Yet another factor—the power of supermarkets—is also being recognized as an important element in Australia's neoliberal agricultural regime (Burch et al., 2012).

Australian supermarket concentration is among the highest in the world, with three chains controlling 70 percent of the sale of grocery items and 60 percent of fresh foods (ACCC, 2008). So as to ensure the foods they sell (especially own-brand products) are of the highest quality, the two largest chains—Coles and Woolworths—have imposed increasingly stringent environmental, safety, and quality standards on their suppliers. This has resulted in a reduction in the number of farmers who are able and willing to adopt the standards; those supplying are the largest, most capitalized, farmers capable of providing bulk supplies that meet the necessary standards (Burch and Lawrence, 2007; Vorley, 2007). That is, through private regulation, supermarkets are imposing standards that farmers are compelled to accept—or face losing contracts (Richards et al., 2012). Globally, as in Australia, financially burdensome requirements are marginalizing the small- and medium-sized producers and undermining their ability to compete (Bienabe et al., 2007), thereby fostering the further concentration and centralization of production in agriculture.

But it is not only through the sourcing of supplies that the supermarkets are fostering the restructuring of Australian agriculture, it is also in the pricing of farm products on supermarket shelves. In 2011, the two major chains, engaging in so-called "price wars," reduced the price of milk to below one Australian dollar per liter—cheaper than the per liter costs of soft-drinks, fruit juice, and bottled water. For many producers, the price they were receiving was below the cost of production (ABC News, 2011; Millward, 2011). Milk became a "loss leader" for the supermarkets and was subsidized from profits made on other grocery products and from the stores' petrol and liquor sales (Cox and Chicksand [2007] explain how this operates). This had a direct effect on farm viability. In a series of "crisis" meetings held in Queensland at the end of 2012, hundreds of farmers protested about the tactics of the supermarkets, indicating that farmers were being forced from the industry as a consequence (Rego, 2012). It also had a direct effect on the companies supplying branded milk. The two supermarkets were discounting their own milk, resulting in branded milk sales slumping by some 26 percent within the year, while the supermarkets' own-brand sales increased by between 10 and 20 percent (Cook, 2012).

The "milk wars" were yet another significant challenge for the dairy industry, which was deregulated in 2000 and saw the end of the monopoly marketing of table milk under the auspices of former state-based statutory marketing bodies (Edwards, 2003). This was accompanied by a A$1.5 billion structural adjustment package to assist unviable farmers to leave the industry. Basically, under neoliberal-inspired National Competition Policy, dairying was viewed as a protected industry and in need of urgent restructuring (Edwards, 2003). This is despite the fact that market forces had reduced the number of dairy farmers from some thirty thousand in 1974/5 to around fourteen thousand in 1999/2000 (Martin et al., 2000). Under deregulation, it has since halved again to less than seven thousand producers (Dairy Australia, 2013). While it has been claimed that the increased productivity that has flowed from the amalgamation of farms and the geographical repositioning of the industry (from states like Queensland and New South Wales to those of Victoria and Tasmania) has benefitted remaining producers, milk manufacturers, and consumers (ACCC, 2001; Edwards, 2003), there is evidence that this is not the case. Large conglomerates like the transnational Kirin beverage company (which controls some 81 percent of the nation's fresh milk market) have such power that they can force producers to accept lower prices if they want contracts with Kirin (Durie, 2011). The continuing milk wars of the retailers are keeping prices low, which although beneficial for consumers of own-brand milk, reduces the viability of cooperatives and other manufacturers of branded milk. Corner stores supplying the higher-priced branded milk are likely to find themselves with fewer customers, undermining viability and further strengthening the supermarket oligopoly in Australia (Kruger, 2013).

A final issue of concern is that of the "financialization" of food and farming in Australia. Neoliberalism has loosened controls over the flow of finance capital entering the nation, resulting in the purchase of food businesses and agricultural properties by overseas interests. Overseas-based companies currently control 60 percent of Australia's raw sugar production, 50 percent of the wheat export industry, 50 percent of Australia's milk processing, and 40 percent of lamb and beef processing (Bita, 2012). Sovereign wealth funds, pension funds, and other financial entities from the Middle East, China, Brazil, Singapore, the UK, and Canada have been investing in prime farmlands both for short- and long-term capital gains, as well as for the repatriation of food. For example, Qatar's Hassad Food has acquired some 750,000 ha for sheep and wheat production; Canada's Alberta Investment Management Company purchased 250,000 ha for forestry and agriculture; Southern Agricultural Resources—a UK finance firm—has identified some twenty-four properties (covering some 100,000 ha), which it has targeted for acquisition for cotton and wheat production; China's state-owned Beidahuang, the country's largest agricultural company, has made offers on some 80,000 ha of Western Australian farmland for the export of dairy products and wheat to China; and UK's Terra Firma Capital—another firm from the finance sector—currently controls more than 2.6 million ha of grazing lands in the state of Queensland (GRAIN, 2012). It is estimated that these purchases/intended purchases amount to approximately US$2.7 billion (GRAIN, 2012).

Although prepared to lend capital to farmers, financial firms have generally eschewed the ownership and control of farm lands—leaving the risks with the owner-operators. However, in an era of global population expansion, the demand by the burgeoning middle classes of India and China for Western-style diets, the loss of farmland worldwide as a result of desertification, expected climate change impacts that are set to reduce global food production, the advent of carbon markets in agriculture, and the financial attractiveness of biofuel production (McMichael, 2012; Burch and Lawrence, 2013) land acquisition is viewed as providing an excellent return on capital. Globally, farm values have increased by some 400 percent during the last decade, outperforming residential and commercial property markets and providing reliable gains at a time of continuing economic instability (Savills Research, 2012). Between 2002 and 2010, average cropland values increased by 300 percent in Australia and 262 percent in New Zealand—making farm lands in these countries some of the most attractive in the world for investors (Savills Research, 2012, p. 5). The Labor government and Liberal-National opposition endorse foreign investment and applaud the free-flow of finance into Australian farming; in keeping with neoliberalism, foreign direct investment is viewed as the lifeblood of economic expansion.

Protesting the Neoliberal Trajectory in Australia

While now largely hegemonic, neoliberal policy has not proceeded without contestation over its impacts—although these contests have been limited in comparison to the New Zealand case study that will be examined next. As suggested above, farmers have complained bitterly about the tactics of supermarkets and have forced enquiries to be held in relation to unfair practices of the retailers. When the reductions in tariffs occurred in the 1980s and cheaper vegetables were being sourced from abroad by companies such as McCain's and Simplot, contract growers blockaded factories demanding increased prices for home-grown produce (Rickson and Burch, 1996). When banks began to rationalize their regional branches and governments reduced their commitments to regional health, education, and legal services, there were quite vocal rallies and protest meetings throughout the nation (Argent and Rolley, 2000; Gerritsen, 2000), leading some scholars to label rural Australia a "land of discontent" (Pritchard and McManus, 2000).

Dismay at what was happening in the regions led, in 1997, to the formation of a new protest party—One Nation. It had particular appeal for rural and regional Australians who felt that they had lost ground under both Labor and Liberal-National Party governments. One Nation was a populist party, which argued that immigrants and Aboriginal Australians were being pampered by urbo-centric governments that were out of touch with "real" Australians. It blamed free-market policies for having undermined the economic prosperity of workers in the cities and regions, and it advanced protectionist policies as a solution to the problem of jobs leaving Australia's shores (Lockie, 2000; Gray and Lawrence, 2001). By reducing migrant intake and reinstating tariffs on imports, jobs would be created for Australian workers. This would herald the reemergence of the manufacturing sector, providing widespread benefits to struggling small businesses and to rural communities that had been damaged by neoliberalism—particularly through agricultural trade liberalization (Pritchard, 2000).

Hit by a series of internal brawls and scandals, the party imploded in 2001, and in subsequent state and federal elections, it was clear it had become a spent force. However, during its brief time, it mobilized the sentiments of a variety of citizens in industries that were waning under a more globalized trade regime along with others who saw few personal or community benefits arising from neoliberalism. As Pritchard (2000, p. 96) notes for the farm sector:

> Although the principles of comparative advantage indicate that net national benefits can be generated from the pursuit of free trade, the realization of those gains is contingent upon the distributional flow of trade benefits through the economy. In agriculture's case, the benefits of free trade increasingly are being captured by larger farm enterprises and globally mobile agribusiness traders with the scale and scope to take advantage of shifts in factor prices.

With the demise of One Nation, it is only the Greens who are offering an alternative vision for Australia. Yet, their stance is about intervention for the environment not for fundamental changes to trade policy. Capturing in the vicinity of 12–15 percent of the federal vote, their policies appear to have limited political appeal. Therefore, at present, Australia's current economic trajectory is one wedded to bipartisan neoliberalism.

Part 2: Uneven Neoliberalization? Contested and Uncontested Pathways in New Zealand

The Australian case becomes even more compelling when contrasted to its near neighbor—New Zealand. While the narrative of neoliberalization moves down parallel tracks in the early stages of the reform process in the 1980s, New Zealand has experienced a more uneven and contested process of neoliberalization of land-use and of agricultural industries. This section will briefly describe this slightly different trajectory and then seek to illuminate why key differences emerged between the two countries.

The Neoliberal Moment in New Zealand

Like Australia, New Zealand experienced a major shock from the dismantling of what had been nearly a century of stable colonial trading relationships. In the case of New Zealand, the entry of Britain into the Common Market in 1973 brought these long-term relationships to a very abrupt end with a collapse in trade in sheepmeat and dairy products with the UK (Le Heron and Pawson, 1996; MacDermott et al., 2008).

Stable trade relations were accompanied by strong state involvement in agriculture and the business of exporting. Large monopoly export boards controlled the main agricultural products—that is, sheepmeat/beef, wool, and dairy products (Blunden et al., 1996). A range of other boards controlled less visible exports (like fruit) or local supply of "essentials," like milk, eggs, and bread. The state also provided strong investment into agricultural research and development (R&D) and intervened regularly into the political conflicts of the agricultural sector by siding with farmers against unionists in meat processing and waterfront disputes, operating favorable tax regimes for productivist expansion of farming, and subsidizing lending for the purposes of purchasing farmland. When the 1973 crisis struck, the state intervened once more to support farm incomes by introducing a price support scheme for sheepmeat—New Zealand's first ever foray into direct subsidization of agricultural production (Le Heron and Pawson, 1996).

The nature of the intensifying crisis and subsequent neoliberal reform period in the early 1980s has been the subject of considerable scholarship in New Zealand (for a summary, see Le Heron and Pawson, 1996). What has generally been noted in the New Zealand situation (including by Campbell and Lawrence, [2003]) was the purity with which reforms were pursued in New Zealand with the complete removal of tariff protection, agricultural subsidies, favorable tax regimes, currency management, and progressive privatization of prior state activities, like agricultural extension and lending for rural land purchasing (Sandrey and Reynolds, 1989; Johnson, 1991). Wider economic reforms were also instigated. Seen through this lens, New Zealand was described as a paragon of neoliberal intent, with the *Economist* magazine in 1987 dubbing the country "Adam Smith's other islands" (Campbell and Rosin, 2008, p. 11).

Diverging Industry Trajectories

Almost all the early dynamics reviewed in the Australian case are also observable in New Zealand. Le Heron and Pawson (1996) provide the most comprehensive review of the impacts of the reforms in declining terms of trade for meat and wool exporters, changes in the structure of farming, and the rise of new and decline of old industries (much of this review being substantiated by Campbell and Lawrence, [2003]). However, the situation differs in two important respects. First, the extreme export orientation of New Zealand agriculture has exposed farmers to the changing dynamics and strategies of supermarket chains in the global North, much more so than retail structures within the New Zealand economy itself. The broad trend with these supermarkets has been quite different from those manifested by the duopoly in Australia with a drive towards quality, sustainability, food safety, and the elaboration of complex systems of production and supply chain audit (Campbell and Rosin, 2008). While these dynamics have not been entirely absent in Australia (see Campbell et al., 2006), producers in Australia are much more exposed to domestic retail monopolization with few alternatives. Second, in New Zealand, the complex dynamics of trading in quality foods in the world market are evident—with some highly successful (dairy and kiwifruit) and some very unsuccessful (wool and sheepmeat) outcomes of the neoliberal transition "meeting market requirements" in agricultural products. This differs from Australia where the terms-of-trade of the agricultural export sector seemed to have generally declined across the board. Those sectors in New Zealand that have actually prospered under neoliberalization provide a significant departure from the narrative of Campbell and Lawrence (2003) and will become the focus for the rest of this chapter.

This mixed outcome for exporters demonstrates what might be termed the "diverging trajectories" explanation for the outcomes of neoliberalization in New Zealand (Campbell and Rosin, 2008; Rosin et al., 2012). This has been discussed in numerous parallel studies, including an examination of farming cultures between dairy and sheep farming (Stock and Peoples, 2012) and pastoral and horticultural sectors (Rosin, 2008); and the politics and contestation of deregulation by both dairy farmers (Muirhead and Campbell, 2012) and kiwifruit producers (Rosin, 2008; Rosin and Campbell, 2012).

While these diverging trajectories are evident in the way in which different export sectors have become part of reconfigured global trading arrangements, this is strongly influenced by changes occurring within New Zealand's farming sectors themselves. Here, a second line of difference with the Australian experience warrants examination. If the Australian case is characterized by what Cheshire (2005) and Cheshire and Lawrence (2005) describe as an ideological shift or congruence between farming culture and neoliberal subjectivity, then New Zealand demonstrates a much more contested arena of farm politics, particularly the implicit understandings of the role of collectives and cooperatives-versus-individualism in farming. This contested realm of farming culture sits at the heart of a divergence between wider industry sectors in terms of strategic responses to, and new trajectories under, neoliberalism. The best examples of this are the collective industries—dairy and fruit—versus the highly individualistic and competitive meat and wool sectors. This difference became apparent in the earliest stages of deregulation of the monopoly export boards.

Contested and Uncontested Deregulation of the Monopoly Export Boards

In 1989, only five years after the process of deregulating agriculture commenced in New Zealand, a Ministry of Agriculture publication (Sandrey and Reynolds, 1989) evaluated and praised the success of the removal of agricultural subsidies and deregulation of state interventions and governance structures in agriculture. The rapid deregulation of the local milk supply, egg supply, and wheat (for domestic bakers) towards unregulated, competitive domestic supply was hailed as an unparalleled success. While the politics of transition from a unitary structure for meat exporting (The Meat Export Board) and wool (The Wool Board) were complex, Le Heron and Pawson (1996) describe a process whereby large entities were either sold into international ownership, or, when an entity seemed less desirable or likely to survive the deregulation process (some large meat companies), they were reluctantly sold into farmer cooperatives. Even when cooperatives were formed to take over particular elements of the sector, the transition for the meat and wool export sectors was from a monopoly controlled "single desk" export market agency into multiple, competing export entities (both cooperatives and companies). This new direction, however, was not emerging in all aspects of export infrastructure, with commentators like Evans (2004) suggesting that the remnants of the large producer boards were the last blot on the otherwise stainless neoliberal record of New Zealand agriculture.

Despite this confidence in the preferred model of the neoliberal project, Rosin and Campbell (2012) argue that: (1) the outcomes of producer board deregulation have been highly contrasting and (2) that these outcomes were strongly influenced by the degree to which a prevailing culture of competitive or cooperative farming subjectivity dominated each sector (a claim that is pursued in greater depth by Stock and Peoples, [2012]).

Two trajectories emerged. One trajectory was comprised of those industries that experienced (and cooperated with) the full deregulation of monopoly export boards. In particular, the meat industry and the wool export sector cooperated with the state in dismantling the Meat Board and Wool Board. The Wool Board, in particular, had been the subject of grower hostility since the early 1970s through perceived abuse of statutory powers, inefficiency, and limited effectiveness. The subsequent decades since deregulation have not been kind to the wool industry, with a dominant model of competitive selling still seen as the correct way to market wool by both a majority of producers and the various actors in the agri-fiber chain that profit from an ongoing supply of cheap wool. Terms of trade for wool have continually declined with the sector shifting from contributing 19 percent of primary product export earnings in 1980 to only 3 percent in 2007 (MacDermott et al., 2008). One small breakaway—the fine-wool merino sector—has overcome difficulties and moved into new profitability. Despite this alternative model, the rest of the wool sector has fitfully attempted to coordinate and collectivize without success. Stock and Peoples (2012) attribute much of this to the prevailing existence of a strongly competitive and individualistic industry culture.

The second, alternative, trajectory is that demonstrated by those sectors that contested producer board deregulation or sought to capture those powers through grower-owned corporate monopolies. The kiwifruit sector had a regulated, multiple exporter structure at the outset of the reform period but entered a period of such catastrophically declining terms of trade that they successfully lobbied the government to establish the NZ Kiwifruit Marketing Board in 1988. This board, at best, stopped the decline of revenues for a few years before a second major crisis around fruit quality and residue levels in 1991 led to technical insolvency for the board in 1992 and further calls for deregulation. Campbell et al. (1997) reviewed how a combination of determined negotiation by sector leaders and very strong support by the wider group of kiwifruit producers saw the government approve a "grower owned corporation" model that retained (with some small exceptions) a monopoly exporting single desk for New Zealand kiwifruit. This model was quite different from the usual deregulationist scenario of the time and represented a major victory for kiwifruit growers and their highly empowered negotiators (Olssen, 2011). The new entity was established in 1997 as Zespri International Ltd and used a combination of new sustainable production techniques, direct contracts with elite retailers in the global North, a new variety of kiwifruit, and a new branding strategy around the Zespri label to oversee a period of massive expansion in the scope and profitability of kiwifruit production (Campbell and Rosin, 2008).

The other significant success story since the neoliberal reform period has been the dairy export sector. Emerging from the NZ Dairy Board (NZDB), the dairy industry has dramatically expanded both in terms of its political influence and economic output to become New Zealand's largest earner of export revenue (Gray et al., 2007; Gray and Le Heron, 2010). The average size of dairy herds has increased and—driven by its profitability and the productivist orientation of both farmers and their representatives—dairy farms are now expanding into increasingly unsuitable climatic zones along the drier East Coast of New Zealand on the back proliferating access to irrigation (Rosin and Campbell, 2012). Clearly, the dairy success story has been achieved partly through farm intensification, with attendant environmental problems (see Burton and Wilson, 2012), but the overall narrative is one of success and expansion in an era when the dominant neoliberal model oversaw the opposite for almost every other farm sector in Australasia. One important element of this success was that dairy industry leaders resisted deregulation and competition at every turn—supported by the strong cooperative ethos of the dairy industry. No farmers had entered dairying outside of some kind of cooperative structure, and when the pressure arose to deregulate the NZDB, cooperatives engaged in a defensive set of mergers until the final "mega-merger" in 2001, which created the fully farmer-owned and controlled Fonterra—an industry configuration that did not conform to any of the expectations of neoliberal advocates of structural reform in agriculture (Gray et al., 2007; Gray and Le Heron, 2010). As Muirhead and Campbell (2012) argue, to further deregulate the New Zealand dairy industry, the government would have had to purchase the shareholding of every dairy farmer in New Zealand and then manufacture an entire private market for the sale of dairy quota; there is little enthusiasm for such a radical reform.

After fifteen years of contestation, a compromise position was reached in which the full government assets of the NZDB were transferred into the ownership of the large cooperatives, which then merged to form Fonterra. The Dairy Industry Restructuring Act 2001 did impose a number of conditions to promote competition in the sector so as to avoid the accusation that the New Zealand dairy sector didn't meet World Trade Organization (WTO) requirements, but the conclusion of Gray et al. (2007) and Gray and Le Heron (2010) is that Fonterra represents an expression of the ongoing power of cooperative politics to underwrite a highly successful model of success in the global dairy trade. At the time of writing, Fonterra had become well entrenched as the largest dairy export organization in the world.

The Diverging Pathways: What Made the Difference in New Zealand?

The highly contrasting outcomes of the calamitous sheepmeat and wool trajectory in comparison to the success of Fonterra and Zespri can be argued to reside in four key differences between the sectors:

1 *Enduring Cooperative Structures*: in those sectors that mobilized against deregulation, the existence of a prior cooperative structure or, in the case of kiwifruit, the success in demanding that government legislate such a structure into existence, provided an institutional basis from which the political claims of the sector could be mobilized. Both kiwifruit and dairy had strong cooperative structures at key moments in the reform process that enabled effective push-back against efforts at reform. In contrast, sectors like wool had seen the dismantling of the large boards supported by producers as the first order of business in the reform process.
2 *Cooperative industry cultures*: the existence of the institutional framework for cooperatives was not enough in itself but tended to be aligned with strong producer adherence to cooperative politics. The sheep sector (competitive) and the dairy sector (collaborative) have very different industry cultures, something that has persisted for much of New Zealand's history (Stock and Peoples, 2012). Even when the meat sector introduced cooperatives as a crisis measure during the 1980s, this was arguably only to boost the number of competitors in the meat export sector at a time when the industry was so depressed that there were not enough private buyers to take over collapsing meat companies. In contrast, the kiwifruit and dairy sectors were characterized by a wider industry culture that celebrated cooperation as a pathway to shared prosperity and fought any attempts by government to set producers against each other.

3 *Elaboration of industry skills*: it is no coincidence that the sectors that retained strong, collaborative, industry structures also retained a higher level of R&D capacity, capability to direct research, and the continuation of extension services. Part of the dairy and kiwifruit success story has been the dramatic elaboration of new production techniques and management systems—albeit in diverging directions with dairy taking an intensificationist path, while kiwifruit has targeted a higher-quality, audited pathway to greater profitability.
4 *The destiny of ecologies*: what is less recognized about the New Zealand case is that the established deregulationist trajectory of farm amalgamation and corporate ownership does not translate well into New Zealand's highly ecologically diverse landscape. From the earliest moments of reform, rural sociologists like Fairweather (1989) did not observe a disappearing middle of family-sized farms or rapid elaboration of corporate farming. It is arguable that this is partly due to the particular landscape of farming in New Zealand. It is highly variegated with multiple land management techniques often required on one family-sized farm property. This does not translate well into large-scale corporate aggregation of farms of the kind that have consolidated farming in parts of the United States and Australia.

The combination of these four factors assists in explaining the difference in trajectory of the successful and unsuccessful industry sectors post-deregulation in New Zealand.

Discussion and Conclusion: What has Happened in the Last Ten Years?

Writing in 2003, we recognized that agriculture in both Australia and New Zealand had entered a structural crisis relating to various contradictory aspects of capital accumulation. These included polarization within the farm sector, expanding output accompanied by severe environmental consequences, and social and economic instability. These continue to remain concerns today. However, in the period since 2003, two key dynamics have come into play, both of which elaborate the particular trajectories of liberalization in Australia relating to the dynamics of marketization and financialization as well as the extent to which the 2000s represented a period of greater contestation of deregulation in some sectors in New Zealand.

Financialization in Australia

For Australia, neoliberalism can be viewed as having altered the conditions under which farming operates, enabling powerful actors (supermarkets, globally linked producers, and overseas firms) to improve their positions of wealth and influence.

While supermarket power has always been concentrated in Australia, what was not apparent a decade ago was the extent to which neoliberal policies would foster the financialization of the farm and food sectors. Finance capital is now prominent in the purchase of farmlands and irrigation licenses. Agriculture is a new "asset class" for a variety of financial actors (including hedge funds, merchant banks, sovereign wealth funds, private equity consortia, and pension funds), which are seeking, inter alia, food for export to water- and land-scarce nations, new markets in carbon trading, profits from biofuel production, and opportunities for speculation in land (Lawrence et al., forthcoming 2014). Quite simply, there is now much more to farming than food production. The so-called fungibility (such as the ready substitution of foods for biofuels) of farm crops gives investors flexibility in production, while the bypassing of traditional trading mechanisms (in terms of food repatriated by sovereign wealth funds) changes market dynamics. Globally, these changes have the potential to undermine the present multilateral systems of agricultural and energy trading (Cotula, 2012; McMichael, 2012).

Whereas futures markets in past decades helped farmers to transfer risk to those who hoped to make profits from anticipated price rises, deregulation has encouraged significant speculation (Lagi et al., 2011). Commodity Index Funds (CIFs) combine futures contracts in products as disparate as coal, oil, pigs, and wheat into a single bundle, which is then traded with investors seeking to make profits when the contracts reach fruition (IATP, 2011). Globally, CIF purchases doubled in value to four hundred billion US dollars between 2005 and 2008, which, in turn, helped to drive up the price of commodities (Clapp, 2012). But while these food price hikes are known to have had a major impact on the world's poor (see Lawrence et al., 2010), they have not translated to additional profits for the average Australian producer. From 2008 to 2009, some 69 percent of Australia's broadacre (wheat, sheep, cattle) properties posted a negative farm business profit, while the figure for dairy properties was 62 percent (Lagura and Ronan, 2009). Profits appear to have gone to the supermarkets and speculators, not the food producers.

There have been other worrying signs for farming in Australia. Neoliberal policies have reduced the government's commitment to agricultural R&D, which has resulted in a fall in farm-based productivity growth from 1.8 percent per annum to around 1.3 percent during the ten years to 2010 (PMSEIC, 2010, p. 16). Despite calls by the National Farmers' Federation and Australia's Chief Scientist for increased levels of investment (NFF, 2012), the states and federal governments—bound by fiscal constraints and wanting the farmers and farm industries, themselves, to take responsibility for their own fate—have been reluctant to act.

Contested Deregulation in New Zealand: An Antipodean Agrarian Question?

At the outset of this analysis, it was recognized that neoliberalization of Australian agriculture has had two distinguishing features: marketization and the ideology of "self-help" (Cheshire, 2005; Cheshire and Lawrence, 2005). While financialization is the defining feature of the first of these, the second warrants some closer consideration. As Guthman (2008) has argued, the pervasiveness of neoliberalism has limited the capacity of people to think and act. It "limits the conceivable because it limits the arguable, the fundable, the organizable, the scale of effective action, and compels activists to focus on putting out fires" (Guthman, 2008, p. 1180). That there are no strong and societally endorsed challenges to neoliberal hegemony from the left, the Greens, or populist politicians in Australia is indicative of the power of corporate business not only to shape the debates about the nation's economic future but also to determine the trajectory of Australia's farming and wider economy.

The contrast between this and the situation in New Zealand demonstrates that such subjectivities of neoliberalism are not totalizing but arise from situated politics, cultures, and contests in farming sectors. In Guthman's (2008) terms, alternatives to deregulation were unthinkable for the competitively minded sheep sector in New Zealand. However, the dairy and horticulture sectors—with their long history of cooperative and collaborative industry structure—did not find alternatives to deregulation unthinkable. The cases of Zespri and Fonterra reviewed here are even more helpful in recognizing where such alternatives are sited. In contrast to Australia, where there was a brief and controversial expression of rural protest in the national political scene through the One Nation party, New Zealand's main political parties since 1984 have had the same kind of broad commitment to liberal economic policy to that which can be seen in Australia. If dairy and kiwifruit sectors earned the consent of national political actors to engage in their compromise or alternative trajectories, they did it against the tide of national political sentiment.

Digging down to a deeper level, while the national farming lobby group—Federated Farmers of NZ—did everything in its power to support deregulation, other grower organizations, like the Fruit Growers Federation of NZ, did contest the deregulation process. Alongside them, the cooperatives (with significant farmer representation on their governing bodies) and wider farmer participants in farmer ownership supported those political leaders in each sector who had contested and negotiated the processes of industry restructuring.

The above analysis provides three important conclusions. First, in the last twenty-five years in both Australia and New Zealand, the marketization of agriculture, in the form of the unfettered pursuit of free markets, has proceeded apace. The recent "financialization" of farming—particularly for Australia—is simply a logical extension of marketization. Second, while marketization has become entrenched, the economic trajectory has been uneven. In fact, it is more fruitful to speak of "trajectories" rather than any coherent trajectory. In New Zealand, the cooperatives provided a site of strong resistance and subsequently faced neoliberalism on slightly more favorable terms. Given the ideological support for individualist strategies and solutions under neoliberalism, it is indeed ironic that the fiercely independent and competitive meat and wool sectors in New Zealand largely failed under neoliberalism, while the cooperative structures that underpinned dairy and fruit appear to have allowed those sectors to thrive. Third, the resistance shown in New Zealand agriculture notwithstanding, there has been overwhelming agreement in both countries, at the political level, for the continuation of neoliberal policy formation, including the embedding of economic rationalist settings. This is, perhaps, the most compelling finding in this review. The farm leaders, politicians, and opinion leaders have been captured, and farmers, themselves, accept the inevitability of an uncertain life under free-market forces. This is the most profound legacy of neoliberal hegemony. What also remains clear is that the contradictions, crises, and uncertainties of capitalist farming do not appear to have disappeared during an era of neoliberalism; in fact, they have become more apparent. By sketching out some sites of resistance and contestation that have emerged over the last decade, we have, in effect, demonstrated the opposite—how partial and situated those few points of contestation have been in an era where neoliberalized agriculture in the Antipodes transitioned from a radical proposition to a hegemonic reality.

References

ABC News. (2011) "Woolies admits milk war will hurt farmers," ABC News, February 28, www.abc.net.au/news/stories/2011/02/28/3150349.htm, accessed March 20, 2011.

Australian Competition and Consumer Commission [ACCC]. (2001) *Impact of Farmgate Deregulation on the Australian Milk Industry: Study of Prices, Costs and Profits*, ACCC, Canberra.

ACCC. (2008) *Report of the ACCC Inquiry into the Competitiveness of Retail Prices for Standard Groceries (July 2008)*, Commonwealth of Australia, Canberra.

Almås, R. and Campbell, H. (eds) (2012) *Rethinking Agricultural Policy Regimes: Food Security, Climate Change and the Future Resilience of Global Agriculture*, Emerald Group Publishing, Bingley, UK.

Andrews, N., Buetre, B., Davidson, A., McDonald, D., Jotzo, F., and Fisher, B. (2003) "Agricultural trade reform: Benefits for Australian broadacre agriculture," *Australian Commodities*, vol 10, no 2, pp. 249–59.

Argent, N. (2002) "From pillar to post? In search of the post-productivist countryside in Australia," *Australian Geographer*, vol 33, no 1, pp. 97–114.

Argent, N. and Rolley, F. (2000) "Lopping the branches: Bank branch closure and Rural Australian communities," Pp. 140–68 in B. Pritchard and P. McManus (eds) *Land of Discontent: The Dynamics of Change in Rural and Regional Australia*, UNSW Press, Sydney.

Beck, U. and Beck-Gernsheim, E. (2002) *Individualization*, Sage Publications, London.

Bienabe, E., Boselie, D., Collion, M., Fox, T., Rondot, P., van der Kop, P., and Vorley, B. (2007) "The internationalization of food retailing: Opportunities and threats to small-scale producers," Pp. 140–68 in B. Vorley, A. Fearne, and D. Ray (eds) *Regoverning Markets: A Place for Small-scale Producers in Modern Agri-food Chains?*, Gower, Farnham.

Bita, N. (2012) "Brawl heats up with foreign-owned farms on the rise," *Australian*, January 19, www.theaustralian.com.au/national-affairs/brawl-heats-up-with-foreign-owned-farms-on-the-rise/story-fn59niix-1226247779267, accessed February 26, 2013.

Blunden, G., Moran, W., and Bradly, A. (1996) "Empowering family farms through cooperatives and producer marketing boards," *Economic Geography*, vol 72, no 2, pp. 161–77.

Burch, D., Dixon, J., and Lawrence, G. (2012) "Introduction to symposium on the changing role of supermarkets in global supply chains: From seedling to supermarket—agri-food supply chains in transition," *Agriculture and Human Values*, vol 30, no 2, pp. 215–24.

Burch, D. and Lawrence, G. (eds) (2007) *Supermarkets and Agri-food Supply Chains: Transformations in the Production and Consumption of Foods*, Edward Elgar, Cheltenham, UK.

Burch, D. and Lawrence, G. (2013) "Financialization in agri-food supply chains: Private equity and the transformation of the retail sector," *Agriculture and Human Values*, vol 30, no 2, pp. 247–58.

Burton, R. and Wilson, G. (2012) "The rejuvenation of productivist agriculture: The case for 'cooperative neo-productivism,'" Pp. 51–72 in Almås and Campbell, *Rethinking Agricultural Policy Regimes.*

Campbell, H. (2012) "Let us eat cake? Historically reframing the problem of world hunger and its purported solutions," Pp. 30–45 in C. Rosin, P. Stock, and H. Campbell (eds) *Food Systems Failure: The Global Food Crisis and the Future of Agriculture*, Earthscan, London.

Campbell, H., Fairweather, J., and Steven, D. (1997) *Recent Developments in Organic Food Production in New Zealand: Part 2, Kiwifruit in the Bay of Plenty*, Studies in Rural Sustainability No. 2, Department of Anthropology, University of Otago, New Zealand.

Campbell, H. and Lawrence, G. (2003) "Assessing the neo-liberal experiment in antipodean agriculture," Pp. 89–102 in R. Almås and G. Lawrence (eds) *Globalization, Localization and Sustainable Livelihoods*, Ashgate, Aldershot, UK.

Campbell, H., Lawrence, G., and Smith, K. (2006) "Audit cultures and the Antipodes: The implications of EurepGAP for New Zealand and Australian agri-food industries," Pp. 69–94 in J. Murdoch and T. Marsden (eds) *Between the Local and the Global: Confronting Complexity in the Contemporary Agri-Food Sector*, Elsevier, The Netherlands.

Campbell, H. and Rosin, C. (2008) "Global retailer politics and the quality shift in NZ horticulture," Pp. 11–26 in M. Butcher, J. Walker, and S. Zydenbos (eds) *Future Challenges in Crop Protection: Repositioning New Zealand*'s *Primary Industries for the Future*, NZ Plant Protection Society, Hastings.

Cheshire, L. (2005) *Governing Rural Development: Discourses and Practices of Rule in Australian Rural Policy*, Ashgate, Aldershot.

Cheshire, L. and Lawrence, G. (2005) "Neoliberalism, individualization and community: Regional restructuring in Australia," *Social Identities*, vol 11, no 5, pp. 85–96.

Cheshire, L. and Woods, M. (2013) "Globally engaged farmers as transnational actors: Navigating the landscape of agri-food globalization," *Geoforum*, vol 44, pp. 232–42.

Clapp, J. (2012) *Food*, Polity, Cambridge, UK.

Cook, H. (2012) "Milk wars leave sour taste in farmers' mouths," *Sydney Morning Herald*, www.smh.com.au/business/milk-wars-leave-sour-taste-in-farmers-mouths-20120120-1q9st.html, accessed December 13, 2012.

Cotula, L. (2012) "The international political economy of the global land rush: A critical appraisal of trends, scale, geography and drivers," *Journal of Peasant Studies*, vol 39, no 3–4, pp. 649–80.

Cox, A. and Chicksand, D. (2007) "Are win-wins feasible? Power relations in agri-food supply chains and markets," Pp. 74–99 in Burch and Lawrence, *Supermarkets and Agri-food Supply Chains.*

Dairy Australia. (2013) "Cows and farms," Dairy Australia, www.dairyaustralia.com.au/Statistics-and-markets/Farm-facts/Cows-and-Farms.aspx, accessed February 21, 2013.

Dean, M. (1999) *Governmentality: Power and Rule in Modern Society*, Sage Publications, Thousand Oaks, California.

Department of Foreign Affairs and Trade (DFAT). (2012) *Agriculture and the WTO*, DFAR, www.dfat.gov.au/trade/negotiations/trade_in_agriculture.html, accessed November 21, 2012.

Dibden, J. and Cocklin, C. (2005) "Sustainability and agri-environmental governance," Pp. 135–52 in V. Higgins and G. Lawrence (eds) *Agricultural Governance: Globalization and the New Politics of Regulation*, Routledge, London.

Drummond, I., Campbell, H., Lawrence, G. and Syme, D. (1999) "Contingent or structural crisis in British agriculture?," *Sociologia Ruralis*, vol 40, no 1, pp. 111–27.

Durie, J. (2011) "Big supermarkets gain fresh food market share at the expense of the small guys," *Australian*, March 9, www.theaustralian.com.au/business/opinion/big-supermarkets-gain-fresh-food-market-share-at-the-expense-of-the-small-guys/story-e6frg9if-1226018006708, accessed February 21, 2013.

Edwards, G. (2003) "The story of deregulation in the dairy industry," *Australian Journal of Agricultural and Resource Economics*, vol 47, no 1, pp. 75–98.

Evans, L. (2004) "Structural reform: The dairy industry in New Zealand," paper prepared for the APEC High Level Conference on Structural Reform, Tokyo, September 8–9, 2004.

Fairweather, J. (1989) *Some Recent Changes in Rural Society in New Zealand*, AERU Discussion Paper No. 124, Lincoln University, New Zealand.

Gerritsen, R. (2000) "The management of government and its consequences for service delivery in regional Australia," Pp. 123–39 in Pritchard and McManus, *Land of Discontent.*

GRAIN. (2012) "GRAIN releases data set with over 400 land grabs," GRAIN, February 23, www.grain.org/article/entries/4479-grain-releases-data-set-with-over-400-global-land-grabs per cent0D, accessed November 8, 2012.

Gray, I. and Lawrence, G. (2001) *A Future for Regional Australia: Escaping Global Misfortune*, Cambridge University Press, Cambridge.

Gray, S. and Le Heron, R. (2010) "Globalising New Zealand: Fonterra Co-operative Group, and shaping the future," *NZ Geographer*, vol 66, no 1, pp. 1–13.

Gray, S., Le Heron, R., Stringer, C., and Tamasy, C. (2007) "Competing from the edge of the global economy: The globalising world dairy industry and the emergence of Fonterra's strategic networks," *Die Erde*, vol 138, no 2, pp. 1–21.

Guthman, J. (2008) "Neoliberalism and the making of food politics in California," *Geoforum*, vol 39, pp. 1171–83.

IATP. (2011) *Excessive Speculation in Agricultural Commodities*, Institute for Agriculture and Trade Policy, Minneapolis, MN.

Johnson, R. (1991) "Current changes in New Zealand agriculture: A review," *Review of Marketing and Agricultural Economics*, vol 59, no 2, pp. 130–48.

Kruger, C. (2013) "Supermarkets war heads for the corner," *Northern Daily Leader*, August 19, www.northerndailyleader.com.au/story/1259934/supermarkets-war-heads-for-the-corner/?cs=9, accessed 21 February 2013.

Lagi, M., Bar-Yam, Y., Bertrand, K., and Bar-Yam, Y. (2011) *The Food Crises: A Quantitative Model of Food Prices Including Speculators and Ethanol Conversion*, New England Complex Systems Institute, Cambridge, MA.

Lagura, E. and Ronan, G. (2009) *How Profitable Is Farm Business in Australia? An Interpretation of ABARE Broadacre and Dairy Industries Farm Performance Data and Some Implications for Public Policy*, www.agrifood.info/connections/2009/Lagura_Ronan.html, accessed February 27, 2013.

Lawrence, G. (1987) *Capitalism and the Countryside: The Rural Crisis in Australia*, Pluto Press, Sydney.

Lawrence, G., Lyons, K., and Wallington, T. (2010) *Food Security, Nutrition and Sustainability*, Earthscan, London.

Lawrence, G., Richards, C., and Burch, D. (2013) "The impact of climate change on Australia's food production and exports," Pp. 173–86 in Q. Farmar-Bowers, J. Miller, and V. Higgins (eds) *Food Security in Australia: Challenges and Prospects for the Future*, Springer Science, New York.

Lawrence, G., Richards, C., Gray, I., and Hansar, N. (2012) "Climate change and the resilience of commodity food production in Australia," Pp. 131–46 in C. Rosin, P. Stock, and H. Campbell, *Food Systems Failure.*

Lawrence, G., Sippel, S., and Burch, D. (forthcoming 2014) "The financialization of food and farming," in G. Robinson and D. Carson (eds) *Handbook on the Globalization of Agriculture*, Edward Elgar, Cheltenham, UK.

Le Heron, R. and Pawson, E. (eds) (1996) *Changing Places: New Zealand in the Nineties*, Longman Paul, Auckland.

Le Heron, R. and Roche, M. (2009) "Rapid reregulation, agricultural restructuring, and the reimaging of agriculture in New Zealand," *Rural Sociology*, vol 64, no 2, pp. 203–18.

Lockie, S. (2000) "Crisis and conflict: Shifting discourses of rural and regional Australia," Pp. 14–32 in Pritchard and McManus, *Land of Discontent.*

Lockie, S., Lawrence, G. and Cheshire, L. (2005) "Reconfiguring rural resource governance: The legacy of neoliberalism in Australia," Pp. 29–43 in P. Cloke, T. Marsden, and P. Mooney (eds) *Sage Handbook of Rural Studies*, Sage, London.

MacDermott, A., Saunders, C., Zellman, E., Hope, T., and Fisher, A. (2008) *Sheep Meat: The Key Elements of Success and Failure in the NZ Sheep Meat Industry from 1980–2007*, Agribusiness Research and Education Network, Lincoln, New Zealand.

Martin, P., Riley, D., Lubulwa, M., Knopke, P., and Gleeson, T. (2000) *A Report to the Australian Dairy Industry Survey: Australian Dairy Industry 2000*, ABARE, Canberra.

McMichael, P. (2012) "Biofuels and the financialization of the global food system," Pp. 60–82 in C. Rosin, P. Stock, and H. Campbell, *Food Systems Failure.*

Millward, J. (2011) "Milk price war will dry up competition warns Choice," *West Australian*, March 15, http://au.news.yahoo.com/thewest/regional/gascoyne/a/-/news/9012966/milk-price-war-will-dry-up-competition-warns-choice/, accessed March 22, 2011.

Muirhead, B. and Campbell, H. (2012) "The worlds of dairy: Comparing dairy frameworks in Canada and New Zealand in light of future shocks to food systems," Pp. 147–68 in Almås and Campbell, *Rethinking Agricultural Policy Regimes.*

Munro, P., Tippet, G., and Hyland, T. (2013) "Dark side of the dollar," *Age*, www.theage.com.au/national/dark-side-of-the-dollar-20120211-1sylg.html, accessed February 27, 2013.

NFF. (2012) "Chief scientist backs calls for renewed R&D investment," National Farmers' Federation, December 11, www.nff.org.au/read/3763/chief-scientist-backs-calls-for-renewed.html, accessed February 24, 2013.

Olssen, E. (2011) *The Fruition: NZ Fruit Growers Federation 1991–2005*, New Zealand Fruitgrowers Federation, Wellington, New Zealand.

Peters, J. (2011) "Neoliberal convergence in North America and Western Europe: Fiscal austerity, privatization and public sector reform," *Review of International Political Economy*, vol 19, no 2, pp. 208–35.

PMSEIC. (2010) *Australia and Food Security in a Changing World*, The Prime Minister's Science, Engineering and Innovation Council, Canberra.

Pritchard, B. (2000) "Negotiating the two-edged sword of agricultural trade liberalization: Trade policy and its protectionist discontents," Pp. 90–104 in Pritchard and McManus, *Land of Discontent.*

Pritchard, B., Burch, D., and Lawrence, G. (2007) "Neither 'family' nor 'corporate' farming: Australian tomato growers as farm family entrepreneurs," *Journal of Rural Studies*, vol 23, pp. 75–87.

Pritchard, B. and McManus, P. (eds) (2000) *Land of Discontent: The Dynamics of Change in Rural and Regional Australia*, UNSW Press, Sydney.

Productivity Commission. (2005) *Trends in Australian Agriculture*, Commonwealth of Australia, Canberra.

Rego, F. (2012) "Supermarket 'milk wars' continue to push farmers out," ABC News, September 25, www.abc.net.au/news/2012–09–25/supermarket-milk-wars-continues-to-push-farmers-out/4278824, accessed December 13, 2012.

Richards, C., Lawrence, G., Bjorkhaug, H., and Hickman, E. (2012) "Retailer-driven agricultural restructuring—Australia, the UK and Norway in comparison," *Agriculture and Human Values*, vol 30, no 2, pp. 247–58.

Rickson, R. and Burch, D. (1996) "Contract farming in organizational agriculture: The effects upon farmers and the environment," Pp. 173–202 in D. Burch, R. Rickson, and G. Lawrence (eds) *Globalization and Agri-food Restructuring: Perspectives from the Australasia Region*, Avebury, Aldershot.

Rose, N. (1999) *Powers of Freedom: Reframing Political Thought*, Cambridge University Press, Cambridge.

Rosin, C. (2008) "The conventions of agri-environmental practice in New Zealand: Farmers, retail driven audit schemes and a new *Spirit of Farming*," *GeoJournal*, vol 73, no 1, pp. 45–54.

Rosin, C. and Campbell, H. (2012) "The complex outcomes of neoliberalization in New Zealand: Productivism, audit and the challenge of future energy and climate shocks," Pp. 191–210 in Almås and Campbell, *Rethinking Agricultural Policy Regimes.*

Rosin, C., Stock, P., and Campbell, H. (eds) (2012) *Food Systems Failure: The Global Food Crisis and the Future of Agriculture*, Earthscan, London.

Sandrey, R. and Reynolds, R. (1989) *Farming Without Subsidies*, GP Books, Wellington.

Savills Research. (2012) *International Farmland: Focus 2012*, Savills, UK.

Schneider, M. and McMichael, P. (2010) "Deepening, and repairing the metabolic rift," *Journal of Peasant Studies*, vol 37, no 3, pp. 461–84.

Stock, P. and Peoples, S. (2012) "Commodity Competition: Divergent Trajectories in New Zealand Pastoral Farming," Pp. 263–84 in Almås and Campbell, *Rethinking Agricultural Policy Regimes.*

Vorley, B. (2007) "Supermarkets and agri-food supply chains in Europe: Partnership and protest," Pp. 243–67 in Burch and Lawrence, *Supermarkets and Agri-food Supply Chains.*

15

CONCLUSION

The Plasticity and Contested Terrain of Neoliberalism

Steven A. Wolf and Alessandro Bonanno

The Paradox of Interpretive Flexibility

As an ideology and a discourse, neoliberalism informs and sustains practice, investment, education, and regulation. It has been dominant for the last three decades and has real power. Yet, as a social construct interpreted and invoked in very different circumstances, it is incoherent and self-contradictory. Going further, we note that it is adaptive to be variable and flexible in terms of expression, both material and linguistic. In other words, the plasticity of concepts such as neoliberalism explains their power in technocratic and popular domains (Allaire and Wolf, 2004; Wolf and Klein, 2007). Concepts such as sustainability, multifunctionality, or organic have become mainstream references in significant part because of ample space within the concepts into which people and organizations can project variable justifications and prescriptions. The commons, community-based action, and even property rights are concepts characterized by remarkable capacity to integrate the projects of sets of actors occupying very different structural and ideological positions. The strength of these references is their capacity to catalyze and maintain conversations among sets of people who share very little. Their power derives from interpretive flexibility.

Rigidity and fidelity in people's understanding of objects and ideas in global circulation are rare. When it does happen and a standard or technique emerges, institutional change slows down dramatically. These are exceptional circumstances, and it is clear that problems and questions do not, as a rule, get resolved. Generally, there is a measure of sociotechnical coherence in the modern world, but this coherence is uneven and incomplete. Most commonly, we muddle through (Lindblom, 1959), and this muddling is supported by institutions. Conventional understanding that makes it possible to plan, share, and transact emerges through social labor of coordination (Storper, 1998). Order is neither imposed from outside and above nor strictly derived from interactions within communities. Efficiency considerations, political economy, cultural norms, and learning through trial and error are interdependent explanations of emergent institutional forms. The implication here is that coordination is often local and contingent.

More often than not, there is no comprehensive clarity or closure. References circulate, ideas compete, mash-ups happen. Misattributions and appropriations sometimes produce powerful results that initiate a new thread. The imprecision and flexibility of neoliberalism allows it to sustain a far-reaching conversation with material consequences for people and the planet. As an open-ended object capable of orienting people and organizations at great distances and across great divides, neoliberalism is clearly powerful. It is, however, an ad hoc social construction, and its power derives from its openness and flexibility. This claim is, of course, perverse in the sense that Margaret Thatcher's position that "there is no alternative (TINA)" and the Washington macroeconomic and strategic "consensus" do not evoke images of flexibility and accommodation. Here we want to stress the paradox. Concepts anchoring some of the most critical political debates are often poorly specified and weakly supported empirically.

As Busch points out in this volume, neoliberalism is mythical. It is not true, yet its power is real. It is performed in varying circumstances, on an ongoing basis, in a dizzying array of forms. Because its meaning, practices, outcomes, and justifications are varied, it is easy to imagine alternative performances. This volume demonstrates the flexibility and organic character of neoliberalism and its status as a heterogeneous amalgamation. If standards and techniques imply institutional lock-in (Ellul, 1964), lack of agreement and openness implies institutional flux—perhaps even abrupt change (True et al., 1999). This does not mean that change will result in progressive developments. We can imagine regressive outcomes just as easily. But change is certain.

The plasticity of neoliberalism is often accompanied by theorizations that aim to systematize its characteristics in coherent analytical constructs. Despite warnings (Harvey, 2005, p. 21), Kotz and McDonough (2010, p. 93) write: "we view neoliberalism as a coherent, multi-leveled entity whose core features include political-economic institutions, policies, theories, and ideology."[1] These types of interpretations of neoliberalism as a *regime* are not simply periodizations of the growth of mature capitalism, but they are meant to be theories of the transformation of society. One of their major characteristics, but also grave weaknesses, is their tautological tendency to assume, and also conclude, that the end of one regime is followed by the emergence of a new one. Accordingly, neoliberalism as the response to the crisis of Fordism is understood as a confirmation of the theory that regimes follow each other. Underscoring the complexity and diversity of the forms of domination, but also the internal contradictions of neoliberalism, the chapters in this book suggest that the notion of regimes could be better employed in reference to historical conditions (most notably, see chapters in this volume by Patel-Campillo, Wolf, and Lawrence and Campbell) rather than theoretical constructions.

In their diversity, the chapters underscore the simultaneous crisis and resilience of neoliberalism, its domination and the resistance that it engenders, and the discursive power along with its myths and inconsistencies. These chapters point out the inadequacy of the instruments advocated to solve neoliberalism's many crises, as well as the difficulties associated with the availability of alternatives (Buck and Som Castellano in this volume). As at the end of the crisis of Fordism, Claus Offe (1985) spoke of ***disorganized capitalism***, we may find ourselves in a world in which an organized regime and a dominant discourse may not necessarily emerge. As this hypothesis is often dismissed for the search of the new emerging regime and acts of opposition to neoliberalism continue, the contested terrain of global capitalism assumes historical and epistemic relevance. Its historical dimension centers on both episodes of domination and the struggles against them that constitute contemporary agri-food, governance, and capitalism. Epistemologically, its open-endedness allows the inclusion of alternative explanations to the one-sided proposal of the inevitable emergence of a new dominant regime.

Because the book defines and surveys the variable, contingent, and contested terrain in which neoliberalism unfolds, we want to avoid a reification of the concept. The plasticity discussed above and demonstrated in the chapters of this volume highlight our insistence that neoliberalism is a socially constructed process generated by the actions of social groups. It is not a single, unified, coherent project endowed with its own life and energy, and on this basis, we chose not to offer a definition of neoliberalism in the Introduction of this volume. Neoliberalism is the process and outcome of contested actions that find their origins in historically, cognitively, organizationally, and geographically situated human agency. It is a general reference, and thus rupture is a partially misleading metaphor. Neoliberalism draws on various ideologies, evidence bases, and material practices. When any single element of neoliberalism loses its power to enroll and discipline actors, the ship does not sink.

Crisis of Neoliberalism

The crisis of neoliberalism is not only a structural crisis with its economic and social downturns, it is also a crisis of legitimacy (Bonanno, this volume). Within the institutional tradition (Meyer and Rowan, 1977), legitimacy is the core resource that lends power to ideas, actors, and actions. In this sense, it is fundamental. The effectiveness and efficiency of prescriptions deriving from neoliberal thought are increasingly questioned. Critique and contestation continue to derive from civil society as it has for many years, but we observe sets of state and commercial actors questioning the objectives and justifications of policy orthodoxy with increasing frequency and conviction. The IAAKSTD report is an excellent example of critique emanating from near the core. Peschard's account in this volume of Brazilian soy farmers' withdrawal from the pro-GMO coalition is another. The cause and effect relationships asserted by neoliberal thought are increasingly unconvincing. Claims regarding the wisdom of more atomization, expanded reliance on self-organization, and fewer social controls are losing their power. Neoliberalism does not deliver the goods, even for an increasing segment of the elites.

The crisis of legitimacy also derives from the inability of the state to brook the cracks that emerge from neoliberal practices and the human condition more generally. The inability of neoliberal prescriptions to produce new knowledge and new resources for development and problem solving is increasingly perceived by its subjects and its many local architects. The nation-state has been the subject of a double transformation. First, it has been restructured from within through the neoliberalization of its agencies and policies. The socially oriented interventions and policies of the old Fordist state have been significantly reduced and/or eliminated under neoliberalism. Second, the ability of the state to regulate socioeconomic processes has been reduced through extensive diffusion of popular myths about the ineptitude of public bureaucracies, on the one hand, and the transnationalization of economic and social relations on the other. In this context, the governance of processes that constitute the essence of agri-food as well as other sectors of the economy has been increasingly placed outside the sphere of action of the state. Yet, the state is continuously asked to intervene when crises emerge (witness the 2008 global financial collapse). It follows that the ability of the state to address the unwanted consequences of the expansion of the economy and the aspirations of the civil society cannot materialize. The nation-state does not have the instruments and does not control the conditions that affect current socioeconomic relations. This core internal contradiction—neoliberalism has undermined the standing of the state, which is a necessary component for sustaining structural coherence of governance and sociotechnical arrangements—continues to erode the legitimacy of neoliberalism. Lock-in to a set of ideas, promises, and practices that are untenable constitutes a death spiral.

Beyond the structural and cognitive dimensions of the crisis of neoliberalism, we identify material and institutional contradictions. Food riots, land/natural resource grabs, biodiversity decline, climate change, and grinding poverty for billions of people represent core material problems. Economic, social, and ecological indicators point to tipping points. Many people of many persuasions hold views that we confront significant social problems that require major changes in economy and our relationships to the natural world on which our lives and our prosperity depend (Jackson, 2009). To the extent that scarcity is a technical problem, this does not necessarily constitute a crisis. The crisis stems from a lack of institutional alternatives. What is scarce is coordination capacity. We do not have proven, practical institutional responses to what ails us, and the switching costs and the learning challenges make transition frightening. On this basis, transitions under democratic rule are not easy to imagine.

In the past, Milton Friedman ([1962] 1982) and like-minded neoliberals eased social and economic fears through the promise that relying on the functioning of the free market and eliminating the distortions caused by an overbearing interventionist state would generate a period of sustained economic growth, social well-being, and stability. These promises have not been met. The major alternative proposal rested on a return to greater state intervention. Summarizing this intellectual current, the left-leaning economist Paul Krugman (2013, 2012, 2008) stressed that Fordist-style Keynesian deficit spending and the reintroduction and strengthening of welfare state expenditures would end the neoliberalism-induced great recession. However, this neo-Fordist option has not yielded convincing results, as bailouts, state control of interest rates, and other measures reminiscent of Fordist-style state intervention have produced controversial outcomes at best (Lopez, 2013). Accordingly, claims that we should not expect a transition to a new equilibrium and that implosion is likely are difficult to contemplate. Hard landings are bad outcomes ideally left to producers of poorly made dystopian movies. But we must contemplate such events. It is not self-evident that the "system" is self-correcting or correctable in relevant time steps. Crisis implies disjuncture and irreversibility, a trajectory opposite to learning and adaptation.

While authors in this volume are not always explicit about their future expectations, mobilization of the state by civil society (Otero in this volume), fragmentation of corporatist advocacy coalitions (Peschard in this volume), institutional hybridity (Wolf in this volume), reimagining of the commons (Buck in this volume), pragmatic entrepreneurial action in solidarity with local people (Iba and Sakamoto in this volume), enlisting actors in a reflexive assessment of their role in performance of neoliberalism (Busch in this volume) are, fundamentally, optimistic notions. They suggest learning and a progressive dynamic in which we will realize a democratic surplus and material payoffs in terms of equity and social, economic, and ecological risk reduction. There are reasons to believe that we are not locked into a neoliberal paradigm.

At the same time, authors highlight clear evidence of continual unfolding of neoliberalism through financialization (Fairbairn in this volume), cultural acceptance of inequities in labor markets (Harrison in this volume), mobilization of competitiveness as a means of discipline (Patel-Campillo and Lawrence and Campbell in this volume), and dissolution of national commitments to, for example, marketing cooperatives (Magnan in this volume). There is ample evidence that neoliberalism is healthy. The conclusion we draw from what was expected to be contradictory evidence is that reports of rupture, morphing, continuity, and deepening of neoliberal modes of operating and reasoning are valid. But more generally, our analysis leads us away from any effort to make a global statement about neoliberalism. It is not an "it." It is a variable process that is realized differently at various scales and in specific contexts, and it's a process that emerges from the contested terrain of domination and opposition.

The crisis of neoliberalism can be understood as angst and uncertainty regarding development pathways during a historical transition. While it is quite likely that all peoples in their time perceive epochal shifts to be underway, we are between periods. We are moving away from a world based on international relations to a world dominated by global relations. A focus on economic growth is, haltingly, giving way to a focus on the implications of ecological scarcity and social inequality. The coordination strategies and the narratives regarding the coordination strategies that we rely on to support development and to respond to social and ecological problems must change. What is behind us is, of course, more clear than what lies ahead. The crisis stems from institutional stickiness—the costs and risks of departure and the politics of unseating incumbents and embedded ideologies.

Neoliberalism grew out of insecurity. Whether we emphasize problems of capital accumulation and recall of the welfare state (Harvey, 2005) or we emphasize concerns about totalitarian states in the middle of the twentieth century (Busch this volume), liberalization, deregulation, and celebration of rights of individuals have produced a response. We have lived through a period of unprecedented economic expansion and a time in which nation-states have come to find themselves subject to a range of new controls from above and below. By some accounts, democracy, civil society, and accountability have expanded, but the rise and significance of transnational corporations and their often destabilizing power cannot be disputed. This is what lies behind us.

We now find ourselves confronting very different kinds of insecurity. The problems of economic inequality and disunity at global and domestic levels threaten us all. The prosperity of our children and their life prospects are undermined by inequality. Second, ecological scarcity and degradation constitutes a grave risk. The problems and constraints of our age are not the problems and constraints of the previous period. A new specification is required, and new institutional arrangements are needed. While some would argue that neoliberalism was a misguided project from the start, we can take a more generous view. It has outlived its useful life. The problem is where to go from here and how to get there. While abstract, Polanyi's ([1944] 2001) analysis remains central. Social protection by the state from an excess of market rule, as specified by civil society, is a good place to start.

Institutional Hybridity

While Polyani's metaphorical pendulum suggests that the institutional orders of state and market are discrete and that we move through time from dependence on one or the other pole of governance, it is essential to recognize their interdependence and even their co-construction. The classical notion of political economy contains within it the understanding that politics and law are embedded in market relations and vice versa. Quite frequently, this conception is lost. While academic analyses of neoliberalism reference the relational nature of state and market processes, theories and empirical analyses of interconnections are needed. We do not have a vocabulary, let alone a solid understanding of how institutional orders are combined, layered, and put into productive tension with one another. Tapping into the power of public (state), private (market), and collective (community) modes of coordination and accountability rather than focusing on sweeping claims about the virtues of one over the others could yield institutional innovations leading to improved material welfare and a democratic surplus. Evans (1996), Allaire and Wolf (2004), Lockie and Higgins (2007), and Hodge and Adams (2012) may be useful references in this regard.

The programmatic suggestion that the analytical project of advancing understanding of how combinations and admixtures of public, private, and collective coordination strategies perform in various contexts could unlock new governance capabilities and move us beyond a stale political debate invites critical reflection. As the nation-state adopts neoliberal postures but, simultaneously, remains the locus of resistance and the private sphere emerges as the new site of governance, blurring and blending of public and private logics becomes one of the characteristics of current socioeconomic arrangements. The chapters show the complex, uneven, and ultimately contradictory dimensions of institutional hybridity. They provide evidence that the crude contraposition between state and market does not capture the multifaceted neoliberalization of the state and privatization of governance. The nation-state has been an active agent of the deregulation and/or reregulation of socioeconomic processes. The adoption of neoliberal policies transformed it into a vehicle for the implementation of ideas and practices that reinforce neoliberalism. Simultaneously, it has also been the locus of resistance in terms of promoting conscious efforts in opposition to the introduction of market-oriented policies but also reproducing an intricate web of entrenched practices that retard and, in some instances, also defeat, the move toward the market and individual entrepreneurship. The privatization of regulatory mechanisms and the introduction of market calculus into administrative routines have been presented as a fruitful and democratic form of organization of agri-food. While this evaluation continues to be supported by some, the criteria of evaluation employed in these assessments often assumes the inequitable and ruling-class-supporting connotations that are characteristic of neoliberalism.

Conclusion: Ways Forward

The chapters in this book provide evidence that insecurity, a turn toward more power of corporate entities, but also hopes for democratic changes, characterize this period of neoliberalism. As intellectual efforts to analyze the characteristics and evolution of agri-food continue, the contributions in this volume offer relevant rationales about research areas for the immediate future. While the stimuli contained in the chapters are multiple, we would like to stress the relevance of three areas of research: public policy and the state, resistance and its individualization, and the class dimension of neoliberalism.

A number of chapters (Roy, Peschard, Wolf, Patel-Campillo, Magnan, Som Castelano in this volume) call for attention to public policy and the working of the state in its various and/or emerging forms. Along with concerns about the availability of effective institutional instruments (Bonanno, Busch in this volume), these contributions offer reasons for further investigation of the intersection of market, the state, and agri-food. As the transition toward a more integrated global society continues, the instruments to operate in this emerging social, economic, and political environment are difficult to identify and therefore to be discussed with analytical but also political clarity. Additionally, the common proposals for an increased dose of neoliberalism on one side, and a return to more state interventionism on the other, beg for the discussion of alternatives. As nation-states find it increasingly difficult to deliver security and material benefits through reliance on the old repertoire of economic growth and pushing costs off onto peripheral territories and future generations, there would seem to be an incentive for state actors to participate constructively in a shift toward accountability founded on different metrics. Here, more holistic, more localized, more subjective measures of well-being gain traction as public purposes and sources of legitimation for the state. In this context, the characteristics, meanings, and potential uses of public policies along with the identification of other available instruments of intervention and action are questions that should be addressed.

Agri-food corporations developed new agendas to enhance their power in the evolving global society. Their control of new technology and the resistance that this action entailed have been discussed in the book (Otero, Peschard and Roy in this volume). While this interest in the social and political dimensions of, and opposition to, the diffusion of biotechnology is reminiscent of older debates of the 1980s and 1990s, it opens the question of resistance and the forms through which it is carried out. In this regard, we note that no chapters in this volume address consumption as a form of resistance. Recognizing that the power but also limits of ethical and/or reflexive consumption are relevant and popular themes of analytical investigation in the contemporary setting, we should also understand that the propensity to introduce individualistic solutions to crises represents an overt pattern. In emphasizing tendencies toward individuation in analysis and prescriptions regarding consumption vis-a-vis neoliberalization, we recognize a strong parallel with emphasis placed on entrepreneurship on the supply side (Iba and Sakamoto wrestle with this issue explicitly in this volume). One of the problematic dimensions of attention to an individual's agency rests on its neglect of the collective and structural dimensions of domination. While it can be argued that individual-oriented actions can be—and in some cases are—translated into collective movements and/or successful outcomes, it is equally arguable that placing opposition exclusively in the hands of individuals narrows the range of opportunities for resistance. It also fosters the adoption of dominant discourses that devalue collectivity and stress individual initiative and merit (Szasz, 2007). In this context, the capacity to be reflexive can become an element that favors domination rather than emancipation. Reminiscent of left-leaning critiques of modern arrangements based on the problematic emancipatory power of science and reason (i.e., Horkheimer and Adorno, [1944] 1969), the claimed liberation of the individual and the power of individuality associated with consumption and other forms of opposition should be the subjects of further scrutiny.

Neoliberalism increased class polarization. The creation of a precarious class of immigrants and overexploited workers is accompanied by the existence of an impoverished and disillusioned middle class. The development of financialization (Fairbairn, and Lawrence and Campbell in this volume), the global exploitation of labor (Harrison, this volume), and processes of violent dispossession (Buck, this volume) that characterized contemporary patterns of capital accumulation in agri-food call attention to the class dimension of the evolution of neoliberalism.

Yet, as the class dimension of neoliberalism is overtly visible, discussions about class and class relations have been marginalized in academic, but also political, circles. While the exploitation of immigrant workers powers the growth of agri-food production globally, immigration is discussed mostly as a matter of national security, social control, social stability, individual attributes, national identity, and demographic trends. A similar critique can be made regarding public health and international trade as they relate to agri-food. Attention to class issues has become indirect if not silenced. In this context and recognizing the validity of some of these approaches, sociological efforts that would reintroduce class in the analysis of contemporary phenomena are sorely needed.

The chapters contained in this volume offer evidence of the plasticity and the contested dimensions of neoliberalism. They suggest the limits of unified and organic views of neoliberalism as an ideology and a model of public policy. As the debates on the present characteristics and future developments of agri-food continue, these observations could be relevant instruments for the understanding and the formulation of proposals for change of agri-food.

Note

1 These types of pronouncements are not limited to works produced by authors who subscribe to the Social Structure of Accumulation Theory (e.g., Gordon et al., 1982) but can be found in works generated by members of the Regulation Theory School (e.g., Aglietta, 1979) and World Systems Theory (e.g., Chase-Dunn, 1998). In the Sociology of Agri-Food, the tradition of the "Food-Regimes" approach represents a pertinent instance. Pioneered by the seminal writings of Harriet Friedmann and Philip McMichael (1989), the "Food Regimes" approach includes works that describe changes in agri-food in terms of the continuous existence of socioeconomic systems or regimes.

References

Aglietta, M. (1979) *A Theory of Capital Regulation*, New Left Books, London.

Allaire, G. and Wolf, S. (2004) "Cognitive representations and institutional hybridity in agrofood innovation," *Science, Technology & Human Values*, vol 29, no 4, pp. 431–58.

Chase-Dunn, C. (1988) *Global Formation: Structure of the World-Economy*, Rowman & Littlefield, Lanham, MD.

Ellul, J. (1964) *The Technological Society*, translated by J. Wilkinson, Knopf, New York.

Evans, P. (1996) "Government action, social capital and development: Reviewing the evidence on synergy," *World Development*, vol 24, no 6, pp. 1119–32.

Friedman, M. ([1962] 1982) *Capitalism and Freedom*, University of Chicago Press, Chicago, IL.

Friedmann, H. and McMichael, P. (1989) "Agriculture and the state system: The rise and decline of national agriculture," *Sociologia Ruralis*, vol 29, no 2, pp. 93–117.

Gordon, D., Edwards, R., and Reich, M. (1982) *Segmented Work, Divided Workers: The Historical Transformation of Labor in the United States*, Cambridge University Press, Cambridge.

Harvey, D. (2005) *A Brief History of Neoliberalism*, Oxford University Press, Oxford.
Hodge, I. and Adams, W. (2012) "Neoliberalisation, rural land trusts and institutional blending," *Geoforum*, vol 43, pp. 472–82.
Horkheimer, M. and Adorno, T. ([1944] 1969) *Dialectic of Enlightenment*, Continuum, New York.
Jackson, T. (2009) *Prosperity without Growth—Economics for a Finite Planet*, Earthscan, London.
Kotz, D. and McDonough, T. (2010) "Global neoliberalism and the contemporary structure of accumulation," Pp. 93–120 in T. McDonough, M. Reich, and D. Kotz (eds) *Contemporary Capitalism and Its Crises: Social Structure of Accumulation Theory for the 21st Century*, Cambridge University Press, Cambridge.
Krugman, P. (2008) *The Return of Depression Economics and the Crisis of 2008*, W. W. Norton, New York.
Krugman, P. (2012) *End this Depression Now*, Melrose Road Partners, New York.
Krugman, P. (2013) "Austerity doctrine benefits only the wealthy," *Houston Chronicle*, April 26, http://www.chron.com/opinion/outlook/article/Krugman-Austerity-doctrine-only-benefits-the-4466944.php, accessed June 19, 2013.
Lindblom, C. (1959) "The science of 'muddling through,'" *Public Administration Review*, vol 19, no 2, pp. 79–88.
Lockie, S. and Higgins, V. (2007) "Roll-out neoliberalism and hybrid practices of regulation in Australian agri-environmental governance," *Journal of Rural Studies*, vol 23, no 1, pp. 1–11.
Lopez, R. (2013) "UCLA Anderson forecast paints dismal picture of economic recovery," *Los Angeles Times*, http://www.latimes.com/business/la-fi-ucla-forecast-20130605,0,7676874.story, accessed June 5, 2013.
Meyer, J. and Rowan, B. (1977) "Institutionalized organizations: Formal structure as myth and ceremony," *American Journal of Sociology*, vol 83, no 2, pp. 340–63.
Offe, C. (1985) *Disorganized Capitalism*, MIT Press, Cambridge, MA.
Polanyi, K. ([1944] 2001) *The Great Transformation: The Political and Economic Origins of Our Time*, Beacon Press, Boston, MA.
Storper, M. (1998) "Conventions and the genesis of institutions," unpublished paper, available at http://webu2.upmf-grenoble.fr/regulation/Journees_d_etude/Journee_1998/Storper.htm.
Szasz, A. (2007) *Shopping Our Way to Safety: How We Changed from Protecting the Environment to Protecting Ourselves*, University of Minnesota Press, Minneapolis.
True, J. L., Jones, B. D., and Baumgartner, F. R. (1999). "Punctuated-equilibrium theory: Explaining stability and change in American policymaking" Pp. 97–115 in P. Sabatier (ed.) *Theories of the Policy Process*, Westview Press, Boulder, CO.
Wolf, S. and Klein, J. (2007) "Enter the working forest: Discourse analysis in the northern forest," *Geoforum*, vol 38, no 5, pp. 985–98.

INDEX

Aboriginal Australians 269
ABRANGE *see* Brazilian Association of Non-Genetically Modified Grain Producers
ABRASEM *see* Brazilian seed Producers Association
ABSPII *see* Agricultural Biotechnology Support Group II
Access to Local Foods: Farm to School Program 121
Acharia, Basudeb 162
adaptation (ability of economy to distribute wealth) 27
additionality principle 194
ADM *see* Archer Daniels Midland Company
Administrative Council for Economic Defense (Conselho Administrativo de Defesa Econômica (CADE)) (Brazil) 181, 183
AEP *see* agri-environmental policy
Africa 214–16, 245, 254
Agency for International Development (U.S.) 43
Agreement on Technical Barriers to Trade (TBT) 185n4
Agreement on the Application of Sanitary and Phytosanitary Measures (SPS) 185n4
Agreement on Trade-Related Aspects of Intellectual Property Rights (TRIPS) 172, 185n4, 229
agribusiness transnational corporations (ATNCs): development and production of agrochemicals 230; goal of maximizing their own profits 240; new agendas of 291–2; patenting ethics 171–2, 177–80, 182, 185n8, 228; state and biotechnology interdependence with 7, 226–39; *see also* transnational corporations
Agricultural Biotechnology Support Group II (ABSPII) (India) 161
Agricultural Commission of the Brazilian Chamber of Deputies 175
Agricultural Commission of the Senate (Brazil) 184
Agricultural Federation of Rio Grande do Sul (FARSUL) (Brazil) 173, 174
Agricultural Investment Summits 251
Agriculture, Food & Human Values Society 4
Agriculture and Human Values (Wolf and Harrison) 4
agri-environmental policy (AEP): contemporary efforts to rationalize 193–6; efforts to introduce more market discipline 196–8; failure to impose more market discipline 199–203; as set of national programs legislated by Congress 204n1
agri-food *see* neoliberal food regime
agrochemicals 41, 232–3, 236; Brazilian consumption of 180; five ATNCs dominating development and production 230; U.S. consumption of 180
Alberta, Canada 87n1
Alberta Investment Management Company 267
Allen, P. 114
American Enterprise Institute 33, 44
American Legislative Exchange Council 44
ANA *see* National Articulation of Agroecology
apomixis 47n4
APROSMAT *see* Mato Grosso seed Producers' Association

APROSOJA-RS *see* Soybean Producers Association
Archer Daniels Midland Company (ADM) 76, 84
Argentina 76, 229, 236–8
Aron, Raymond 33
AS-PTA *see* Consultancy and Services for Projects in Alternative Agriculture
ATNCs *see* agribusiness transnational corporations
Austin, J. 44
Australia 76, 84, 132, 234, 275; dairy industry 267–8, 273; neoliberalism's distinguishing features in 265–8; New Zealand's trajectory compared to 266, 269–70, 276–8; protests against neoliberal trajectory 268–9; PSE for 264
Australian Wheat Board 84

Bacillus thuringiensis 166n2
bananas 221n6
Bancada Ruralists 174
barley 41, 80–6, 87n1
BASF Global 41
Batie, S. S. 197
Baumer, Meghan 4
Beck, Ulrich 27n5
Becker, Gary 37, 39
Beidahuang 268
Belgium 216
biodiversity: competitive productionist agriculture versus 265–6; decline 288; in Guatemala's maize 229; in India's brinjal 163; land and ecosystem services impacting 201; peasants' preservation of 234; U.S. agriculture and 193, 195; U.S. non-regulation of polluters 202
biotechnology: Bt crops in India 153–6, 166n2; Green Revolution and Gene Revolution 171; India's ABSPII 161; MAHYCO Monsanto Biotech Limited 160, 162; paradoxes of neoliberal capitalism illustrated by 171–3; of poor 167n4; as response to Green Revolution era 56; state and ATNCs' interdependence with 7, 226–39; Truman on sharing technical knowledge 155; U.S. free trade rhetoric versus support of 171–2; *see also specific corporation*; *specific nation or state*
Boltanski, L. 43
Bonanno, Alessandro 58, 113, 210, 221n5
Border Patrol 100
Boxer, Barbara 123
BP Gulf of Mexico oil spill 25
Bracero Program, U.S. 100
BRASPOV *see* Brazilian Plant Breeders Association
Brazil: anti-GMO activism 61; food-price inflation 236–8; National Campaign for a GM-Free Brazil 174, 175, 182, 185; resistance to transgenic crops 170–86, 237, 238, 254, 267
Brazilian Agricultural Research Corporation (Empresa Brasileira de Pesquisa Agropecuária) (EMBRAPA) 173–4, 180
Brazilian Association of Non-Genetically Modified Grain Producers (Associação Brasileira de Produtores de Grãos Não Geneticamente Modificados) (ABRANGE) 174, 175, 181
Brazilian Institute of Consumer Protection (IDEC) (Brazil) 174
Brazilian Institute of Environmental and Renewable Natural Resources (Instituto Brasileiro do Meio Ambiente e Recursos Naturais Renováveis) (IBAMA) 174, 180
Brazilian Plant Breeders Association (Associação Brasileira dos Obtentores Vegetais) (BRASPOV) 173, 174
Brazilian seed Producers Association (Associação Brasileira de Sementes e Mudas) (ABRASEM) 173, 174, 182; soy growers' class action filed against Monsanto 175–6
Bretton Woods system 248
brinjal (*Solanum melongena*) 154, 161–3, 166n2
Britain 33, 55, 208, 211, 250, 265, 271; hegemony in food staples 58, 74, 226; joining European Union 234
British Columbia 87n1
Brown, Ralph 4
Brown, S. 46
Bt brinjal (eggplant) 154, 161–3, 166n2
Bt cotton hybrid plants 154, 160–3, 167nn10–11
Bt crops 153–6, 160–3, 166n2, 167nn10–11
Bt gene 166n2
Buffet, Warren 246
Bunge 76, 84
Burch, D. 250, 268
Busch, Lawrence 286
Butler, J. 44
Buttel, F. H. 4

CADE *see* Administrative Council for Economic Defense
CAFOs *see* concentrated animal feeding operations
Callon, M. 45
Campbell, Hugh 264, 271, 272, 273
Canada 73, 214, 232, 238, 267; Customs Act 87n3; government neoliberalization of prairie grains sector 81–7
Canadian Food Inspection Agency 230
Canadian Wheat Board (CWB) 41, 73, 87nn1–6; government dismantling of single-desk marketing 81–7; single-desk grain marketing 74–81, 86–7
Canadian Wheat Board Alliance 83
Capitalism and Freedom (Friedman, M.) 16–19, 22–3
Cardoso, Fernando Henrique 173
Cargill 76, 84, 87n5, 250
Carrefour 41, 251–2
Cartagena Protocol on Biosafety (2000) 172, 185n4
Castells, S. 232
CDs *see* certificates of deposit
CEAP *see* Conservation Effectiveness Assessment Project
Central American immigrants 91, 95, 99
Centre for Advanced Studies in Applied Economics (Centro de Estudos Avançados em Economia Aplicada) (CEPEA) (Brazil) 180
Centre for the Renewal of Liberalism (France) 33
CEPEA *see* Centre for Advanced Studies in Applied Economics
Cerny, P. G. 210, 217
certificates of deposit (CDs) 255
CFEs *see* Community Farming Enterprises
CFSC *see* Community Food Security Coalition
CFTC *see* Commodity Futures Trading Commission
Chayanov, Alexander 2
Cheshire, L. 271
Chief Scientist for Australia 276
Child Nutrition Act (1966) 120, 123
Child Nutrition Forum (CNF) 123
China 214, 232, 245, 267; Beidahuang 268; PSE for 264
Christie, J. 233
Chu, Winnie 4
CIFs *see* Commodity Index Funds
City of London 259
civil society organizations (CSOs) 154, 158
classical agrarian question 2
class polarization: Friedman's inaccurate predictions 13, 287; McMichael on political and economic elite 166n1; racial segregation case study 91–108; TNCs and ATNCs creating 291–2
Clean Air Act 195
climate change: constraints on PES and MES 201; crop loss due to weather-driven events 257; desertification from 269; markets created through environmental regulation 195; modern agriculture's contribution to 236; as sales pitch 253
CNA *see* National Agricultural Confederation
CNBS *see* National Biosafety Council
CNF *see* Child Nutrition Forum
coal, CIFs and futures contracts 276
Coase, Ronald 36
Cocklin, C. 265
Coles 266
Colombia, cut-flower agro-industry 208–9, 213–16, 218–19, 220n3
Commodity Futures Trading Commission (CFTC) 249
Commodity Index Funds (CIFs) 276
Common Market 270
the Commons 63–5; air, sea, and soil as global resource commons 62; manipulation of, toward capitalist ends 62
Communist Party of India 162
Community Farming Enterprises (CFEs) (Japan) 135–8, 146n3, 147n4; advantages and strengths of 142–4
Community Food Security Coalition (CFSC) 114
community supported agriculture (CSA) 114
competitiveness: Australia's competitive productivism 265–6; Colombia's cut-flower agro-industry 208–9, 213–15; Netherlands's cut-flower agro-industry 208–9, 215–17
competitive sheep sector (New Zealand) 274–5
CONAB *see* National Food Supply Company
concentrated animal feeding operations (CAFOs) 197
conditionality principle 194
Congress 23, 123, 179, 183, 193, 195, 197, 199, 201, 202, 204n1
CONPES *see* Consejo Superior de Política Económica y Social

Consejo Superior de Política Económica y Social (CONPES) (Columbia) 218, 219
Conservation Effectiveness Assessment Project (CEAP) 202
Conservation Reserve Program (CRP) 196, 197
Constance, D. H. 113
constructivism 2
Consultancy and Services for Projects in Alternative Agriculture (AS-PTA) (Brazil) 174
The Consumer Goods Forum 42
consumption process 1, 2, 45
CONTAG *see* National Confederation of Agricultural Workers
Convention of the International Union for the Protection of New Varieties of Plants (UPOV) 179, 180
Convention on Biological Diversity (1992) 185n4
Cooperative Extension Service (U.S.) 45
cooperatives 45–6; Brazilian 181, 183; Dutch grower 216–17; farmer-owned 78; Japan's FCC 140; in milk wars 267–8, 271; New Zealand's dairy sector 274–5
CPT *see* Pastoral Land Commission
Croatia 216
CRP *see* Conservation Reserve Program
CSA *see* community supported agriculture
CSOs *see* civil society organizations
CTNBio *see* National Technical Commission on Biosafety
Cuba 236–8
Customs Act 87n3
CWB *see* Canadian Wheat Board
CWB-Warburtons contract 79

dairy industry: Australia's 267–8, 273; milk wars 267–8, 271; New Zealand's 274–5; Wisconsin's 5–6, 91–108
Dairy Industry Restructuring Act (2001) (New Zealand) 274
Daniel, S. 259
Dean, M. 265
De Angelis, M. 62, 64
Declaration of Principles 179
De Goede, M. 250
democratic society: efforts toward more 5; neoliberal myths and opportunities for more 32–47, 47n4; return of democratic developmentalist state 156–7; state's inability to represent its people 211; TNCs creating class polarization 291–2; *see also* class polarization; state
Department of Agriculture, U.S. (USDA) 37, 118, 120; breaking monopoly of 200–2; continuity despite marketization pressures 203; internal friction from shift in organizational culture of 202, 203; nation-state and neoregulation via 228–30; Natural Resources Conservation Service 199–203, 204n1; neoliberal policy entrepreneurship impacting 193–9
Department of Biotechnology (India) 159
Department of Transportation, U.S. 102
Descartes, R. 33
development, Roy's definition of 167n6
Dibden, J. 265
direct marketing abroad 266
disorganized capitalism 286
DNA 171, 173, 185n2, 225, 231
Dosi, Giovanni 231
Dow Chemical Company 41
Dryzek, J. 200
DuPont 41

Eastern Europe 245
EBI *see* Environmental Benefits Index
ecocertification standards 221n6
Ecuador 214, 216, 221n6; food-price inflation 236–8
elderly population 129; Law for Stable Employment of Seniors 146n2; Silver Human Resource Centre 139–40, 146n2
EMBRAPA *see* Brazilian Agricultural Research Corporation
employment: in Australia 268–9; CFEs and 143–4; conditions 20–1; creation of 22, 215; desirable 16; developing countries' flexibility of employment practices 208, 215, 220; high levels of un- 18–19, 32; Law for Stable Employment of Seniors 146n2; NRCS employees' relationship with farmers 202; peasants' lack of alternative 235–6; Wisconsin's racially segregated workplaces 91–108
enclosures: as both local and global 57–60; from, toward reclaiming and reinventing of the Commons 63–5; as ongoing phenomenon 52–5; policing and violence backing up 5, 61–3; three new types of 55–7
Environmental Benefits Index (EBI) 197
Environmental Protection Agency (EPA) 198
Environmental Quality Incentives Program (EQUIP) 108n2, 197

EPA *see* Environmental Protection Agency
EQUIP *see* Environmental Quality Incentives Program
Ethiopia 216
Europe 21, 74, 213
European Court of Justice 178
European Trading System 195
European Union 19, 77, 81, 236, 264; Britain's membership impacting Australia 234; U.S. trade war with 75–6
Evans, L. 272
Evans, P. 210, 217

Fairweather, J. 275
FAMATO *see* Mato Grosso Agriculture and Livestock Federation
Family Farming Workers Federation (Federação dos Trabalhadores na Agricultura Familiar (FETRAF) (Brazil) 174, 175
FAO *see* Food and Agriculture Organization
Farm Bill, U.S. (1985) 197–8, 204n1; on subsidies to farmers' incomes 193
Farm Bill, U.S. (1996): as alteration of Fordist policy 98, 193, 196, 197, 204n1; as safety net 23–4
Farm Bill, U.S. (2008), EQUIP and market-defying logic 108n2, 197
farm cooperative corporation (FCC) (Japan) 140
farmers' markets 42, 45–6, 112–15, 122, 139; *see also* supermarket chains
farmland investment 245–9; discursive construction of 254–9; strategies and emerging institutions 250–4
farms/farmers: Australia's compared to New Zealand's trajectories 266, 269–70, 276–8; Brazilian farmer's organizations 174–5; classical agrarian question 2; decline in number of 76–7, 265–6; displaced and expelled 232; family farmers compared to capitalist farmers 233–4; FCC 140; financialization of agriculture 247–9; Fordism applied to 2, 3, 96–7, 131; FTS program 6, 113–25; Japan's CFEs 135–8, 142–4, 146n3, 147n4; Japan's Green Work case study 138, 140–1, 142; Japan's neoliberal reforms impacting its 133–4; Japan's Tsukinoya Healing case study 138–40, 142; Mexican immigrants in U.S. 5–6, 91–108; monocropping 231, 233; multifunctional values of 147n4; on-farm seed selection 166n2; plant biological diversity maintained by 233; privatization and commodification of land impacting 57–60; production agriculture 97; PSE for Australia, U.S., China, and Norway 264; resilience of 144–5; shortage of workers 136; suicides by Indian farmers after neoliberal reforms 156–7; *see also* peasants/rural communities; *specific nation*; *specific state or region*
Farm to School Network (FSN) 114
farm-to-school (FTS) program 6, 113–25
Farr, Sam 123
FARSUL *see* Agricultural Federation of Rio Grande do Sul
FCC *see* farm cooperative corporation
FDIC *see* Federal Deposit Insurance Corporation
Federal Court of Canada 83
Federal Deposit Insurance Corporation (FDIC) 258
Federal Reserve 249, 256
Federated Farmers of NZ 277
Federation of Agricultural Workers (FETAG-RS) (Brazil) 177–8
Federici, S. 61, 62
Ferguson, J. 59
FETAG-RS *see* Federation of Agricultural Workers
FETRAF *see* Family Farming Workers Federation
Fifth National Policy Document ("Fifth Plan") (Netherlands) 217–18
The Fight over Food (Wright, W., and Middendorf) 4
financialization 20; of agriculture 247–9; of Australia's food and farming 267–8; as new form of enclosure 56–7
Fitting, Elizabeth 229
Floramérica (Columbia) 214
FNS *see* Food and Nutrition Service
Fonterra (New Zealand) 273, 274, 277
Food and Agriculture Organization (FAO) 232
Food and Nutrition Service (FNS) 125n3
food-price inflation 236–8
food regime *see* neoliberal food regime
Food Research and Action Council (FRAC) 121
food riots 60, 288
Fordism 2–4, 96–7; -Keynesian regime 131, 288
Foucauldian approach 130–2, 145–6
Foucault, M. 44

FRAC *see* Food Research and Action Council
France 213, 236–8, 237; Centre for the Renewal of Liberalism 33
Free Soy Program (Programa Soja Livre) (Brazil) 182
FRETAF *see* Family Farming Workers Federation
Friedland, William 4
Friedman, Milton: *Capitalism and Freedom* 16–19, 22–3; neoliberal agenda 33, 35; on social class and disadvantaged groups 13, 287
Friedman, Thomas 27n7
Friedmann, Harriet 221n6, 226, 292n1
From Columbus to Con-Agra (Bonanno, Busch, Friedland, Gouveia, Mingione) 4
Fruit Growers Federation of NZ 277
FSN *see* Farm to School Network
FTS *see* farm-to-school program
functionalist theory 1–2
futures contracts 276

Gandhi, Indira 167n9
Gandhi, Rajiv 167n9
GATT *see* General Agreement on Tariffs and Trade
GATT Uruguay Round agreements 134
GCC *see* Global Commodity Chain
GE *see* genetically engineered content
GEAC *see* Genetic Engineering Approval Committee
General Agreement on Tariffs and Trade (GATT) 229
Gene Revolution 171
genetically engineered (GE) content: apomixis compared to 47n4; lack of labeling for 230; *see also* transgenic crops
genetically modified organisms (GMOs) 61, 167n2, 286; access to conventional seeds versus 181–2; Bt crops in India 153–6, 166n2; controversy over royalties to Monsanto 176–80; diminishing profitability of transgenic varieties 180–1; National Campaign for a GM-Free Brazil 174, 175, 182, 185; Smith, N., on 56
Genetic Engineering Approval Committee (GEAC) (India) 160, 162
Germany 213, 236–8, 237
Getz, C. 46
GHG *see* greenhouse gas
Glass-Steagall Act 248
Glencore International PLC 84
Global Commodity Chain (GCC) 212
GlobalGAP 42
global land grabs 245–60
global North: neoliberalization of agricultural sector 98; Zespri's contracts with retailers in 273, 274, 277; *see also* supermarket chains
global positioning satellite systems (GPS) 231
global South: commoners' struggles against enclosure in 63–4; debt crisis in 76–7; neoliberal reforms impacting 99–100; peasants' preservation of biodiversity in 234; *see also specific nation*
glyphosate 176, 180
GMOs *see* genetically modified organisms
Goldberger, Jessica 4
The Good Society (Lippmann) 33
Gouveia, L. 4
GPS *see* global positioning satellite systems
Grantham, Jeremy 257
Gray, S. 274
Great Depression 74, 248
Greece 46
greenhouse gas (GHG) offset: potential for fraud and fake claims 195, 205n4; problem of accounting procedures and contract design 201; unwillingness to regulate polluters 202
Greenpeace 174
Green Revolution 41, 56, 228, 230; biotechnology exports to India 153–6, 166n2; Gene Revolution and 171; social polarization brought about by 235–6; as technological paradigm 231
Greens (Australia) 265, 269, 270, 277–8
Green Work (GW) case study (Japan) 138, 140–1, 142
Guatemala 229, 236–8
Gujarat, India 160–6, 167n11
Guthman, J. 46, 114, 132, 277
GW *see* Green Work case study

Habermas, Jürgen 13–28, 27nn1–5, 45
HAIG *see* Hancock Agricultural Investment Group
Hall, P. A. 210, 211
Hall, S. 52, 62, 63
Hancock Agricultural Investment Group (HAIG) 252
Hardwick, Tim 4
Harvey, D. 26, 246; on contradictory nature of neoliberalism and neoliberalization 131–2, 172; on land grabbing 250; on neoliberal claims of

meritocracy 105–6; on NGO claims of protecting disenfranchised groups 27n8
Hassanein, N. 114
Hatch Act 40
Hayek, F. A. 33, 35, 38
Healthy, Hunger-Free Kids Act (2010) 121
Helleiner, E. 15
herbicides 180
Heritage Foundation 33, 44
Hiemstra, N. 101, 104
Higashimura area, Izumo City, Shimane prefecture 140
Higgins, V. 132, 145
Higham, J. 100
high value export commodities 207–20, 220n1
high yielding varieties (HYVs) 158
Hinrichs, Clare 4
Hispanic *see* Mexican immigrants
Hobbes, T. 36
Holland *see* Netherlands
Holt, Rush 123
Hood, C. 198
hybrid assemblages 130
hybridity, institutional 289–90
HYVs *see* high yielding varieties

IAAKSTD *see* International Assessment of Agricultural Knowledge, Science and Technology for Development
IBAMA *see* Brazilian Institute of Environmental and Renewable Natural Resources
IDEC *see* Brazilian Institute of Consumer Protection
identity-preserved (IP) shipments 79
illegality: of Bt cotton hybrids 154, 160–1; concept of 5–6, 91–108; contract of mutual authorization and 60; Federal Court of Canada on dismantling of CWB 83; introduction of RR soy into Brazil 175; Monsanto's collection of royalties 179
IMEA *see* Mato Grosso Institute for Agricultural Economics
IMF *see* International Monetary Fund
"In Defense of 'Extreme Apriorism'" 35
India 167n4, 167n9, 214, 232, 268; anti-transgenic crop activism 61; central and regional government 167n3; as neoliberal globalization project in crisis 153–6, 166n1; return of democratic developmentalist state 156–7; typology of seed markets in 157–9, 167n7
Indonesia 79
Integrated Triple Fiscal Reform (Sanmi-ittai no kaikaku) (Japan) 135
Internal Review Board (IRB) 125n4
International Assessment of Agricultural Knowledge, Science and Technology for Development (IAAKSTD) 45, 286
International Journal of Sociology of Agriculture and Food 27n8
International Monetary Fund (IMF) 58, 64, 76, 92, 99, 207, 248
International Rural Sociological Association 4
International Sociological Association 4
IP shipments *see* identity-preserved shipments
Iran 79
IRB *see* Internal Review Board
Ireland 46
Italy 46
Izumo City, Japan 140–2

Japan 21, 79, 81, 214; CFEs 135–8, 142–4, 146n3, 147n4; Green Work case study 138, 140–1, 142; MAFF 134, 137, 145; rural social service enterprises 129–38, 146nn2–3; Tsukinoya Healing case study 138–40, 142
Jasanoff, S. 45
Journal of Rural Social Sciences 27n8

Kaldor-Hicks efficiency 36
"*Kan kara min e*" (from bureaucracy to private sector) 133
Katz, Cindi 53, 57, 65n2
Kaup, B. Z. 92
Kautsky, Karl 2, 266
Kelly, N. J. 18
Kenya 214, 216
Kingdon, J. W. 116, 117
Kirin beverage company 267
Klepek, James 229
Kloppenburg, J. 114
Koizumi, Jun-ichiro 133–5, 142
Kotz, D. 285
Krugman, Paul 287
Kuhn, Thomas 231
Kumar, N. 156
Kyoto Protocol 195, 205n4

Labor Coalition 265, 268, 269
Labor Party (Australia) 265
Lahiff, E. 56

Land, Water and Environmental Conservation (LWEC) (Japan) 137, 142, 143
land grabbing: as dimension of neoliberalism 8, 55–6, 288; global 245–60; institutional violence to enforce 60
Land Grant Universities 228
Land Investment Expo 251, 258
Landless Rural Workers Movement (Movimento dos Trabalhadores Rurais Sem Terra) (MST) (Brazil) 61, 174
Langley, P. 39
Latin America 93, 214, 226, 232, 245
Latour, B. 45
Law for Stable Employment of Seniors (Japan) 146n2
Lawrence, Geoff 250, 264, 268, 271
Leahy, Patrick 123
legitimacy crisis 3–4, 153–6, 166n1, 182–5, 286–7; Beck on weakened nation-state and 27n5; Habermas on 13–28, 27n4, 27nn1–2; Hall, S., on 52, 62, 63; policing and violence contributing to 5, 61–3
Legitimation Crisis (Habermas) 14
Le Heron, R. 270, 272, 274
Lenin, Vladimir 2
Leyshon, A. 254
Li, T. 257
Liberal-National Coalition (Australia) 265, 268, 269
Lin, K. H. 8, 248
Linebaugh, P. 64
Lippmann, Walter 32, 33
Lithuania 216
Lockie, S. 132, 145
Louis Dreyfus 76, 84
Lula da Silva, Luiz Inácio 173, 229
Lund, C. 60
Lupel, A. 15
LWEC *see* Land, Water and Environmental Conservation

McCain's 268
McCarthy, Thomas 27n3
McDonough, T. 285
McMichael, Philip 112, 153, 166n1, 230, 292n1; on corporate food regime 226–8; on land rush 245–6, 250
MAFF *see* Ministry of Agriculture, Forestry and Fishery
Magdoff, H. 249
Maggi, Blairo 184, 186n10
Mahtani, M. 105
MAHYCO (Maharashtra Hybrid seed Company) 160, 161, 162
Mahyco gene construct 167n10, 167nn10–11
MAHYCO Monsanto Biotech Limited (MMB) 160, 162
Malthusian crisis 155, 257
Manitoba, Canada 87n1
MAPA *see* Ministry of Agriculture
market, Roy's definition of 167n7
market conditioning 18
marketization 131; AEP and constraints to 192–204; of all spheres of life 61; in Australia compared to New Zealand 266, 269–70, 276–8; Australia's food and farming industries 265–8; competitiveness as justification for 207–8, 217–18; Japanese CFEs in response to 143; USDA continuity despite marketization pressures 203
markets for provision of ecosystem services (MES) 195, 201
Marx, Karl 2, 56, 57, 265
Marxism 2, 131, 132; on metabolic rift between people and nature 265
Mato Grosso, Brazil 176, 178–9, 180–3, 186n10
Mato Grosso Agriculture and Livestock Federation (Federação de Agricultura e Pecuária de Mato Grosso) (FAMATO) (Brazil) 178, 179
Mato Grosso Institute for Agricultural Economics (Instituto Mato-grossense de Economia Agropecuária) (IMEA) (Brazil) 180
Mato Grosso seed Producers' Association (Associação de produtores de sementes de Mato Grosso) (APROSMAT) (Brazil) 182
MCT *see* Ministry of Science and Technology
MDA *see* Ministry of Agrarian Development
Mead, G. H. 38
Meat Export Board (New Zealand) 272
meat industry, Australia's 267, 273, 276
Medema, S. G. 36
Melo, C. J. 221n6
Mercuro, N. 36
Merton, Robert 2
MES *see* markets for provision of ecosystem services
Mexican immigrants in U.S. 5–6, 91–108

Mexico 24, 99, 235; aggressive neoliberal policies 231–2; food-price inflation 236–8; maize exports 229
MIC *see* Ministry of Internal Affairs and Communications
Middendorf, G. 4, 112
Midnight Notes 55
milk wars 267–8, 271
Miller, M. J. 232
Minaya, Jose 254
Mingione, E. 4
Ministry of Agrarian Development (Ministério do Desenvolvimento Agrário) (MDA) (Brazil) 172, 173
Ministry of Agriculture (Ministério da Agricultura, Pecuária e Abastecimento) (MAPA) (Brazil) 172, 173; Brazilian soy growers' class action filed against Monsanto 175–6, 183; soy growers' class action filed against Monsanto 272
Ministry of Agriculture, Forestry and Fishery (MAFF) (Japan) 134, 137, 145
Ministry of Agriculture, New Zealand 272
Ministry of Environment and Forests, India 162
Ministry of Health (Ministério da Saúde) (MS) (Brazil) 172, 173
Ministry of Internal Affairs and Communications (MIC) (Japan) 135
Ministry of Science and Technology (Ministério da Ciência e Tecnologia) (MCT) (Brazil) 172, 173
Ministry of the Environment (Ministério do Meio Ambiente) (MMA) (Brazil) 172, 173
Mirowski, P. 34
Mises, Ludwig von 33, 35
MMA *see* Ministry of the Environment
MMB *see* MAHYCO Monsanto Biotech Limited
MMC *see* Peasant Women's Movement
modernization theory 2
Mol, A. 45
monocropping 231, 233
Monsanto 41, 42, 81; Brazil's controversy over royalties to 176–80; control over seed technologies 172–3; diminishing profitability of transgenic varieties 180–1; lobbying efforts of 229–30; MAHYCO Monsanto Biotech Limited 160, 162; *see also specific products*
Mont Pelerin Society 33, 44
MPA *see* Small Farmers Movement
MS *see* Ministry of Health
MST *see* Landless Rural Workers Movement
Muirhead, B. 273
multifunctional values of agriculture 147n4

NAFTA *see* North American Free Trade Agreement
Nagaraj, K. 157
Nakasone, Yasuhiro 133
NANA *see* National Alliance on Nutrition and Activity
National Agricultural Confederation (Confederação Nacional da Agricultura e Pecuária do Brasil) (CNA) 173, 174, 179
National Alliance on Nutrition and Activity (NANA) (U.S.) 121, 123
National Articulation of Agroecology (ANA) (Brazil) 174
National Biosafety Council (Conselho Nacional de Biossegurança) (CNBS) (Brazil) 173, 174
National Campaign for a GM-Free Brazil 174
National Competition Policy (Australia) 267
National Confederation of Agricultural Workers (Confederação Nacional dos Trabalhadores na Agricultura (CONTAG) (Brazil) 174, 175
National Farmer's Federation (Australia) 276
National Food Supply Company (Companhia Nacional de Abastecimento) (CONAB) (Brazil) 180
National School Lunch Program (NSLP) 6, 113–25, 125nn1–3, 125nn5–6
National Technical Commission on Biosafety, Brazil (Comissão Técnica Nacional em Biossegurança) (CTNBio) 173, 174
Natural Resource Conservation Service (NRCS) 199, 200, 202, 204n1
Navbharat 151 (NB 151) 160, 167n10, 167nn10–11
NB 151 *see* Navbharat 151
neoliberal food regime 125n1; AEP failure to impose market discipline 199–203; CWB and history of 41, 73, 74–87, 87nn1–6; India's seed sector unraveling of 153–66; internal faltering of 52–65; key drivers of 228; national policy and transformation of 113–25; neoliberal model 1; plasticity of 4,

284–7, 292; racial segregation case study in 91–108; state, biotechnology, and ATNCs' interdependence in 7, 226–39; suicides by Indian farmers after economic reforms 156–7; TNCs as key component in 186n11; *see also specific nation, state, or region*
neoliberalism/neoliberalization: agricultural biotechnology and paradoxes of 171–3; civil conflict and war as integral to 62–3; combinations of public, private, and collective coordination strategies moving beyond 289–90; competitiveness and haphazard expansion of 207–20, 220n1; continuing unfolding of 288–9; creating ruptures in, through policy change 112–25; crisis of material and institutional contradictions 52–65, 230–8, 287–8; financialization and 247–9; food riots 60, 288; Foucauldian approach to 130–2, 145–6; Harvey on contradictory nature of 131–2, 172; land grabbing 8, 55–6, 60, 245–60, 288; legitimacy crisis of 3–5, 13–28, 27nn1–2, 27nn4–5, 52, 61–3, 153–6, 166n1, 182–5, 286–7; multifacetedness and contradictions of 144–6; opportunities provided by 27n7; ruptures in 1–2, 6–7, 54, 107, 265, 287; as socially constructed process 32–47, 47n4, 285–6; transgenics and paradoxes of 171–3; *see also specific nation, state, or region*
Netherlands 208–9, 213, 215–19, 220n3, 221n6
New Public Management (NPM) 40, 191, 192, 198
New Sociology of Agriculture movement 1–2
New Zealand 8, 40, 76, 86, 263–4; Australia's neoliberal trajectory compared to 266, 269–70, 276–8; dairy and kiwi success story 270–5
Ngai, M. M. 100
NGO *see* nongovernmental organization
nongovernmental organization (NGO) 58, 62, 171, 173, 193, 197, 202, 229; Harvey on 27n8; National Campaign for a GM-Free Brazil 174, 175, 182, 185
noninnocent topography 65n2
nonprofit organization (NPO), TS in Japan 138–9
North American Free Trade Agreement (NAFTA) 24, 99, 101
Norway 264
NPM *see* New Public Management
NPO *see* nonprofit organization
NRCS *see* Natural Resource Conservation Service
NSLP *see* National School Lunch Program
NZ Dairy Board (NZDB) 273, 274
NZDB *see* NZ Dairy Board
NZ Kiwifruit Marketing Board 273

Obama, Barack 123
Obama, Michelle 122
obesity, farm-to-school program and 6, 113–25
Occupy Wall Street 18, 25, 259
OECD *see* Organization for Economic Cooperation and Development
OEM *see* Office of Ecosystem Services and Markets
Offe, Claus 286
Office of Ecosystem Services and Markets (OEM) 195, 204n3
oil: CIFs and futures contracts 276; dependency 236; prices 74, 264; spill 25
One Nation 269, 277
open pollinated varieties (OPVs) 161
OPVs *see* open pollinated varieties
Organization for Economic Cooperation and Development (OECD) 237
Otero, Gerardo 58, 74, 113, 125n1, 171–2, 186n11

Paraná, Brazil 173
Parenteau, R. 249
Parliamentary Standing Committee (PSC) Report, India 162
Parsons, Talcott 2, 27n1
Pastoral Land Commission (CPT) (Brazil) 174
patenting ethics 171–2, 177–80, 182, 185n8, 228
Pawson, E. 270, 272
payment for ecosystem services (PES) 193–202; PES-like schemes 203
PE *see* private equity funds
peasants/rural communities: alternative job opportunities 235–6; biotechnology of poor in India 167n4; Brazil's Landless Rural Workers Movement 174; Brazil's Peasant Women's Movement 174; displaced and expelled 232; Japan's neoliberal reforms impacting its 133–4; Mexican immigrants in U.S. 5–6, 91–108; on-farm seed selection 166n2; plant biological diversity maintained by

233; privatization and commodification of land impacting 57–60; resilience of 144–5; *see also* farms/farmers
Peasant Women's Movement (MMC) (Brazil) 174
Pechlaner, G. 58, 74, 113, 125n1
Peck, J. 59, 131
Peoples, S. 272
Peoples Company 251
PES *see* payment for ecosystem services
Peschard, Karine 230, 286
Plant Variety Protection Act (PVPA) 177, 179
Plehwe, D. 34
Poland 216
Polanyi, Michael 33
policing: to buttress neoliberal food system 52–3; from, toward reclaiming and reinventing of the Commons 63–5; local, regional, national, and global 53, 57–60; neoliberal legitimation crisis and 5, 61–3; U.S. immigration and practices of 93, 100–2, 104–6
Polyani, Michael 289
Porter, Michael 212, 221n6
Portugal 46, 211, 237
Posner, Richard 37
Potter, C. 196–7
price pooling 78, 84–5, 87n1
primitive accumulation: as both local and global 57–60; from, toward reclaiming and reinventing of the Commons 63–5; as ongoing phenomenon 52–5
Pritchard, B. 269
private equity funds (PE) 251, 253, 260
producer cars 85
Producer Payment Options 81
Producer Support Estimate (PSE) 264
production agriculture 97, 265–6
production process 1
Prudential Agricultural Investments 252
PSC *see* Parliamentary Standing Committee
PSE *see* Producer Support Estimate
PVPA *see* Plant Variety Protection Act

Qatar's Hassad Food 267

The Rainforest Alliance 42
Rajan, K. S. 171
Rajiv Gandhi era 167n9
Ramanna, A. 160–1
Ramesh, Jairam 162, 163
Randeria, S. 172
Raulet, Gérard 15–16
R&D *see* research and development
Readers Digest 33
Reagan, Ronald 23, 99, 131, 133, 232
real estate investment trust (REIT) 251, 253
Regulation Theory School 292n1
REIT *see* real estate investment trust
Requião, Roberto 173
research and development (R&D) 275, 276
Research Committee on Sociology of Agriculture and Food (U.S.) 4
responsibilization 265
Ricardo, David 211
rice 23, 133–41, 144
Richardson International 84
Rio Grande do Sul, Brazil 173–5, 177–8, 180, 183
The Road to Serfdom (Hayek) 33
Roberts, D. J. 105
Rose, Carol 57
Rose, N. 44, 265
Rosin, C. 272
Rothbard, Murry 35
Rougier, Paul 32, 33
Roundup Ready (RR) soybeans 171, 175; access to conventional seeds versus 181–2; controversy over royalties to Monsanto 176–80; diminishing profitability of 180–1
Rousseff, Dilma 173
Royal Ahold 41, 251–2
RR *see* Roundup Ready soybeans
Rudy, A. 4
Rural Sociological Society 3–4
"Rural Sociology" (Friedland) 1, 4
Russia 214, 216

SAFRIG *see* Sociology of Agrifood Research Interest Group
Sanitary and Phytosanitary Measures Agreement (SPS) 185n4, 229
"*Sanmi-ittai no kaikaku*" (Integrated Triple Fiscal Reform) 135
Saskatchewan wheat pool, Canada 78, 87n1
Saudi Arabia 79, 245
School Nutrition Association (U.S.) 121
Schütz, Alfred 33, 38
Scott, J. C. 144
Second Green Revolution 155
seed companies: ABRASEM 173, 174, 175, 182; Brazilian Plant Breeders Association 173, 174; genetic engineering compared to

apomixis 47n4; India's return to democratic developmentalist state 156–7; MAHYCO 160, 161, 162; Mato Grosso seed Producers' Association 182; Monsanto's control over seed technologies 172–3; on-farm seed selection versus 166n2, 167n8; typology of seed markets in India 157–9, 167n7; *see also* genetically modified organisms; transnational corporations; *specific company*
self-help entrepreneurship 132, 136, 144–5, 265–6, 276
Selznick, P. 200
"*Shijô ni dekiru koto wa shijô ni*" (let the market do what the market can) 133
SHRC *see* Silver Human Resource Centre
Sikor, T. 60
Silver, Beverly 61
Silver Human Resource Centre (SHRC) 139–40, 146n2
Simons, H. C. 35
Simplot 268
single-desk grain marketing agency 74–86, 87; *see also* Canadian Wheat Board
Small Farmers Movement (MPA) (Brazil) 174
Smith, Adam 211, 212
Smith, Neil 56
Social Structure of Accumulation Theory 292n1
Sociology of Agrifood Research Interest Group (SAFRIG) 3, 4
Soros, George 246
Soskice, D. W. 210, 211
South Africa 214
Southeast Asia 245
Southern Agricultural Resources 267
South Korea 245
South Sea Bubble 248
Soyatech 251
Soybean Producers Association (Associação de Produtores de Soja do Rio Grande do Sul) (APROSOJA-RS) (Brazil) 177, 179, 181; class action filed against Monsanto 175–6, 183
Spain 46, 214, 237
SPS *see* Agreement on the Application of Sanitary and Phytosanitary Measures
Sreenivasan, G. 233
state: Beck on legitimation crisis and 27n5; biotechnology and ATNCs' interdependence with 7, 226–39; captured by narrow interests of economic actors 211; centrality of, in neoliberal regime 226–39; changing modes of intervention 131; disciplinary and punitive function of 54–63; as domain of innovation 115–18; financialization versus regulation by 248–9; India's state-led versus corporate-controlled capitalism 156–9; Japan's curtailing of national fiscal supports 129–35
State Agricultural Experiment Stations 40
Stock, P. 272
sugar, Australia's export of 267
suicides 156–7
Superior Court of Justice (Brazil) 178
supermarket chains 24, 38–9; Australian agriculture and 266–7; Colombian flower business supplying 214; consumers governed by 42; as key driver of neoliberal food regime 228; New Zealand dairy and kiwi success story 270–4; Walmart, Tesco, Carrefour, and Royal Ahold 41, 251–2
Supreme Court of India 162
Swaminathan, M. S. 163
Sweezy, P. 249
Switzerland 213
Szasz, A. 93

Tanaka, Keiko 4
TBT *see* Agreement on Technical Barriers to Trade
Teachers Insurance and Annuity Association-College Retirement Equities Fund (TIAA-CREF) 247, 255, 258
TEC *see* Technical Expert Committee
Technical Expert Committee (TEC) (India) 162
Technology Stewardship Agreements (1996) 172–3
Terra Firma Capital 268
Terrapinn 251
Tesco 41, 251–2
TH *see* Tsukinoya Healing
Thailand 76, 214
Thatcher, Margaret 61, 131, 164, 232, 285
"There is No Alternative" (TINA) 61, 164, 285
Thévenot, L. 43
Third Food Regime 75–83
Thomas, W. I. 32
Thrift, N. 254
TIAA-CREF *see* Teachers Insurance and Annuity Association-College Retirement Equities Fund

Tickell, A. 131
TINA *see* "There is No Alternative"
TNCs *see* transnational corporations
Tomaskovic-Devey, D. 8, 248
Toward a New Political Economy of Agriculture (Friedland, Busch, Buttel, Rudy) 4
transgenic crops 235; access to conventional seeds 181–2; Brazil's controversy over 173–5, 185n2, 185nn4–8; Bt brinjal case study 154, 161–3, 166n2; Bt cotton case study 160–3, 166n2; conflicts between producers and companies 185n7 (*see also* genetically modified organisms); controversy over royalties to Monsanto 176–80; diminishing profitability of transgenic varieties 180–1; in India 154–9; unavailability of official statistics on 185n6; *see also* genetically modified organisms
transnational corporations (TNCs): biotechnology and crisis of neoliberal food regime 230–8; as component of neoliberal regime 186n11; India's seed sector unraveling of neoliberal policies 153–66
TRIPS *see* Agreement on Trade-Related Aspects of Intellectual Property Rights
TRIPS Agreement (1994) 172
Truman, Harry S. 155
TS *see* Tsukinoya Shinkoukai
Tsing, A. 257
Tsukinoya area, Un-nan City, Shimane prefecture 138
Tsukinoya Healing (TH) case study 138–40, 139t
Tsukinoya Shinkoukai (TS) 138
tulipmania 248

UBS Agrivest 252
Uganda 214
UN Cartagena Protocol on Biosafety 172, 185n4
Underhill, G. 15
U.N. Food and Agriculture Organization 237
United Kingdom 40, 79, 133, 213, 214, 267; food-price inflation 236–8
United Nations 58, 61, 172, 237
United States (U.S.) 20, 33, 74, 77, 133, 208, 213, 232; agri-environmental policy in 191–205, 204n1, 205n4; food-price inflation 236–8; free trade rhetoric versus support to biotech industry 171–2; housing bubble 245–6; neoregulation via USDA 228–30; policing and deportation of immigrants 93, 100–2, 104–6; PSE for 264; Wisconsin's racially segregated workplaces 91–108
United States Patents and Trademarks Office (USPTO) 172
Un-nan City, Japan 138–40, 142
UPOV *see* Convention of the International Union for the Protection of New Varieties of Plants
U.S. *see* United States
USDA *see* Department of Agriculture, U.S.
USDA Natural Resources Conservation Service 199–203, 204n1; *see also* Department of Agriculture, U.S.
USPTO *see* United States Patents and Trademarks Office

Valbo, Mio 198, 199
Van der Linde, C. 221n6
Van der Ploeg, J. D. 234
Varsanyi, M. W. 100
Vasavi, A. R. 167n5
Vía Campesina. 112, 227, 239
violence: commitments to enclosure backed up by 5, 52–65; contract of mutual authorization and 60; Mexican-U.S. border deaths and 100; in policing and neoliberal legitimation crisis 5, 61–3; real and symbolic 57–60, 65n1
Viterra 78, 84, 87n5
Volscho, T. M. 18

Wacquant, L. 103
Walmart 32, 38, 41, 42, 251–2
Walzer, M. 43
war: Cold War thinking 42; India post-World War II 156; as integral to neoliberal expansion 62–3; U.S. versus European Union trade war 75–6; U.S. war on terror 100; World War I 74; World War II 1, 33, 156, 226, 263, 264; World War II and single-desk marketing system 74
Warburtons 79
water resources: AEP policy designs for improving 194; Australia's 265–6; competitive productionist agriculture versus 265–6; decline and pollution of 288; ecosystem services impacting 201; Japan's LWEC 137, 142, 143; peasants' preservation of 233; U.S. agriculture and 193, 195; U.S. non-regulation of

polluters 202; water-and-land-scarce nations 276
Watts, Michael 65n1
The Wealth of Nations (Smith, A.) 211
Wellington West 258
Western Grain Marketing Panel (Canada) 80
wheat: Australian Wheat Board 84; Australia's deregulation 273; CIFs and futures contracts for 276; financialization of Australian 267–8; quality-differentiated 79–80; world wheat markets 75; *see also* Canadian Wheat Board
Winders, Bill 4
Wisconsin 5–6, 91–108
Wisconsin Governor's Council on Migrant Labor 102
Wolf, Steven A. 196–7, 221n6
Wood, Spencer 4
wool industry 272–5; CIFs and futures contracts 276
Woolworths 266
Workers Party, Brazil 173
World Bank 56, 58, 76, 92, 99, 207, 209, 213, 217, 220, 232, 236; on naturally efficient management of resources 61
World Systems Theory 292n1
World Trade Organization (WTO) 38, 58, 92, 98, 185n4, 229; The Agreement on Agriculture 76; agricultural biotechnology and paradoxes of neoliberal capitalism 171–3; Canada's free trade agreements through 77, 81; impasse at 75; members of 227
Wright, Ashley 4
Wright, Wynne 4, 112
WTO *see* World Trade Organization
WTO Doha Round 227
WTO TRIPS Agreement (1994) 172
WTO Uruguay Round 227, 229

Zambia 214
Zapatistas 61
Zespri International Ltd 273, 274, 277
Zhang, X. 15
Zimbabwe 214
Žižek, Slavoj 47